THE URBAN POLITICS READER

The Urban Politics Reader draws together classic and contemporary writings that best illuminate the basic questions of urban politics – how interests contend for power over the distribution of resources and why some win while others lose. Contributions from Martin Shefter, Clarence Stone, Rufus P. Browning, and Saskia Sassen are included among the thirty-two generous selections.

The Reader juxtaposes the main theoretical approaches to urban power with vivid accounts of actual political experiences on such key themes as the urban crisis, the politics of race, ethnicity and gender, national urban policy, suburbanization, and globalization. Strom and Mollenkopf illustrate how thinking about cities is central to our understanding of democracy and citizenship and how although the parties to urban politics may change over time, the struggle of new groups to gain access and representation is a constant theme.

Editors' introductions suggest topics and questions for class discussion, demonstrate the significance of urban politics, and suggest directions for further reading and thinking, while the associated bibliography enables deeper investigations. By drawing together important but widely dispersed writings, *The Urban Politics Reader* provides an essential resource for students of urban politics. The volume will also have widespread appeal for students of urban sociology, urban affairs, urban planning, and public policy.

Elizabeth A. Strom is Associate Professor at the University of South Florida.

John H. Mollenkopf is Distinguished Professor of Political Science and Director of the Center for Urban Research at the CUNY Graduate Center.

THE ROUTLEDGE URBAN READER SERIES

Series editors

Richard T. LeGates

Professor of Urban Studies, San Francisco State University

Frederic Stout

Lecturer in Urban Studies, Stanford University

The Routledge Urban Reader Series responds to the need for comprehensive coverage of the classic and essential texts that form the basis of intellectual work in the various academic disciplines and professional fields concerned with cities.

The readers focus on the key topics encountered by undergraduates, graduates and scholars in urban studies and allied fields. They discuss the contributions of major theoreticians and practitioners and other individuals, groups, and organizations that study the city or practice in a field that directly affects the city.

As well as drawing together the best of classic and contemporary writings on the city, each reader features extensive general, section and selection introductions prepared by the volume editors to place the selections in context, illustrate relations among topics, provide information on the author and point readers towards additional related bibliographic material.

Each reader will contain:

- Approximately thirty-six *selections* divided into approximately six sections. Almost all of the selections will be previously published works that have appeared as journal articles or portions of books.
- A *general introduction* describing the nature and purpose of the reader.
- Two- to three-page *section introductions* for each section of the reader to place the readings in context.
- A one-page *selection introduction* for each selection describing the author, the intellectual background of the selection, competing views of the subject matter of the selection and bibliographic references to other readings by the same author and other readings related to the topic.
- A plate section with twelve to fifteen plates and illustrations at the beginning of each section.
- An index.

The types of readers and forthcoming titles are as follows:

THE CITY READER

The City Reader: Third edition – an interdisciplinary urban reader aimed at urban studies, urban planning, urban geography and urban sociology courses – will be the *anchor urban reader*. Routledge published a first edition of *The City Reader* in 1996 and a second edition in 2000. *The City Reader* has become one of the most widely used anthologies in urban studies, urban geography, urban sociology and urban planning courses in the world.

URBAN DISCIPLINARY READERS

The series will contain *urban disciplinary readers* organized around social science disciplines. The urban disciplinary readers will include both classic writings and recent, cutting-edge contributions to the respective disciplines. They will be lively, high-quality, competitively priced readers which faculty can adopt as course texts and which will also appeal to a wider audience.

TOPICAL URBAN ANTHOLOGIES

The urban series will also include *topical urban readers* intended both as primary and supplemental course texts and for the trade and professional market.

INTERDISCIPLINARY ANCHOR TITLE

The City Reader: third edition
Richard T. LeGates and Frederic Stout (eds)

URBAN DISCIPLINARY READERS

The Urban Geography Reader
Nicholas R. Fyfe and Judith T. Kenny (eds)

The Urban Sociology Reader
Jan Lin and Christopher Mele (eds)

The Urban Politics Reader
Elizabeth Strom and John Mollenkopf (eds)

The Urban Design Reader
Michael Larice and Elizabeth Macdonald (eds)

Forthcoming:

The Urban and Regional Planning Reader
Eugenie Birch (ed.)

TOPICAL URBAN READERS

The City Cultures Reader, second edition
Malcolm Miles, Tim Hall with Iain Borden (eds)

The Cybercities Reader
Stephen Graham (ed.)

The Sustainable Urban Development Reader
Stephen M. Wheeler and Timothy Beatley (eds)

The Global Cities Reader
Neil Brenner and Roger Keil (eds)

For Further Information on The Routledge Urban Reader Series
please visit our website:

www.geographyarena.com / geographyarena / urbanreaderseries

or contact:

Andrew Mould
Routledge
Haines House
21 John St
London WC1N 2BP
UK
andrew.mould@routledge.co.uk

Richard T. LeGates
Urban Studies Program
San Francisco State University
1600 Holloway Avenue
San Francisco, California
94132
(415) 338-2875
dlegates@sfsu.edu

Frederic Stout
Urban Studies Program
Stanford University
Stanford, California 94305-6050
(650) 725-6321
fstout@stanford.edu

The Urban Politics Reader

Edited by

Elizabeth A. Strom

and

John H. Mollenkopf

Routledge
Taylor & Francis Group

LONDON AND NEW YORK

First published 2007
by Routledge
2 Park Square, Milton Park, Abingdon, Oxon OX14 4RN

Simultaneously published in the USA and Canada
by Routledge
711 Third Avenue, New York, NY 10017

Routledge is an imprint of the Taylor & Francis Group, an informa business

© 2007 Elizabeth A. Strom and John H. Mollenkopf for selection and editorial matter; the contributors for individual readings

Typeset in Amasis Mt Lt and Akzidenz Grotesk by Graphicraft Limited, Hong Kong

British Library Cataloguing in Publication Data
A catalogue record for this book is available from the British Library

Library of Congress Cataloging in Publication Data
A catalog record for this book has been requested

ISBN10: 0–415–31995–1 (hbk)
ISBN10: 0–415–31996–X (pbk)
ISBN10: 0–203–62611–7 (ebk)

ISBN13: 978–0–415–31995–9 (hbk)
ISBN13: 978–0–415–31996–6 (pbk)
ISBN13: 978–0–203–62611–5 (ebk)

Contents

List of Plates, Tables and Figures

PLATES

TABLES

FIGURES

Acknowledgments

We would like to thank Frederic Stout for the opportunity to contribute to this remarkable series. We are also grateful to Andrew Mould and the Routledge editorial staff for their patience and good cheer through this Reader's gestation. Thanks are also due to our students at Rutgers and the City University Graduate Center, for their help in shaping our approaches to studying and teaching urban politics.

This collection owes its largest debt to the authors whose works we have reprinted. Several were especially helpful in assisting us to secure permission from publishers, allowing us to edit their work to meet the space requirements of this volume, and offering valuable suggestions. Alice O'Connor, Clarence Stone, and Raymond Wolfinger, most notably, have extended themselves to enable us to include their work.

Elizabeth Strom would like to thank Janice Bockmeyer, Jill Gross, Lorraine Minnite, and Mara Sidney for their continued inspiration. She has benefited from the warm support of her colleagues at Rutgers-Newark and the University of South Florida. Martin and Benjamin Muschol have now lived amid stacks of scanned urban politics articles in three states over four years; we thank them for their forbearance. John Mollenkopf is grateful to the CUNY Graduate Center for providing him with many wonderful colleagues who have a strong interest in the comparative analysis of urban dynamics. In addition to Frederic Stout, with whom he co-taught Stanford's urban studies core seminar for many years, he would like to thank Sophie Body Gendrot, Manuel Castells, Peter Dreier, Susan Fainstein, Ian Gordon, Chris Hamnett, Hartmut Häussermann, Michael Jones-Correa, Toshio Kamo, Edmond Preteceille, Clarence Stone, and Todd Swanstrom for years of stimulating friendship.

We are grateful to Ron Hayes for the photograph "Homeland security billboard". The photographs at the beginning of each section are courtesy of John Mollenkopf.

INTRODUCTION

This volume brings together what we regard as the best thinking that social scientists have done about politics, power, and governance in American and European cities. These essays reflect a central tension in how we think about cities – we know that the larger social, economic, and political contexts shape and constrain what city dwellers and their political leaders may do, yet we also know that their choices nonetheless remain critically important for the quality of urban life and urban democracy. In one way or another, all the writings in this volume explore how the interplay between social, economic, and political structure and human agency shapes what kinds of challenges city governments take on, how they respond to these challenges (or fail to do so), and whose interests are best served by their actions.

At one time, urban politics could be studied as if we remained in the time of the Greek *polis*, where each city was a free-standing entity seeking to determine its own fate. (Indeed, the concepts of city, citizen, civic, politics, public, and democracy can all be traced back to closely related roots.) As late as the 1950s, the great political scientist Robert Dahl could study New Haven as if it were a miniature republic and American politics simply urban politics writ large. Today, however, we are increasingly aware that all localities are interdependent on a global scale. A fisherman in Maine may decide what to catch based on the prices prevailing in Tokyo's Tsukiji market; the changing population of a small town in North Carolina reflects the need for Mexican workers in its chicken-processing plants; an Irish software firm sells its products to companies across Europe, and locates its service center in India. Across North America and Europe, new transportation and telecommunication technologies have shifted new office and housing construction out of central cities; urban fringes and suburbs have taken on new economic significance. These major trends are all well beyond the reach of any given city to influence and then shape the environment in which all of them operate. Equally important, cities are not politically or legally independent. Different national systems grant cities various degrees of "home rule," but nowhere are cities sovereign. As both a legal and a practical matter they face unique fiscal constraints, with less freedom to borrow or spend compared to higher levels of government.

At one time, as well, American urbanists seldom looked beyond their borders to understand how cities function. The American metropolis was *sui generis*, it seemed, with its ethnic political machines and racial tensions. In the past two decades, a new interest in comparative urban research has injected welcome energy into this subdiscipline. In part, the new focus on cross-national comparison stems from the new awareness of global interconnectedness. If, indeed, steel workers in Pittsburgh, Germany's Ruhr Valley, and the South Korean port of Gwangyang are all producing for the same market, it seems natural to inquire whether the workings of that market lead to similar policy responses in all three places. Indeed, some analysts have argued that policies and governing processes have shown signs of convergence across national borders. Devolution of responsibilities from national to local/regional governments; privatization of services and increasing reliance on public–private partnerships; and greater concerns with policies promoting competitiveness are seen as hallmarks of this new regime.

While acknowledging these developments, we remain convinced that governments respond to these changes in distinctive ways, shaped by their histories and institutional structures. There is no "one-size-fits-all" urban governance paradigm that can explain power relations and policy outcomes across all cities, or even all similarly industrialized cities. While we stress the specificity of political responses in each nation, we believe that students and scholars can have important conversations across boundaries. Therefore, although many of the selections chosen for this Reader are grounded in the U.S. urban political experience, they all contribute to the international conversation about local governance in a global age.

Given the importance of these economic, social, and political contexts, we begin by looking at the forces currently shaping the urban experience. Part 1 examines how demographic change, new technologies, and economic restructuring have reshaped American cities, how changing national political dynamics have affected the standing cities within their national settings, and how globalization has altered social stratification in urban areas. It begins with an overview of the most salient trends from a group of urban analysts then at Rutgers. Political scientist Margaret Weir then contrasts the role of cities and left-liberal parties in the welfare states of Europe and America, stressing the distinctive features of American politics. Saskia Sassen presents the essence of her view of how the forces of economic change are propagated through global cities, with far-reaching social and political implications, while the eminent British geographer Chris Hamnett questions whether globalization does in fact lead to social polarization. Although brief, these essays are incisive about the key forces now shaping and constraining the actions of local governments, perhaps enhancing the power of some cities and actors while undermining others.

Part 2 turns to the historical origins of urban politics in the United States. After visiting the United States in 1904, with stops in New York and St. Louis, the great German sociologist Max Weber observed that American cities had produced a new class of professional politicians who made a business out of electoral politics and government – living "off" politics, rather than "for" politics. While many middle-class reformers lamented the way that machine politicians sold off public policy to the highest bidder, traded jobs for votes, and so on, there is no doubt that urban political machines did effectively organize the fractious politics of American cities during their period of rapid urbanization, immigration, and industrialization between 1850 and 1950. Richard Croker, perhaps the most eminent political boss of his time, makes the case that his organization had a powerfully positive impact on New York City. Political scientist Milton Rakove, himself a participant, shows how the Chicago political machine continued to operate effectively a century later. Others would disagree about the benign nature of the political machine. Amy Bridges and Richard Kronick explore how the constriction of the electorate, particularly in certain regions of the country, fostered reform efforts to take politics out of local administration through nonpartisan government and the professionalization of the Civil Service. These efforts made great headway in the South and West, but Ray Wolfinger shows that machine politics, if not the political machines of old, had great staying power, especially in the older cities of the Northeast. Michael Jones-Correa provides a case in point, showing how older ethnic political elites in Queens, New York, resisted attempts by Hispanic immigrants to break into local politics.

Given this array of actors and interests, what can be said about the organization of political power in American cities? Political scientist John Mollenkopf opens Part 3 with an overview of how theorizing about the dynamics of urban power has evolved over the last half-century. Although party politicians sustain the allegiance of ethnic and racial voting blocks in many cities by providing jobs and contracts to their followers, their interactions with grass-roots constituencies do not always determine what policies the leaders of city governments adopt. Indeed, the question of the extent to which the composition of a city's governing coalition – both in terms of whose votes put it into office and what private interests it aligns itself with – drives its actions or its policies remains central to a lively debate in the theory of urban politics. Are such coalitions a coherent force that operates across policy areas, or do multiple coalitions operate in different policy domains? Are they durable or transitory? Do they reflect the values, goals, and strategies of their members, or do systematic imperatives instead largely determine what the coalition does?

The pluralist school descended from Robert Dahl's justly famous study of politics in New Haven in the late 1950s provides one extremely influential answer to these questions. Dahl argues that each policy area is characterized by its own set of interested parties, that no single elite dominates decision-making in all policy areas, and that every constituency can organize to have significant influence over any question of real interest to it.

Paul Peterson, however, has offered the alternative view that competition among cities for people who will invest and pay taxes means that urban political leaders have little choice but to adopt land use policies that promote growth and to forswear expensive social welfare policies that will redistribute those taxes to the poor. Clarence Stone reaches a somewhat similar theoretical destination in his studies of Atlanta, but by a different route. Beginning from the insight that urban political leaders must rely on private sector power holders to accomplish their goals, he develops what has become known as "regime theory."

Part 4 turns to the larger setting of American and European cities in the six decades following the end of World War II. Deindustrialization, economic and demographic transformation, suburbanization, and regional competition have remade central cities in both settings. These forces largely erased the white, blue-collar neighborhoods created by the fusion of urbanization, industrialization, and migration in the late nineteenth and early twentieth centuries, often with a strong push from local and national urban development policies. The crisis of old neighborhoods yielded in some places to new forms of racial and ethnic segregation, often accompanied by neighborhood decline and growing poverty, and gentrification and the emergence of new urban social strata in others, sometimes in close proximity. Martin Shefter shows how the deep tensions between the demand for social spending and the of tax revenues from the economic base in New York City have periodically led to grave fiscal and political crises, most recently in the mid-1970s. John Mollenkopf shows how modernizing political entrepreneurs formed "pro-growth coalitions" that married a base of popular constituencies to elite interests in programs of urban redevelopment that would transform both urban space and urban politics. Political scientists Hank Savitch, Paul Kantor, and Selena Vicari outline the dependence of cities on higher levels of government where their influence has been waning, while Peter Dreier outlines both the achievements of neighborhood-based movements for improvement and the limits to their efforts. Amy Lind highlights the important gender dimension in this form of activism.

If economic transformation has provided one major challenge to cities over the last half century, racial and ethnic transition has provided the other. As white populations moved out of large central cities, African Americans, Puerto Ricans, Mexican-Americans, and more recently waves of newer immigrants from Latin America, Asia, and the Caribbean have moved in. Given the persistence of machine politics dominated by earlier white ethnic groups described in Part 2, it is not surprising that racial conflict and demands for racial empowerment characterized urban politics in America from the late 1960s through the 1990s. In Part 5, political scientists Rufus Browning, Dale Rogers Marshall, and David Tabb analyze the conditions under which minority political empowerment took place in American cities. While they believe it yielded important policy gains, Adolph Reed takes a much more critical view of the constraints preventing black mayors from delivering substantial change for their constituencies. Over time, Rob Gurwitt argues, the racial polarization of the early elections yielding the first black and Hispanic mayors has become less pronounced, at least in some places, while Michael Jones-Correa provides a path-breaking analysis of how gender shapes the forms of political participation among Hispanic immigrants in New York.

The rising influence of national government on big cities has been another hallmark of the postwar period. In the U.S., federal policies have had a huge direct and indirect impact on cities. The indirect effects of many federal policies, such as the impact of military spending, the building of interstate freeways, or federal tax incentives for home-ownership, may have been even more influential than the direct effect of federal urban policies, especially those designed to remediate the increasingly urban concentration of poverty in the 1970s and 1980s. As the center of gravity in the nation's electorate moved away from the older central cities of the Northeast, the ability of urban interest to make claims on state and federal resources has also waned. In Part 6, various aspects of these developments are traced by policy historian Alice O'Connor, journalist Buzz Bissinger, and a group of political scientists. Peter Dreier, John Mollenkopf, and Todd Swanstrom trace the evolution of federal policies toward cities. O'Connor documents the history of urban anti-poverty programs. Swanstrom, and Richard Sauerzopf trace the declining influence of urban voters in state and national politics and the consequent diminution of their influence in the balance of legislative power. Pietro Nivola put the American experience in broader context by examining the situation of European cities. Finally, Peter Eisinger discusses the rise of terrorism as a new threat to urban life.

As we enter the twenty-first century, then, cities are more than ever constrained by events in larger arenas that they cannot control: the relocation of people and economic activities, the flow of governmental resources and fiscal capacities, the growth of influential political constituencies based outside of central cities. Yet even within this changing environment, cities still possess a considerable amount of resources, have legitimate and important claims for attention and support for higher levels of government, and can forge alliances with their neighbors within their metropolitan areas. The volume concludes with a progressive metropolitan agenda outlined by Bruce Katz, one of America's leading thinkers about urban policy.

Over the past century, urban life has evolved from a stage where cities were rapidly becoming bigger, denser, and more industrial, drawing immigrants from the countryside and abroad, to one where many were becoming smaller, less dense, deindustrialized, suburbanized, racially divided, while also often remaining centers of markets, expertise, culture, and politics. Each epoch has yielded challenges to the achievement of democratic equality, neighborhood tranquility, healthy environments, and economic opportunity. Indeed, these challenges are among the most important faced by developing and advanced societies alike. City governments have the ultimate responsibility for reacting to these challenges, and urban politics is the main arena for debating and deciding what they should try to do. As a consequence, today, just as a century ago, urban politics remains central to the fates of nations.

PART ONE

■ The social and economic context of urban politics

Plate 1 Lower Manhattan after 9/11.

INTRODUCTION TO PART ONE

Cities are always in flux. Their populations change as new population groups move to cities in search of opportunity, while earlier groups age, evolve, migrate, or even melt away as distinct groups. Cities may be likened to a social pump that lifts people across their life cycles, as urban sociologist Herbert Gans once observed. They draw youngsters away from their parents, neighborhoods, and schools and into the early stages of their working lives. When that pump works well, young people rise through the early stages of their working lives and begin to form their own families and have children. As their needs and wants change and their family life evolves, they often move from their initial neighborhoods, which they may have chosen to be close to jobs or to other young people, toward the periphery of the city, where they have more space in which to live and raise their kids. As they age and their own children grow up, they may move back toward the center, or to an environment more suited to older people and less focused on families with children.

The city economy plays an integral role in this process. City economies are always in flux as well. Under the impact of changing technologies and competitive conditions some industries grow, while others decline; city economies as a whole also experience many ups and downs. Cities with stronger economies attract more workers and residents; those with declining economies experience greater social and fiscal distress and, ultimately, lose their residents to other places. While the nineteenth-century city fused the processes of industrialization, immigration, and urbanization and central business districts continue to provide a natural home for certain core economic activities today, many industry sectors migrated away from dense, costly, sometimes politically intrusive central city environments during the twentieth century. That so many cities have remained economically strong in the face of such challenges is a testament to their resilience and creativity. Some cities have not done well, especially those whose economies relied on a primary industry that went into long-term decline, such as the textile cities of New England in the 1950s.

The demographic and economic trajectories of cities are deeply intertwined. The socioeconomic trajectory of population groups moving to a city at any given point in time is generally shaped by the occupations and industries that are open to them, how they build on these initial economic niches, and what happens to the industries in which they become concentrated. Groups that can gain entry to higher positions in growing industries, and that can improve their location across generations, do the best, while those that face blocked opportunities or end up clustered in declining industries tend to do the worst. The reverse may also be true: the resources, talents, creativity, and social organization of the labor force working in a particular industry can often be crucial to whether or not it prospers.

Cities may be seen as a container or vessel as well as a pump, as the great urbanist Lewis Mumford often observed. Indeed, considering that the forces contained within cities interact forcefully and sometimes even violently with each other, it must be a vessel with strong walls. The state provides the principal institutional form for this container. In other words, the agencies and powers of government not only provide the public services that are crucial to the functioning of a dynamic and intense urban society and economy, but they regulate the rhythms of public and private life to keep them within bounds and prevent breakdown.

It has already been observed that city and citizenship have common roots. From the decline of the Roman Empire to the rise of the modern nation state, cities as distinct political corporations bestowed rights on their inhabitants and provided for some degree of self-government, and indeed gave rise to the earliest forms of modern democracy. Yet two trends have compromised the autonomy of cities to govern their own affairs in the modern period.

One was the rise of the modern nation state. Beginning in the seventeenth century and culminating in the twentieth century, cities everywhere became increasingly subordinate to their national political and governmental settings. Although regional variation and federalism persist in many nations, nation states successfully asserted their prerogatives over sections, regions, and cities. Even in the highly decentralized federal system of the U.S., cities were clearly the legal creatures of the states. The other trend was that the relatively dense nineteenth-century industrial city exploded far beyond its initial political boundaries during the twentieth century. Given that city governments often could not annex the jurisdictions surrounding them, suburbs, new cities, various new forms of metropolitan regionalism. Seeking to distill the essence of this new urban form, geographer Jean Gottman coined the term "megalopolis" and defined it as a "multi-nucleated urban realm."

Clearly, as the urban political autonomy was subsumed under the nation state and myriad independent jurisdictions fragmented the megalopolitan realm, it diminished the capacity of any individual city government to regulate or reshape the forces interacting within their boundaries. (Indeed, the next section takes up the question of just how much capacity urban governments have.) At the same time, however, some systemic forces may be having the opposite effect. Indeed, some argue that globalization and internationalization, by weakening nation states, have revitalized cities – especially the largest cities with the most important economic functions – as powerful actors on their own.

The readings in Part 1 provide an overview of all these trends. The reading by Wyly, Glickman, and Lahr provides a broad overview of how demographic and economic trends have reshaped the regional and metropolitan terrain in the United States in recent decades. Many were set in motion by basic changes in the global economy, including increased competition from developing countries, international migration, and trade policy. These are forces with which, like it or not, every city must contend and none can fundamentally alter.

Given that the institutional patterns, domestic policies, and political dynamics of the nation state also have a tremendous impact on cities, Margaret Weir's reading explores why social welfare policy is so weak and fragmented in the U.S. compared to Europe. The U.S. stands at one end of a global political spectrum – it grants the most responsibility for domestic policies to states and most encourages competition among local jurisdictions. This, she points out, has pervasive consequences for what urban governments can and will do.

Excerpts from Saskia Sassen's seminal book *The Global City* outline some of the forces operating at a global scale that amplify some of the themes set forth in the first reading. In particular, Sassen argues that the redistribution of industrial production away from the central cities of the advanced world, combined with their continued role as the primary capital markets and corporate headquarters locations, has produced a characteristic pattern of polarization in global cities like New York, London, and Tokyo. This of course would have enormous consequences for urban governance.

Just as Weir underscores the importance of national and local political institutions for explaining patterns of social welfare policies across national settings, however, geographer Chris Hamnett offers a critique of Sassen's analysis that questions both the degree to which polarization has taken place in the large central cities of the leading economies and the extent to which trends in the global economy, as opposed to national social welfare policies and labor market regulation, shape urban inequalities.

Taken together, these four readings show how the interplay among economic, demographic, and political dynamics and institutions at the global, national, and local levels have enormous local consequences. It may be a major challenge to undertake this type of analysis, but all four authors would agree that we are unlikely to understand the nature of urban politics anywhere in the world today unless we rise to meet it.

"A Top 10 List of Things To Know About American Cities"

from *Cityscape* (1998)

Elvin K. Wyly, Norman J. Glickman, and
Michael L. Lahr

Editors' Introduction

Any discussion of urban politics and policy must take account of the economic and demographic trends affecting cities. "A Top 10 List of Things To Know About American Cities" provides this context for the U.S. case. This chapter was written when the three authors were affiliated with the Center for Urban Policy Research at Rutgers University, where the U.S. Department of Housing and Urban Development asked them to devise a user-friendly database of urban economic and demographic statistics, the *State of the Nation's Cities* (SONC). They use these data to identify the key trends shaping American cities in the 1990s. Although their analysis was not able to draw on the 2000 Census, they describe trends that continue to dominate U.S. urban development. As intensified international trade and migration and new labor-saving technologies reshape the global economy, they also reshape cities.

So what does their analysis tell us about the "state of the nation's cities"? It depends. Some urban areas are fairing much better than others. The Sunbelt, for example, is steadily gaining population and jobs (they note that the fastest growing metropolitan areas are all in Arizona, California, Florida, Nevada, and Texas); but many Northeastern and Midwestern urban economies and populations faced stagnation or decline. (After this article was written, the 2000 Census did show that some "Frostbelt cities" had experienced a surprising rebound, with Chicago's population increasing 4 percent and New York City's increasing 9.4 percent over the 1990s.) Generally, cities specializing in producer services and high technology have fared better than where manufacturing predominates.

Within metropolitan areas, suburbs continue to gain jobs and population. Metropolitan areas remain racially and economically segregated, with central cities home to recent immigrants and racial minorities. Interestingly, the number of African-American, Latino, and Asian families living in suburbs has increased, in some cases dramatically, during the 1990s, suggesting at first glance that long-established patterns of racial segregation may be breaking down. Closer scrutiny, however, reveals that much of this statistical integration reflects the economic decline and population shifts of older industrial suburbs. People of color, most notably African-Americans, continue to live in segregated communities, but those communities are, today, more likely to be located outside of the city limits. Cities also have higher poverty and jobless rates, leading to a host of social problems. The uneven dispersal of good jobs and middle-class residents also contributes to chronic fiscal distress in many cities, as tax revenues fail to keep pace with service demands.

This study's findings raise a host of vital policy questions that will be taken up by different authors throughout this book. First, we ask whether the trends producing inequalities between and within metropolitan areas are subject to influence by policy makers. When national policy makers debate the desirability of free trade policies, such as the North American Free Trade Agreement (NAFTA), they are in part asking whether greater protection against global economic forces could shelter some of our industrial areas, including older cities, from the ravages of economic restructuring.

Others may argue that public policy should concern itself more with ameliorating the impacts of these economic forces. There are numerous ways in which the political arena could address the problems accruing from growing spatial inequalities, from supporting a more robust welfare state, to reducing the political fragmentation within metropolitan areas to allow for more comprehensive, regional approaches to economic development.

The State of the Nation's Cities database (http://socds.huduser.org/) continues to be expanded and updated; students can find a wealth of information on U.S. cities contained within it. Most of these data come from the U.S. Census Bureau, from which detailed information on cities and neighborhoods may be obtained. You can explore their offerings at www.census.gov. If you are particularly interested in exploring patterns of racial segregation, the Racial Residential Segregation Measurement Project of the University of Michigan's Population Studies Center offers online databases with segregation indices for US metropolitan areas (http://enceladus.isr.umich.edu/race/racestart.asp).

Elvin K. Wyly, Assistant Professor of Geography at the University of British Columbia, writes about urban housing and labor market issues, including spatial mismatch, mortgage lending, gentrification, and racial and gender segmentation of occupations. Norman J. Glickman, University Professor at Rutgers and former director of the Center for Urban Policy Research at Rutgers University, is an economist who has written more than 100 articles on issues ranging from urban economic development and econometric analysis to the evaluation of community development corporations. Michael L. Lahr, Assistant Research Professor at the Center for Urban Policy Research, is a regional economist and Associate Director of CUPR's Rutgers Economic Advisory Service (R/ECON™), where he measures the economic impacts of events, operations, and construction projects. He has also studied urban stress and measured the cost of doing business and living across U.S. metropolitan areas.

Henderson, Nevada, is not well known to urbanists, but it has the distinction of being the fastest growing city in America: Its population ballooned by fully 88 percent from 1990 to 1996. This desert suburb and the other nine fastest growing cities reported by the Census Bureau shared several characteristics:

- They were relatively small – all but one had a 1990 population of less than 200,000.
- Several were suburbs of larger urban centers.
- Several were cities close to the Mexican border.

All were in five Sunbelt States – Arizona, California, Florida, Nevada, and Texas. The slow-growing end of the census bureau list was dominated by large Frostbelt/Rustbelt cities: St. Louis; Washington, D.C.; Baltimore; and Philadelphia. St. Louis placed at the bottom of the list with a more than 11 percent population loss, continuing its decades-long tumble from nearly 900,000 people in 1950 to about 350,000 today.

Although the census list for the 1990s provides the latest snapshot of how our cities are developing, it also highlights well-known trends of the past quarter century. Inter-regionally, growth occurred in the South and West while the Frostbelt experienced a decline – especially in large cities.

Those areas that grew had well-educated work forces, high-technology industry, lower-than-average wages, a good quality of life, proximity to universities, and agglomeration economies springing up from proximity to other high-technology firms. Within our metropolitan areas (MAs) a seemingly endless decentralization from central cities to the suburbs and exurbs continued.

.... This article use[s] the Center for Urban Policy Research's (CUPR's) comprehensive state of the Nation's cities database on 77 large American cities and 74 MAs (SONC) developed for the U.S. Department of Housing and Urban Development to produce an overview of urban change since 1970. We concentrate on what we believe to be the most important trends in understanding the cities' economies and national urban policy.

Our top ten list is constructed around two main phenomena resulting from recent changes in the

Nation's economy: the uneven growth of MAs and the ever-increasing imbalances within regions.

1. TECHNOLOGICAL CHANGE, ECONOMIC INTERNATIONALIZATION, AND DEMOGRAPHIC TRENDS

The economic, social, and technological context within which cities grow or decline is the starting point for our top ten list. Advances in technology, changes in the international economy, relative shifts of employment and production among the Nation's industries and a variety of other demographic and social transformations affect urban growth. In response to these broad forces of economic and social history, firms and families relocated from city to city, from city to suburb, and from region to region.

Technological change

In the mid-1970s, the introduction and spread of several new technologies altered the nature and organization of work and, hence, the ways that cities function. In particular, advances in computers and telecommunications – the speed of information processing made it possible for firms to produce and market goods and services practically anywhere on the globe. In addition, new production processes could be quickly introduced worldwide. Information technologies, in particular, permitted the spread and separation of functions. As a result, management, design, production, research, and the other parts of the production process were freed from traditional spatial constraints, particularly from their ties to the central business districts (CBDs) of big cities. Consequently, corporations gained far more freedom to relocate jobs to areas of lowest cost and maximum market potential. Internationalization, facilitated by new production technologies, thus changed the face of work in, and the primary functions of, U.S. cities.

Information technologies facilitated intra-metropolitan change too, reinforcing other elements that promoted suburbanization. These technologies allowed firms to decentralize operations within metropolitan areas as well as across regions and around the world. Corporations no longer needed to locate many of their clerical functions near the head office. For instance, information technologies have allowed many back office functions associated with the rapid expansion of security sales and other activities on Wall Street to relocate to the suburbs. This phenomenon has been replicated in most large cities and in other industries. Thus the vertical disintegration of large firms has not only propelled a rise in producer services but also industrial suburbanization. Growing firms in high-technology industries also tend to locate in suburbs.

International and national economic forces

Technological change eased the way for information and capital to transcend international borders. But it was just one of a myriad of factors that helped to disperse economic activity around the globe and profoundly alter the nature of work and urban development. From 1970 to [1996], international trade doubled its share of the Nation's gross domestic product. At the same time, foreign direct investment swelled enormously. Foreign investors and traders continued to come to this country to tap its $7 trillion market for goods and services, its technology, and its skilled labor. Similarly, U.S. firms went abroad in search of new markets and, in some industries, low-wage workers to reduce production costs. As a result, Americans increasingly worked for multinational corporations (and large, multi-site U.S.-based companies) that had fewer ties to localities than did home grown companies. Thus communities had less economic security, as plants and offices opened and closed more rapidly than previously.

In particular, economic internationalization brought about the process of deindustrialization, the widespread effects of which rippled throughout the economy. Deindustrialization of the U.S. economy resulted in two opposing trends. One was the dispersion of manufacturing and many services to medium-size cities, rural areas, and other countries. The other trend was an increasing spatial centralization of two types of industries: selected service activities oriented toward businesses and manufacturing industries that depend on rapidly evolving technologies.

These trends resulted in the declining share of wealth produced through manufacturing. The descent of manufacturing resulted from many factors, including the increased international competition that reduced the national demand for U.S. workers. In 1970 manufacturing employment boasted 19.4 million workers – 27 percent of all non-farm employment. By 1997 only 15 percent of Americans (18.6 million workers) had jobs in manu-facturing. As a result, cities formerly dependent on manufacturing – many in the Northeast and Midwest – suffered debilitating employment losses from the decline in this once robust sector.

Simultaneously, service industries grew rapidly, from 67 to 80 percent of nonfarm employment. Services include a broad array of jobs, primarily in retail trade and personal services firms, which tend to pay low wages and offer few advance-ment prospects. An increasing proportion of the service industry's employment is part-time with few, if any, employee benefits. Further, these consumer-oriented service jobs continue to move to the suburbs to meet the needs of households with high disposable incomes.

Although the splintering of old corporate struc-tures into more flexible arrangements has threatened job security in many areas, some central cities have prospered. In particular, those with a suitable mix of skilled workers, entrepreneurial capital, and inter-firm linkages have been able to profit from the surge in certain business-oriented services. In particular, many firms in large cities at the top of the Nation's urban hierarchy suddenly needed the administra-tive and financial services required to manage their expanding global trade networks. These kinds of service jobs are what Robert Reich called "symbolic-analytic services" and are a far cry from the jobs that dominate consumer-oriented services. These "pro-ducer service industries" include problem-solving and strategic brokering activities and employ a high number of research scientists, design engineers, bio-technical engineers, management information specialists, public relations executives, and other technologists and scientists. These jobs are high paying and compete with foreign producers based in the United States and abroad.

The business districts of many cities with high concentrations of producer services have been grow-ing quickly, leading the Nation's transformation to a service-oriented economy. Important centers for advanced corporate and producer services include the large banking and corporate headquarters cities – New York, Chicago, and San Francisco. These cities gained employment in producer services while losing jobs in manufacturing. At the same time, grow-ing clusters of producer services emerged in office parks throughout U.S. suburbs. As a consequence, the growth of services generally helped the suburbs as much as, if not more than, it did central cities.

Key demographic forces

Economic changes over the past quarter century were accompanied by dramatic demographic shifts. Nationally, the white population share declined from 88 to 83 percent between 1970 and 1995, whereas the Hispanic share grew to 10.3 percent by 1995. Family and household arrangements have changed radically since 1970: The proportion of the adult population that is married declined from 72 to 61 percent between 1970 and 1995, while the pro-portion of both divorced and single adults tripled (to 9.2 percent). Households conforming to the "traditional" norm – married-couple families with children present – now account for only one-fifth of all households.

The share of the population that is foreign born is currently at its highest level since World War II due to a resurgence in immigration during the 1980s. In addition, the rise in female labor-force par-ticipation – begun in the 1960s – continues apace, reflecting changes in cultural attitudes as well as a response to declining real earnings per worker.

[. . .]

Many of these differences affected the relative demographic composition of cities and their sub-urbs. Immigrants tended to select the Nation's largest central cities as places to live and work. In the general population, married-couple families with children made up 28 percent of suburban house-holds but only about 20 percent of all households in the central cities. Additionally, although the share of all households headed by single parents rose in both cities and their suburbs, the suburban rises tended to be somewhat smaller on lesser starting shares. Hence, it is clear that a disproportionate burden of demographic shifts is being shouldered by the Nation's larger cities.

[. . .]

It is due to these recent phenomena – the resurgence in the immigration of low-skilled workers and a rapid rise in part-time retail trade jobs – that recent national economic expansions have proved less able to shrink the number of poor people than they had historically. This heralds a change in the basic workings of the economy.

Income distribution and poverty

The persistence of poverty remains a difficult problem for many people in our cities. From 1960 to 1973, the rate of poverty fell by nearly one-half, continuing a decline that had begun with the birth of the New Deal during the Great Depression. Then, a substantial turnaround occurred: Urban poverty rates rose sharply in the early 1980s and again in the early 1990s when national recessions occurred. Indeed, from the 1981–82 recession to the present, the poverty rate has fallen only slightly. In particular, cities with significant concentrations of poverty improved little, if at all, even during sustained periods of economic growth. The inability of policymakers to reduce poverty substantially even in flush economic times is perplexing to economists and society as a whole.

Particular groups, most notably children, suffered the greatest amounts of poverty: 23 percent of all children and 46 percent of black children lived poverty in 1993. Blacks (33 percent) and Hispanics (31 percent) experienced far higher incidences of poverty than whites (10 percent). Importantly, poverty remained heavily concentrated in central cities – 43 percent of all the poor in 1993 – especially large industrial cities, and in some neighborhoods. Finally, the "safety net" of social supports, taxes, and transfers – Aid to Families with Dependent Children, food stamps, the Earned Income Tax Credit, and others – proved less able to keep families out of poverty in the 1980s and 1990s. Many analysts believe that the 1996 changes to Federal welfare laws will likely make life more difficult for poor families, especially those who cannot get from their inner-city housing to suburban job locations.

Increasing income inequality partially reflects demographic changes. From 1970 to 1994, the dual-earning, married-couple family was the only household category to gain income (22 percent in real dollars), while all others lost ground. Where these families choose to live influences the nature of fiscal disparities among different units of local government in sprawling fragmented metropolitan areas. Cities losing middle-class families are left with more households in poverty; more single-parent families; declining property values; and under-funded, deteriorating infrastructure and schools. By contrast, suburbs gaining upper-income families enjoy low poverty rates, rising property values, and well-funded infrastructure and schools. In addition, continued decentralization has widened disparities from suburbs, from affluent municipalities with lucrative tax bases of commercial and industrial development to poorer, aging municipalities with static or declining tax bases.

Nearly three decades of rapid income growth and narrowing income differentials ended in the 1970s. Then, incomes not only grew more slowly, but also became more unequally distributed. The dimensions of the increase in interpersonal inequality in the 1970s and 1980s have been substantial. . . . The increasingly suburbanized middle class declined as the proportion of both lower income and upper income families increased! Unfortunately, our major cities experienced the brunt of the Nation's increasingly unequal distribution of income and the slow growth of personal income due to the continued suburban flight of middle-class families and jobs.

2. UNEVEN METROPOLITAN GROWTH

. . . [Metropolitan Areas] have grown steadily in population and spatially over the last quarter century. This growth, especially in central cities, has been highly uneven geographically. Population growth in MAs in the South (2.3 percent annually) and West (2.6 percent) far outpaced that in the Northeast (no growth) and Midwest (0.3 percent) from 1970 to 1996. Cities with big service bases, State capitals, and high-technology and producer services industries were among the winners. The losers were the usual suspects: mostly large cities in the old industrial North and Midwest. Employment in central cities of the Northeast averaged an increase of 0.6 percent annually from 1969 to 1995, compared with much more rapid growth in the South and West (2.8 percent per year).

[. . .]

Large cities and urban distress

Throughout the past twenty-five years, many large cities have confronted serious social, economic, fiscal, and environmental conditions. Although these conditions are not uniquely urban, their concentration in urban areas has induced further problems for the cities and their residents, propelling out-migration of population and jobs. To understand urban change better, we used twenty urban stress indicators of 1990 to construct a single index. We then applied this index across the seventy-seven cities in the SONC database to rank areas by their levels of distress. Income inequality, pollution, housing affordability, and fiscal problems contribute heavily to the distress of older cities. Detroit and Newark were the most distressed cities in 1990, whereas Seattle and Sioux Falls were the least distressed.

[. . .]

3. INCREASED INCOME INEQUALITY AMONG METROPOLITAN AREAS

For the past thirty years, secular changes in employment structure, technology, and international trade have widened income inequality among individuals as well as regions. Interregionally, the ratio of average income earned by the top one-tenth of central-city families to that earned by the bottom one-tenth widened in seventy of seventy-seven cities from 1969 to 1989. In 1969 families in the richest tenth earned more than ten times that of the poorest in only one city (New Orleans). By 1979 this ratio appeared in nine cities, and by the end of the 1980s, nearly one-half of the SONC cities (32) exceeded this threshold. . . .

Historically, inequality has been most pronounced in older cities and those in the South with the highest ratios appearing in Atlanta, New Orleans, Cincinnati, and Washington, D.C. The uneven growth of income polarization across cities in the 1970s and 1980s, however, highlights the effects of nationwide industrial shifts. Although inequality moderated in several small or distinctive settings, it worsened considerably in older goods-producing centers as high-paying industrial jobs vanished and middle-class families moved to the suburbs or beyond. Rapid growth in the service sector magnifies inequality, increasing the process of bifurcation in the job market between well-paying, white-collar positions and poorly paid, insecure, or temporary jobs with few or no benefits.

4. EFFECTS OF INTERNATIONAL MIGRATION

New immigrants have always helped mold the character of our cities. Yet after the elimination of national origin quotas in the 1960s, immigration levels increased dramatically and the source countries of arrivals changed. Thirteen million legal immigrants entered the United States from 1981 to 1996 – more than 80 percent of them from Asian and Latin American countries. This immigration is in addition to the 4.5 million legal immigrants entering from 1970 to 1980. As a result, the proportion of the foreign-born U.S. population (8.8 percent in 1995) stands at its highest since World War II.

The concentration of immigrants in five gateway cities is striking: In the 1980s nearly one-half of all international migrants to cities in the SONC database went to Chicago, Los Angeles, Miami, New York, and Washington, D.C. By the end of the decade, at least 5 percent of all residents in thirteen SONC central cities had immigrated from abroad during the previous five years. Moreover, a majority of central-city Miami and Santa Ana residents had been born abroad. Fully one-half of all immigrants in 1995 settled in California, New York, and Florida. What is more striking is that one out of six settled in the New York metropolitan area while one of twelve went to Los Angeles. This destination selectivity is likely to continue for the remainder of this decade since the family reunification and immediate relative provisions of the U.S. immigration code account for 60 percent of migrants admitted each year.

Immigration policy is commonly discussed as a national issue, yet the most dramatic consequences – the revival of inner-city neighborhoods and the formation of vibrant enclave economies – are uniquely urban. On balance, the benefits of immigration for the U.S. economy – for instance, immigrants' employment in sectors that pay low wages, a reality that helps the Nation remain competitive internationally – appear to be distributed throughout the Nation through interstate trade in goods and services. The costs, however, are borne primarily at the local level. Where immigrants cluster, schools must provide multilingual curricula, while police, emergency services, and government agencies require translators and related services. Moreover, the arrival of large numbers of low-income immigrants widens inequality in gateway cities. Whether this disparity will narrow as immigrants assimilate into the U.S. economy is uncertain because the economic forces described earlier have drastically reduced the need for relatively well-paying, low-skilled labor. . . .

5. DECENTRALIZATION OF METROPOLITAN AREAS

MAs continued to spread out, furthering trends set in place at the beginning of the twentieth century. . . . While four of five Americans now live in MAs, the share of central city dwellers declined rapidly, from 43.0 percent in 1970 to 33.8 percent in 1994. In the 1970s, eleven SONC central cities registered absolute job losses. In the 1980s, nineteen cities experienced job declines. By contrast, suburbs in nearly all SONC MAs gained jobs during this period. While the dispersal of manufacturing and retail functions has been under way for most of this century, the decentralization of sectors traditionally tied to the agglomeration economies of CBDs is a phenomenon unique to the last two decades.

As a result, suburban growth in producer services – particularly finance, insurance, and real estate – has overshadowed that of [many] central cities. . . . These trends reflect the maturation of suburban employment and the potential emergence of agglomeration economies at the urban fringe – variously dubbed suburban downtowns, urban subcenters, or edge cities. These areas enjoy superior access to new infrastructure and highway interchanges, relatively inexpensive land, less-restrictive land-use controls, reduced taxes, and other incentives. As residential growth continues to expand at ever-lower densities, however, these suburban job centers have endured worsening traffic congestion. Mean commuting times lengthened by more than 20 percent in the 1980s for suburbanites in New York and Chicago. One-third of SONC MAs now suffer from system-wide traffic congestion, up

from only about one-seventh in 1982. Nationally, however, average travel times remained stable despite the longer distances and greater congestion of suburb-to-suburb commuting. The continuing shift to the private automobile is a main factor in making this possible, but the movement of population and jobs to smaller cities also plays a role. Even in central cities, the convenience and speed of solo commuting is reducing carpooling and transit ridership.

[. . .]

6. INTERREGIONAL MIGRATION AND CONTINUED SUBURBAN FLIGHT

That Americans continually seek new places to live and work is a widely held perception. Indeed, the fluidity of home locations is a main societal attribute that has enabled changes in the Nation's urban network as well as in the internal structure of urban areas. Each year, three of ten renters and one of ten homeowners move their place of residence. Similarly, many workers leave their current jobs (or are laid off) to search for new opportunities elsewhere.

Across the Nation's urban system, migration currents continue to favor the South and West. From 1970 to 1994, the Northeast experienced a net outflow of 6.7 million people. The Midwest lost 3.8 million people (mostly in the 1970s and early 1980s), and net inmigration flowed into the West (3.3 million) and the South (7.2 million). These interregional shifts have been under way for much of the postwar period, however, as evidenced by SONC migration data. Southern and Western MAs consistently have higher proportions of residents who have moved from out of State. In each decade since the 1960s, for example, at least one-fifth of all metropolitan residents of Anchorage; Cheyenne; Las Vegas; Phoenix; Tampa; Virginia Beach; and Washington, D.C., were out-of-State newcomers. In contrast, older industrial centers are now bypassed by interregional migration streams: In 1990 fewer than one in twenty metropolitan residents of Buffalo, Cleveland, Detroit, New York City, and Pittsburgh were from out of State. Interregional migration streams, however, are diverging for different population groups: The Nation's gateways for international immigration continue to grow more racially and ethnically diverse, while older retiring whites and less-skilled native-born workers gravitate to more homogeneous centers in the U.S. heartland.

Within MAs, the flight outward to the suburbs continues. From 1985 to 1990, the 77 SONC cities saw a net loss of 5.4 million people to their suburbs, corresponding to a 13 percent loss of the population. For every suburbanite moving to the central city, three city residents had left for the suburbs.

7. INCOME INEQUALITY AND SOCIAL POLARIZATION WITHIN CITIES

Decentralization of MAs and continued middle-class flight to the suburbs interact to widen differences in economic opportunity and demographic composition between different parts of MAs. Central-city neighborhoods became more racially and ethnically diverse and poor as suitable employment shifted to the suburbs and as nearby office districts provided professional jobs for affluent suburban commuters.

As a result of the forces that we have discussed, the ratio of city-to-suburban poverty rates widened from 1.9 to 2.3 for the SONC MAs. By 1990 suburban per capita income exceeded city levels in all but one-tenth of the SONC MSAs, and these exceptional MSAs contained relatively new, actively annexing central cities of the South and West that were surrounded by largely rural suburbs. In the Nation's 100 largest MAs, the dissimilarity index for the poor population, which measures the proportion of the population that would have to move to achieve an integrated spatial distribution, increased from thirty-three in 1970 to thirty-six in 1990.

In aging industrial MAs, segregation of the poor is substantially increasing. MSAs posting increases of at least 10 points in the dissimilarity index from 1970 to 1990 include Allentown, Pennsylvania; Buffalo; Detroit; Hartford; Milwaukee; Orange County, California (SONC's Santa Ana MSA); Springfield, Massachusetts; and Syracuse, New York. Increasing income inequality is also evident within central cities. . . . [O]ur inequality index registers the widest disparities in family income in Atlanta, Cincinnati, Hartford, New Orleans, New York, and Washington, D.C.

8. CHANGING CONDITIONS OF INNER-CITY NEIGHBORHOODS

Polarization within MAs is even more pronounced at the neighborhood level. The processes leading upwardly mobile families to the suburbs leave those who remain behind in the cities to face the consequences of concentrated poverty, high crime rates, deteriorating home values, and poor schools. In many instances, these areas of poverty and disinvestment are in the shadow of center-city office districts that make the wealth of the suburbs possible and near high-income residential enclaves close to downtown. The ghettos and barrios are also within commuting distance of suburban jobs for those people with cars although many of the inner-city poor do not have cars. Thus the poor in these areas are isolated from job opportunities that have moved outward.

In 1990 there were about 3,000 neighborhoods with concentrated poverty – those with greater than 40 percent poverty rates. Although blacks dominate two-fifths of these areas, others are either heavily Hispanic (17 percent of these poverty neighborhoods), white (14 percent), or mixed (27 percent). Poverty continues to spread: From 1970 to 1990, the number of census tracts with concentrated poverty and the number of people living in such neighborhoods doubled; the number of Hispanics living in barrios tripled.

In addition, long-term suburbanization of employment continued during the 1990s, widening the spatial mismatch between affordable housing and employment opportunities. In ten large cities from 1990 to 1994, the ratio of city-to-suburban manufacturing jobs declined in 6 MAs, while the ratio for retail employment fell in nine.

[. . .]

Racial segregation remains as deleterious as it was in the 1960s. To achieve a perfectly integrated distribution of blacks and whites in 1990, more than three-fourths of the residents of Atlanta; Baltimore; Chicago; Cleveland; Jackson; Louisville; Miami; Newark; New York City; Philadelphia; St. Louis; and Washington, D.C., would have had to move to a different neighborhood. More ominously, racial segregation interacts with income segregation – both of which are deeply embedded in urban housing markets, given the concentration of affordable housing – to intensify the social polarization described earlier. The magnitude of forces and incentives that encourage people, jobs, and wealth to leave these inner-city neighborhoods is staggering. It includes such seemingly benign policies as the deductibility of home mortgage interest costs, the local financing of school districts, and marginal cost pricing of suburban infrastructure that encourages de facto exclusionary suburban development. The difficulty of solving problems in the inner city with limited resources has thus created an ongoing tension between arguments for dispersal – which reproduces the condition of disinvestment unless done on a massive scale – and those for redevelopment – which has been dubbed gilding the ghetto. This is an ongoing policy and theoretical debate. Trends in the concentration of poverty suggest that it will become more crucial in the remainder of the decade: Even as national poverty rates declined in the recovery of the mid-1990s, the ratio of city to suburban poverty widened from 1.9 to 2.3.

Now that we have traced the effects of national demographic and industrial shifts on income inequality within and across regions, we focus on tracking their effects on the ability of cities to shelter their inhabitants and to effect economic development. We focus on the effects of income inequality as it relates to the fiscal prospects of cities and the localities' ability to provide adequate housing for their poor.

9. WHILE HOUSING QUALITY EDGED UP, AFFORDABILITY FELL

An increase in homeownership is improving housing quality nationwide. After bottoming out at just below 63.9 percent in 1988, the homeownership rate has increased, climbing to 66.0 percent in 1996, a rate that surpassed the previous all-time record set in 1981. Ownership provides an inherent incentive for the rehabilitation, repair, and expansion of homes. In addition, a large share of the new housing that is being added to the existing stock is in single-family units, further improving the amount of housing available per capita – a measure of housing quality. The return to the long-term trend of increasing homeownership reflects two main factors: strong baby-boom demand for home-ownership and a recent improvement in housing affordability due to moderate home mortgage rates,

stable home prices, and greater credit availability predicated on strong economic growth prospects. Since 1990, homeownership rates have increased in suburbs outside the Northeast. They have risen rapidly even in many older cities, such as Boston, Buffalo, Chicago, Cleveland, Detroit, New Orleans, San Francisco, and St. Louis. Since these cities have traditionally had relatively low homeownership rates, their increasing rates may simply be part of a nationwide process of regions converging toward the national average rate.

[. . .]

10. CITY FISCAL CONDITIONS TENUOUS – OUTLOOK, BLEAK

The relative financial condition of cities has not improved over the past twenty-five years. . . . The revenue-raising capacity of cities continues to decline relative to that of their suburban counterparts, as the suburb–city gap in household income noted earlier continues to rise. The New Federalism, which leaves all cities experiencing the effects of more stringent Federal and State fiscal restraint, is forcing urban areas to rely heavily on their own funding sources, which are already strained close to – or even beyond – their revenue-raising capacity.

SONC data reveal that in 1972 smaller, newer cities, such as Charlotte, Santa Ana, and Virginia Beach, were in the best fiscal shape and that larger, older cities, such as Detroit, New Orleans, and New York, were in the worst shape. Ten years later, the financial capacity of 70 major U.S. cities had declined by 10.7 percent on average. In 1982, the average standard expenditure need of cities with more than one million inhabitants was 62 percent higher than their average revenue-raising capacity. Finances of cities such as Baltimore, Cleveland, Detroit, and Newark declined particularly rapidly, [a]lthough financial conditions improved by 1990 for most cities because of a long economic upturn. . . . Although city finances nationwide, including New York City's, have taken a positive turn since 1990, their long-run financial state remains precarious.

[. . .]

The flight of businesses and middle-class households from cities is inducing an economic, social, and fiscal self-aggravating downward spiral. Because poorer households have a greater need for public services, city costs continue to rise on a per capita basis. Unfortunately, the Federal Government has pulled back its direct assistance, and States have not filled the gap. Consequently, local governments often have resorted to increasing property tax rates to make up the shortfall. Partially because of this increasing local burden, the most lucrative portion of the urban tax base is escaping to the suburbs and beyond, giving cities fewer own-source revenues. Hence, although MAs are likely in their best financial shape ever, the flows of revenues to major central cities are more subject to the gyrations of their local economies than they have been in one-half of a century.... Given the lack of potential changes in the growth process and in intergovernmental revenue sharing, the financial prospects of our Nation's cities are bleak. Indeed, when the next recession occurs, cities are likely to find themselves in even worse condition than they had experienced in the midst of the 1990–91 trough.

CONCLUSION

We have highlighted some long-term trends in urban areas. U.S. cities and suburbs have undergone dramatic economic changes during the past quarter century. The transformation of cities has been both the cause and the effect of where people live, the industries in which they work, and the incomes they earn. In this transformation, the forms and functions of cities have been altered, and relations among cities and among neighborhoods have been drastically reshaped.

Cities are the heart of the Nation's economic and social life. They are dynamic and vital concentrations of economic activity and innovation where firms and households make and trade goods, money, and information. They are centers for education, entertainment, cultural enrichment, and artistic creativity. Yet, in the past quarter century, neighborhoods, workers, and families in U.S. cities have endured increasing distress. Trends in global

and national economic restructuring begun in the late 1960s have led to a deepening of inequality in income, wealth, and opportunity. The continued flight of the middle classes to the suburbs has left cities with increasing concentrations of unemployment, poverty, and social dislocation. The selective revival of financial districts and downtown office construction during the 1980s improved the balance sheet of some cities, but most of the long-term benefits went to white-collar workers living in affluent suburbs where most new economic activity continues to gravitate. Taken together, all of these trends have led to a deepening social and spatial polarization of large cities.

These developments carry important implications for national urban policy. U.S. competitiveness remains, as it has been for much of this century, based on the flexibility and adaptability to change on the part of people and businesses in the Nation's cities. Industrial change and the dynamic flow of jobs, people, and wealth have always widened inequality in some places and disadvantaged some cities while others met the challenges of new economic realities and rose to prominence. For much of the post-World War II period, however, prospects for long-term mobility remained optimistic for most groups, and there was general consensus that improving the Nation's development required a strong national commitment to counterbalancing the costs of industrial change in different regions and cities. Although the record of urban policy was certainly mixed, debate focused mainly on the means, not the ends. But in the early 1980s, this commitment was abandoned. Along with the polarization of the world economy and national economic growth has come a division in mobility opportunity for workers and families; and with the devolution of Federal authority to State and local governments, both public and private sectors are competing among themselves and throughout the world. Reconciling the costs and benefits of this dramatic experiment in the geographical structure of urban and regional policy is a challenge to policymakers, cities, and scholars.

"Poverty, Social Rights, and the Politics of Place in the United States"

from Stephan Leibfreid and Paul Pierson (eds), *European Social Policy: Between Fragmentation and Integration* (1995)

Margaret Weir

Editors' Introduction

Margaret Weir's essay comes from a book on social policies in the European Union. It thus examines urban policies in the U.S. through an (implied) comparative lens. She asks, in essence, why the U.S. has failed to develop the comprehensive social safety net found in most other industrialized democracies. Her answer is that the U.S. federal system promotes a great deal of localism – at times what she calls "defensive localism" – that shapes our institutional structures and political practices and makes it difficult to mobilize mass publics behind national entitlement programs.

The New Deal has long been considered the beginning of a new, more centralized era of social policy making in the U.S. Weir notes, however, that this centralization has been uneven. Law and tradition leave key areas, including housing policy, land use, and many aspects of social welfare, to state and local governments. These governments are less able and willing to undertake such efforts than are national governments. As Paul Peterson notes in Part 3, competition among local governments leaves them loath to engage in redistributive policy. Local governments are left with little incentive to cooperate to address regional concerns, while the national government lacks the constitutional authority or the political will to intervene.

Racial antagonisms have exacerbated – and been fueled by – this "defensive localism." Fearing racial and economic integration, suburban communities have defended local control, and for the most part have successfully resisted regional solutions to housing or educational problems. Weir argues that the suburban-urban divide, which has spatial, economic, and racial dimensions, has become the key schism in U.S. politics, surpassing the north–south divide in political significance.

To the extent that US policy makers have addressed social welfare concerns (at least since the 1960s), they have opted for place-based aid, rather than broadly based entitlement policies. This preference again may be attributed to the political salience of localism. Place-based policies can be carried out with little challenge to local political elites, and they satisfy the needs of district-based Congressional representatives to be able to claim credit for benefits flowing to their districts. They have been less successful, argues Weir, in actually alleviating poverty. Moreover, such programs have been susceptible to changes in the political wind. When Ronald Reagan was elected President on a platform of reducing domestic spending, he was most successful in eliminating programs targeted to poor and urban areas. Broad-based entitlements, however, proved far less vulnerable.

The history of the U.S. federal system can offer some insight – and perhaps some causes for concern – to those considering the long-term impact of the European Union on European social policy. As the EU becomes more economically integrated, will similar forces of "defensive localism" serve to undermine programs targeted to poor people and poor regions? Even if European federalism comes to resemble U.S. federalism, Weir believes, there will still be important differences. "Defensive localism" in the U.S. has been shaped by a history of racial antagonism and place-based conflict that is not replicated in Europe. "In the US the possibilities for isolating the poor politically and spatially offered a way to fragment the polity that severely handicapped proponents of a more expansive social policy," she concludes.

Margaret Weir is Professor of Sociology and Political Science at the University of California, Berkeley, and a nonresident Senior Fellow at the Brookings Institution. Her work has focused on U.S. social policies, with a particular interest in how these are shaped by the system of intergovernmental relations. Her numerous books and articles on education, welfare, and urban policies include *The Social Divide* (Washington, DC: Brookings and Russell Sage, 1998).

Works on federalism in the U.S. cited by Weir include: John E. Chubb, "Federalism and the Bias for Centralization," in John E. Chubb and Paul E. Peterson, (eds), *The New Direction in American Politics* (Washington, DC: Brookings, 1985). Other works that consider social and urban policies in the U.S. as a function of the federal system include Richard Franklin Bensel, *Sectionalism and American Political Development, 1880–1980* (Madison: University of Wisconsin Press, 1984); John H. Mollenkopf, *The Contested City* (Princeton, NJ: Princeton University Press, 1983); Michael N. Danielson, *The Politics of Exclusion* (New York: Columbia University Press, 1976); Douglas S. Massey and Nancy A. Denton, *American Apartheid. Segregation and the Making of the Underclass* (Cambridge, MA: Harvard University Press 1983). Chris Pickvance and Edmond Preteceille (eds), *State Restructuring and Local Power: A Comparative Perspective* (New York: Pinter Publishers, 1991), and Douglas Ashford (ed.), *Financing Urban Government in the Welfare State* (New York: St. Martin's Press, 1980) consider shifts in intergovernmental systems from a cross-national perspective.

The New Deal is widely regarded as the pivotal nationalizing episode in American social policy. In addition to inaugurating national social programs, the New Deal enhanced the power of federal institutions and marked the beginning of a long decline in the ability of state governments to resist national initiatives. During the 1930s the undemocratic racial caste system in the South, and the power of the South in national politics, limited social policy in ways that especially disadvantaged African-Americans, the majority of whom lived in deep poverty. But even with the decline in southern influence from the 1960s on and the greater power of the federal government after the 1930s, new federal social policies could not bridge the social and economic gulf separating the black poor from the rest of American society. Indeed, the growing isolation of this group became a central theme in American politics during the 1980s.

The weak response of American social policy to African-American poverty tells us much about the limits of social policy in the United States. It also highlights the limited and peculiar forms of federal government centralization after the New Deal. In contrast to those who emphasize a "bias for centralization," I argue that the failure of efforts to address urban black poverty stems from a reenergized strand of localism. The new racially based territorial divisions that accompanied postwar suburbanization, I stress, have national significance.

[. . .]

RACE, LOCALISM, AND THE POLITICS OF AMERICAN SOCIAL POLICY

Localism has long been a defining feature of American politics. The power that localities have to resist federal government intrusions, and the deference to local interests built into national institutions, have allowed local interests to play an unusually strong role when compared cross-nationally. The federal government had little authority or capacity to impose policies or restrictions on states before the New Deal. But even after the growth of federal capacities and the Supreme Court's affirmation of a broader scope for federal activity in the 1930s and 1940s, the decentralized organization of American political institutions ensured that local interests would continue to play a significant role in national politics and policymaking.

Thus the development of social policy after the New Deal can be understood as a product of two dynamics that emerge from the continuing importance of localities in American political life. The first concerns the rules and practices governing the creation of sub-national political boundaries and the formation of geographically based interests. The struggles among political actors about what level of government should bear which responsibilities are the second. . . .

This perspective contrasts with interpretations that emphasize the centralization in American politics after the New Deal. John Chubb, for example, contends that the emergence of the federal grant system in the 1930s and its expansion in the 1960s and 1970s created a strong force for centralization in American politics. Because congressional careers depend on amassing and distributing resources from the national government, he argues, local governments have declined in importance and a strong federal role, mediated by Congress, has become self-reinforcing. This dynamic may capture the politics of existing grants to individuals, but it does not, I argue, accurately describe the politics of federal aid to places, which has played a central role in poverty policy. Moreover, the focus on congressional politics has little to say about the forces that determine which questions reach the national agenda and what form they take. Both of these issues are more adequately understood by reasserting the enduring importance of decentralization and localities in American politics.

CREATING NEW TERRITORIAL INTERESTS IN POSTWAR AMERICA

For most of American history, region has provided the most significant definition of place-based interests. The division between the economically developed industrial North and the poor rural South marked the central rift in American society and politics. The civil rights movement and the economic modernization of the South over the past forty years, however, attenuated much of the force of the North–South division. In its place, a new kind of geographically based difference has assumed political importance: the split between city and suburb. This division rests on two long-standing features of American politics and society: the laws that govern the formation and autonomy of independent political units and the racial divisions that have been woven into the fabric of American society since the founding of the country.

By cross-national standards, municipalities in the United States enjoy a great deal of autonomy. Although local governments lack formal constitutional recognition (they exist at the discretion of state government), they have, over time, come to wield substantial powers. Among the most important of these is control over local land use. Through zoning and other measures officially aimed at ensuring local health and safety, localities can determine what kinds of people can live within their borders and what kinds of business activities can be conducted there. Another fact contributes to the rift between suburb and city: the formation of new political jurisdictions is relatively easy in most states, but the expansion of older jurisdictions, via annexation or consolidation, is usually a much more difficult process.

A second major impulse toward fragmentation has been racial antagonism. As a major and long-standing social division, race has played a central role in creating the specific urban forms that distinguish the United States from European nations. The salience of race for the metropolitan form grew after World War II, as African-Americans migrated to cities and whites moved to the rapidly growing suburbs.

Although racial zoning was outlawed in 1917, suburbs used a variety of land-use regulations to bar lower-income people, a group in which African-Americans were disproportionately concentrated. These measures were accompanied by informal practices ranging from racial steering to firebombing. Where these tactics failed, the process of white flight was repeated in the suburbs. Black income rose in the 1960s and 1970s, but this did not translate into significant levels of suburban integration. Instead, separate black suburbs formed, often close to inner cities.

Rather than counterbalancing local fragmentation, the federal government reinforced it in several ways. Washington's support for suburban life – building highways, promoting automobile use, and subsidizing private home ownership – is well known. But the federal government also engaged in practices that encouraged racial exclusion. Most significant was federal sanction of discrimination in housing markets by promulgating rules that largely prevented blacks from receiving FHA (Federal Housing Administration) mortgages. Similarly, the federal government bowed to local opposition to subsidized housing that might promote integration; federal housing policies were predicated on local acceptance. As a consequence, subsidized housing was concentrated in cities; very little was located in the suburbs.

The rules governing the creation and authority of separate political jurisdictions combined with federal policies to open numerous exit options for white Americans. In the space of thirty years, a major new territorial division was created. Metropolitan areas took on a distinctive shape, that of a multitude of political jurisdictions sharply segmented by income and race. By the 1980s these patterns had become deeply entrenched.

[. . .]

POLICIES FOR POOR PLACES AND THE POLITICS OF RIGHTS

During the New Deal the major social welfare policies, embodied in the Social Security Act of 1935, provided benefits to individuals who qualified for them on the basis of contributions or a means test. Individually targeted benefits were extended in the 1960s to include medical care for the elderly and the very poor, but the federal government also introduced a new policy approach that focused on poor *places*, not poor *individuals*. At the same time social policy advocates were pressing a newly activist judiciary to elaborate social rights by expanding access to existing individual benefits.

Examination of the political underpinnings of place-based and legal strategies to alleviate poverty clearly shows how fragilely situated they were in the institutional context of American politics. When the distinctive demographic and organizational circumstances that supported these approaches faded in the 1970s, they faltered. As antipoverty strategies, legal and place-based initiatives generally produced narrow and piecemeal measures that not only failed to sustain favorable political circumstances but often actually provoked opposition. Once the immediate political circumstances became less favorable, such approaches could not survive for long, because they had no durable institutional protection.

CITIES AND THE POLITICS OF PLACE

By the 1950s poverty was no longer the widespread generalized condition that it had been in the Great Depression. Instead, poverty appeared to concentrate in particular places. Governors and congressional representatives pointed to the "pockets of poverty" in both rural and urban areas of their states, where industrial change had left whole communities mired in chronic depression. This analysis provided the starting point for the place-based policies adopted in the 1960s.

The first efforts to address these pockets of poverty reflected the logic of congressional politics. In 1961 Congress approved the Area Redevelopment Act (ARA), the first full-fledged "regional" policy in the United States. The administration of the policy attested to the difficulties of distinguishing among regions in American politics: the ARA rapidly became a classic pork barrel program offering benefits to all states to promote economic development.

The more pointed logic of the War on Poverty emerged from the executive branch. Although initially targeted at rural and urban areas, the War on Poverty very quickly came to focus on cities, where poverty had become more visible and more black. And, as riots engulfed city after city, the politics of poverty became inextricably linked with cities and minorities. Reflecting the national concern with cities and an ongoing commitment to place-based policies, President Lyndon Johnson created the Department of Housing and Urban Development in 1965. The department supervised a range of new programs that extended and further institutionalized the idea of place-specific policies.

Why did place-based policies, targeted on cities, become a major route for combating poverty in the 1960s? One reason was congressional resistance to major expansions of social welfare spending in the early 1960s, which ruled out proposals for new income-support programs for poor individuals. By contrast, support for place-based policies in pockets of poverty, both urban and rural, had been building in Congress for years. Moreover, the targeted place-based distribution of benefits that became the hallmark of the War on Poverty was supported by administration economists who viewed geographic concentration of resources as a way to increase the effectiveness of limited funds.

For a Democratic president, providing benefits for cities, in particular, was good politics. The urban vote – along with the South – had constituted the electoral core of the national Democratic party since the New Deal. The strong organization of political parties in many cities, especially in the East and the Midwest, magnified their importance in the electorate and gave city leaders special access to national policymakers. In places like Chicago, where the local party was particularly strong, local party leaders could control delegations to the state legislatures and to Congress. The importance of urban interests within the Democratic party as a whole was clear in Congress, where Democrats from suburban areas in the North voted heavily in favor of measures to aid the cities.

But the drawbacks of place-based attacks on poverty quickly became apparent. Even during the Johnson administration, geographic targeting was limited; both the president and Congress pressed for wider distribution of benefits. Moreover, even those benefits that were targeted to poor places did not automatically flow to the poor; in many cases benefits went to poor communities only after substantial mobilization. The struggle for control of these resources often encouraged urban, and particularly black, political mobilization. In many instances local communities were able to win reforms in federal programs that made them more responsive to the needs of poor people.

When Nixon came to power in 1969, however, targeting became even more attenuated. The new president replaced the urban focus of place-based policies with less targeted general revenue sharing and block grants which made benefits available to a broader range of localities. In particular, benefits became more accessible to smaller localities, from which the Republican party drew its strength. In addition, a newly assertive Congress wrested control over the distribution of funds from the bureaucracy, replacing project grants with congressionally determined formula grants that spread funds even further.

The place-based policies of the 1960s also created political problems. Rather than institutionalizing a broad national responsibility for poverty, the politics of poverty became more narrow and localized. This outcome was somewhat paradoxical because the War on Poverty was clearly national in its origins. It was devised in the executive branch and was pushed through Congress by a president recently elected in a landslide. It was administered by an independent agency in the Executive Office of the President. Yet at the heart of the program were incentives and resources designed to promote local mobilization of the black community. The limited resources devoted to the War on Poverty meant that it served far more effectively as a vehicle for black political empowerment than as a remedy for poverty.

As racial tensions rose in the latter part of the 1960s, black political assertiveness translated into pressures for community control over the institutions that delivered social policy. In the conflicts over community control, the broader national debates over expanding the scope of social policy gave way to narrow, piecemeal squabbles over dividing the shares. Thus, although the War on Poverty started as an odd blend of centralization and localism, by the end of the 1960s its localizing impulses dominated. Government policies had mobilized new locally based interests, but the links between center and locality were not organized to channel or satisfy those interests.

In addition, federal aid to cities was less useful as an antipoverty tool in periods of recession. By the 1970s, when black citizens began to elect mayors, cities faced cross-pressures: they needed both to promote economic development and to assist poor constituents. What they could do for their poor citizens, then, was sharply limited. Moreover, policies aimed at places did not challenge the growth of political boundaries that emerged in the 1950s and 1960s and divided the metropolitan population into separate political jurisdictions by income and race. Once cities began to lose their advantages in population and political organization, place-based policies were doomed.

[. . .]

THE NEW POLITICS OF LOCALISM AND THE AMERICAN POOR

The politics of poverty and the policy options for the poor have narrowed since 1980. In the decades after the War on Poverty, the national will to address African-American urban poverty through broad public means eroded. This change in public sentiment has been fed by and in turn has nourished a "defensive localism" that circumscribes and particularizes the public sphere.

The changes in the politics of urban poverty in the 1980s can be understood by examining the partisan strategies that developed around assigning responsibility for social problems to different levels of government and arenas of policy. Less overt but equally significant was the role that law played in buttressing the legitimacy of localism. Legal rulings treated local jurisdictions as an expression of "natural" community that higher levels of government should leave alone.

THE POLITICAL DECLINE OF AMERICAN CITIES AND THE ABANDONMENT OF THE URBAN POOR

The political weakness of place-based antipoverty policies became evident in the 1980s when cities lost their pivotal political position. Demographic shifts and changes in party organization contributed to the political eclipse of cities, but it was the dynamics of partisan competition that left cities politically isolated: Republicans showed that national elections could be won without urban support. The consequences for poverty policy were significant. Republican policy strategies sought to reduce the national role in social policy, devolving power to the states.

The decline of cities was both political and economic. Federal policies that promoted suburbanization combined with economic shifts and technical advances to hollow out the older cities of the Northeast and the Midwest. In 1960 the nation's population was evenly divided among cities, suburbs, and rural areas; by 1990 both urban and rural populations had declined, leaving suburbs with nearly half of the nation's population. The urban population declined to 31 percent. And as cities lost population, they became poorer and more minority in their racial composition. The shift of population into suburban areas made it possible to win national elections without the urban vote. In 1968 suburbs cast 35.6 percent of the nation's vote for president; by 1988 that figure had risen to 48.3 percent. By the 1992 election, suburbs constituted a majority of the electorate.

Demographic disadvantage was compounded by organizational changes in political parties. One feature of American party organization that had historically benefited cities in national politics was the strong organization of local parties, especially when compared to the weakness of their national counterparts. In the 1960s and 1970s, however, changes in the procedures for nominating candidates and the decline of urban machines turned this relationship on its head. As primaries replaced party nominations as the chief mechanism for selecting candidates, local parties were shunted aside by nationally organized financing sources, such as political action committees. And as minorities took control of city government, racial divisions weakened former strongholds of Democratic politics. With the atrophy of local parties, members of Congress representing urban areas had to assemble their own electoral machines; in the process they became more independent of mayors. These developments significantly muted the voice of cities in national party politics. . . . The growth of the intergovernmental lobby in the 1960 and 1970s, which looked like an increase in the power of localities, really indicated an erosion in the unique political role once occupied by these officials and their party organizations in the political system.

The terms of partisan competition did not take long to register these changes. During the 1980s a sharp disjuncture between urban and suburban voters in the North and the Midwest attested to Republican success in mobilizing a distinctive suburban political identity. In contrast to the pattern prevailing in the sixties and seventies, when central cities and suburbs tended to vote in similar directions, by 1980 there was a marked split between the suburban and urban vote. Central cities remained the only stronghold of the Democratic party, while the majority of suburban areas voted Republican.

This difference reflected a new presidential strategy. In 1968 Richard Nixon had feared the political repercussions of ignoring the cities; in 1980

Ronald Reagan demonstrated that a president could be elected without the support of cities, as well as that urban programs could be curtailed with few political consequences. Running against the trends of the last half-century, Reagan fashioned a successful political appeal around the idea that the federal role in social policy should be minimized. Instead, he argued, initiative should be left with state and local governments, which had shouldered these burdens before the New Deal. Instead of using the presidency to support national policies that spanned local political boundaries, the new Republican strategy sought to reduce federal taxes and devolve social responsibilities to states and localities. The energies of the presidency would be focused on foreign affairs; the domestic responsibilities of the presidency would be reduced as much as possible by giving states more responsibility.

The pattern of success and failure that Reagan's initiatives experienced confirmed the political weakness of cities. The only programs totally eliminated during the 1980s were those that particularly benefited urban areas. For example, in 1981 Congress terminated the Comprehensive Employment and Training Act, which cities had used (unofficially) to bolster the ranks of their employees. General revenue sharing, which provided extra untargeted revenue for localities, ended in 1986. By the end of the Reagan administration, a major urban development program, Urban Development Action Grants lost funding. The area of social policy most severely hit during the 1980s was housing, which disproportionately aided cities. . . . The federal government exacerbated the effects of these cuts by preempting various sources of local revenue. The 1986 Tax Reform Act, for example, extended restrictions on tax-exempt bonds, which localities had increasingly come to rely on to attract business.

The sharp cuts in these place-based policies stand in marked contrast to the relatively minor reductions made in programs targeted at individuals. Most congressional representatives had strong incentives to maintain the programs aimed at individuals, and not simply because policies toward individuals were more likely to be entitlement programs; after all, Congress did eliminate revenue sharing, the largest entitlement program for places. Rather, it was harder for Congress to claim credit for revenue sharing, which disappeared into the coffers of local government. Individually targeted programs, on the other hand, had more direct and tangible meaning for individual constituents. As John Chubb notes, the logic of constituent ties to Congress made these programs important elements in the congressional reelection process. The same is not true of policies to places. Far fewer members of Congress had a comparable stake in these programs.

Cities fared only slightly better at the state level. Long dominated by rural interests, state governments instituted reforms in the 1960s and 1970s that equalized urban representation and modernized administration. By the time the initiative for social policy devolved to the states, however, cities had lost ground to suburbs in state legislatures. Some states made up for federal cuts during the 1980s, but on the whole states did not compensate for the loss of federal supports. To the contrary, by the end of the decade many states, pressed by the recession, made further cuts in programs for the poor.

The political isolation of the cities has had a disproportionately negative effect on the poor because place-based policies had provided a critical supplement to the coverage of social policies aimed at individuals. Poor people and minorities in cities had developed influence and won benefits by their relationship with the federal government. As cities lost federal sources of support, they found themselves shut out of national policy discussions and poorly equipped to compete for resources at the state level.

THE POLITICAL LOGIC OF DEFENSIVE LOCALISM

[Here the author discusses the absence of the federal government from discussions of land use policy, or local government consolidation, and the local administration of housing and education programs in the U.S. This localism is contrasted with experiences in several European nations.]

These features of social policy, together with the ability of localities to impose (or reduce) many taxes, made the costs and benefits of living in one political jurisdiction over another substantial. With the decline in federal aid to places in the 1980s, the trade-off sharpened. American politics and policy can now be described as "defensive localism," in which the aim is to reduce domestic spending by the federal government, push responsibilities down to lower levels of government, and contain the social problems associated with poverty – and their costs – within defined spatial and political boundaries.

Because questions concerning local political boundaries are not constitutionally part of American national politics, the problems that arise from resource fragmentation at the local level rarely enter national political discourse. Once this set of issues fell from the national agenda in the 1970s, the range of explanations and solutions for the problems of cities (and, by extension, urban poverty) was truncated. In this context it became easier to blame local politicians for mismanagement or to point to individual deviant behavior as the cause of the cities' social problems.

Thus the organization of power across political boundaries in the United States made possible the policy strategy first pursued by the Republican party in the 1980s: to devolve social policy responsibility to lower levels of government and to encourage private solutions to social problems. This new approach to policy in turn redefined the problems of poverty and race in the United States. The federal government has abandoned these problems, treating them as inner-city issues that are the proper responsibility of the localities in which they occur. That abandonment accounts for the curious silence about poverty and race in American national politics, unbroken even in the wake of the most violent urban riots in American history in Los Angeles in 1992.

CONCLUSION: MULTI-TIERED POLITICAL SYSTEMS AND SOCIAL POLICY

The American experience highlights some of the ways a multi-tiered political system can affect social policymaking. In the United States the possibilities for isolating the poor politically and spatially offered a way to fragment the polity that severely handicapped proponents of a more expansive social policy. Key to this outcome was the way in which national political institutions were linked to local interests and the different strategies that national political actors could pursue through national institutions. Thus differences in the organization of central institutions create distinct possibilities and perils in social policymaking.

A brief examination of the central institutions of the European Union indicates the ways that differences in decision rules and in the nature of the ties of central actors to the local level could create different political outcomes. As in the United States, the EU has relied heavily on policies for poor places and legalistic extensions of social protections; fiscal constraints make it likely that these will continue to be the main channels for extending European social policy. Yet these policies fit into the politics of the EU quite differently than do their American counterparts.

The political strength of policies for poor places is closely related to the political significance of those places. In the United States, the political importance of poor places (cities) was a temporary phenomenon that disappeared as cities lost population and local party organizations atrophied and lost power at the center. In the EU, poor places – in this case, countries – are better positioned structurally. Because votes in the European Council are allotted by country, and because decisions require either unanimous or qualified-majority votes, shifts in population or party organization will not diminish the power of poor places. Their political importance is much more institutionally rooted than was the power of American cities after the New Deal.

[. . .]

Legal strategies for extending social policy also fit into the European Union quite differently than in the United States. Like the American legal system, the European Court of Justice (ECJ) has played a crucial role in extending social policy when political channels have been blocked or sluggish. But there are crucial differences. In the United States, the constitutional barriers to basing legal arguments on economic inequality led to an increased emphasis on racial inequalities. Yet race is the deepest social cleavage in the United States, and perhaps the most politically divisive issue in American history. Legal decisions extending social policy on the basis of racial discrimination were therefore highly vulnerable politically. By contrast, the employment-based rationale on which the ECJ makes its decisions is far less likely to exacerbate existing cleavages. Indeed, this rationale heightens "the visibility of the benefits bestowed while obscuring the costs." The crosscutting interests of national employers militate against a united employer counterattack on the courts.

The highly centralized welfare states of post-war Europe created a model for social policy that is not likely to be replicated in the more sprawling "federal" institutions of the European Union. But the more fragmented and decentralized organization of the EU does not mean that Europe will follow in the path of the United States; the political and social policy consequences of decentralization depend greatly on the social and political context in which institutions operate. The distinctive combination of race and localism in American history deeply affected the institutions that span sub-national boundaries as they sought to define the balance between local and national responsibilities and authority. Because localities are organized along racial and income lines, the blend of national and local responsibility for social welfare inherited from the 1930s has not only sustained less generous policies for the poor; it has also institutionalized policy and political divisions between the poor and the rest of the nation. The limits of American social policy as a tool for social integration stem not just from its decentralized political structures but also from the localistic and racially divided context in which those institutions are situated.

ONE

"Overview" and "Economic Restructuring as Class and Spatial Polarization"

from *The Global City: New York, London, Tokyo* (1991), excerpts, chs 1 and 9

Saskia Sassen

Editors' Introduction

The ebb and flow of the global economy may seem abstract, but it continually shapes the local political economy. We see, touch, and taste evidence of globalization every day. You may drive a car manufactured in another country, enjoy a cuisine associated with a distant culture, or enjoy movies and CDs produced in another language. You or your family may have migrated to your present home from another land; if not, some of your neighbors and classmates are probably immigrants. You may work for a company with international clients; your employer may be headquartered, or have branches in other nations. Terms such as "out-sourcing" have become part of common political discourse, and debates over trade policies have become the stuff of electoral politics in many nations.

In *The Global City: New York, London, Tokyo*, Saskia Sassen explores the multiple urban dimensions of global economic change. She examines changes in how things are produced, where they are produced, and who produces them in the late twentieth-century global economy, and explores how these shifts are manifested locally. To this end, she examines three major nodes in the global economy: New York, London, and Tokyo.

Her arguments can seem abstract and complex. The investigation of global capital circuits and investment patterns requires a degree of analytical sophistication to understand and can seem far removed from the day-to-day concerns of urban politics, but her arguments have a straightforward relevance to city governance. Since cities are first and foremost sites of production, they will be shaped by larger changes in the nature and location of production. Some cities will lose investment while others gain; some urban neighborhoods will experience decay and abandonment while others fend off new investment and worry about gentrification. Their labor markets will also be affected, as demand for those at the top and bottom ends of the skills hierarchy expands, with fewer jobs for those in the middle. Such changes also have clear political ramifications, as institutions associated with industrial workers (e.g., labor unions, or perhaps certain political parties) shrink and play less of a role in the urban political arena.

In the following reading, Sassen introduces the themes of her book. We then shift to Chapter 9, "Economic Restructuring as Class and Social Polarization," where she discusses the spread of informal work arrangements. Across the developed world, Sassen argues, there has been a bifurcation of wages, with managers and professionals increasing their earnings while the population of low-wage workers becomes larger and more squeezed. These trends, she argues, foster a growing "informal" economic sector, consisting of those who work without formal contracts or benefits.

Some analysts have attributed the rise of informal work to the growth of immigrant populations. Sassen disagrees, maintaining that certain characteristics of advanced capitalism, including the increase in low-wage work, and the severe pressure to keep down labor costs by eliminating workers' rights and benefits, shape the labor market. Immigrants, she claims, are well positioned to "seize the opportunities" presented by informalization, but they are not the driving force behind it.

Sassen documents her claims with fieldwork as well as statistical analysis. She previously co-authored a study of industrial firms in New York City that allowed her to look closely at employment in a few industries likely to depend on a "secondary" labor market, including construction and clothing. In Tokyo, Sassen paints a stark and disturbing portrait of the work halls and living quarters of the city's day laborers she visited. The growth of the informal sector, she notes, has marked all three cities under study, but the differences in the informal sectors of these cities show that "globalization" is not articulated in the same way in every city.

Sassen's work has attracted considerable attention from scholars and policy makers, providing a new vocabulary with which to describe the recent changes in the international political economy. Not surprisingly, her work has also invited its share of critics. Janet Abu-Lughod has suggested, for example, that Sassen is too quick to attribute social and economic polarization to changes in the global economy, when local and national policy decisions have had equal force (see Janet Abu-Lughod, *New York, Chicago, Los Angeles: America's Global Cities*, University of Minnesota Press, 1999). Chris Hamnett similarly criticizes Sassen for understating the importance of local political responses to economic pressures (see the following selection for his critique).

Saskia Sassen is the Ralph Lewis Professor of Sociology at the University of Chicago, and Centennial Visiting Professor at the London School of Economics. Among her many seminal books and articles are: *Denationalization: Economy and Polity in a Global Digital Age* (Princeton University Press 2003) based on her five-year project on governance and accountability in a global economy; *Globalization and its Discontents* (New York: New Press, 1998); *Guests and Aliens* (New York: New Press, 1999); *Cities in a World Economy* (Thousand Oaks, CA: Pine Forge/Sage). Her books have been translated into twelve languages.

OVERVIEW

For centuries, the world economy has shaped the life of cities. This book is about that relationship today. Beginning in the 1960s, the organization of economic activity entered a period of pronounced transformation. The changes were expressed in the altered structure of the world economy, and also assumed forms specific to particular places. Certain of these changes are by now familiar: the dismantling of once-powerful industrial centers in the United States, the United Kingdom . . . Japan; the accelerated industrialization of several Third World countries; the rapid internationalization of the financial industry into a worldwide network of transactions. Each of these changes altered the relation of cities to the international economy.

In the decades after World War II, there was an international regime based on United States dominance in the world economy and the rules for global trade contained in the 1945 Bretton Woods agreement. By the early 1970s, the conditions

supporting that regime were disintegrating. The breakdown created a void into which stepped, perhaps in a last burst of national dominance, the large U.S. transnational industrial firms and banks. In this period of transition, the management of the international economic order was to an inordinate extent run from the headquarters of these firms. By the early 1980s, however, the large U.S. transnational banks faced the massive Third World debt crisis, and U.S. industrial firms experienced sharp market share losses from foreign competition. Yet the international economy did not simply break into fragments. The geography and composition of the global economy changed so as to produce a complex duality: a spatially dispersed, yet globally integrated organization of economic activity.

The point of departure for the present study is that the combination of spatial dispersal and global integration has created a new strategic role for major cities. Beyond their long history as centers for international trade and banking, these cities now function in four new ways: first, as highly concentrated command points in the organization of the world economy; second, as key locations for finance and for specialized service firms, which have replaced manufacturing as the leading economic sectors; third, as sites of production, including the production of innovations, in these leading industries; and fourth, as markets or for the products and innovations produced. These changes in the functioning of cities have had a massive impact upon both international economic activity and urban form: Cities concentrate control over vast resources, while finance and specialized service industries have restructured the urban social and economic order. Thus a new type of city has appeared. It is the global city. Leading examples now are New York, London, and Tokyo. . . .

As I shall show, these three cities have undergone massive and *parallel* changes in their economic base, spatial organization, and social structure. But this parallel development is a puzzle. How could cities with as diverse a history, culture, politics, and economy as New York, London, and Tokyo experience similar transformations concentrated in so brief a period of time? Not examined at length in my study, but important to its theoretical framework, is how transformations in cities ranging from Paris to Frankfurt to Hong Kong and Sao Paulo have responded to the same dynamic. To understand the puzzle of parallel change in diverse cities requires not simply a point-by-point comparison of New York, London, and Tokyo, but a situating of these cities in a set of global processes. In order to understand why major cities with different histories and cultures have undergone parallel economic and social changes, we need to examine transformations in the world economy. Yet the term *global city* may be reductive and misleading if it suggests that cities are mere outcomes of a global economic machine. They are specific places whose spaces, internal dynamics, and social structure matter; indeed, we may be able to understand the global order only by analyzing why key structures of the world economy are *necessarily* situated in cities.

How does the position of these cities in the world economy today differ from that which they have historically held as centers of banking and trade? When Max Weber analyzed the medieval cities woven together in the Hanseatic League, he conceived their trade as the exchange of surplus production; it was his view that a medieval city could withdraw from external trade and continue to support itself, albeit on a reduced scale. The modern molecule of global cities is nothing like the trade among self-sufficient places in the Hanseatic League, as Weber understood it. The first thesis advanced in this book is that the territorial dispersal of current economic activity creates a need for expanded central control and management. In other words, while in principle the territorial decentralization of economic activity in recent years could have been accompanied by a corresponding decentralization in ownership and hence in the appropriation of profits, there has been little movement in that direction. Though large firms have increased their subcontracting to smaller firms, and many national firms in the newly industrializing countries have grown rapidly, this form of growth is ultimately part of a chain. Even industrial homeworkers in remote rural areas are now part of that chain. The transnational corporations continue to control much of the end product and to reap the profits associated with selling in the world market. The internationalization and expansion of the financial industry has brought growth to a large number of smaller financial markets, a growth which has fed the expansion of the global industry. But top-level control and management of the industry has become concentrated in a few leading financial centers, notably New York, London, and Tokyo. These account for a disproportionate share of all financial transactions and one that has grown rapidly since the early 1980s. The fundamental dynamic posited here is that the more globalized the economy becomes, the higher the agglomeration of central functions in a relatively few sites, that is, the global cities.

The extremely high densities evident in the business districts of these cities are one spatial expression of this logic. The widely accepted notion that density and agglomeration will become obsolete because global telecommunications advances allow for maximum population and resource dispersal is poorly conceived. It is, I argue, precisely because of the territorial dispersal facilitated by telecommunication that agglomeration of certain centralizing activities has sharply increased. This is not a mere continuation of old patterns of agglomeration; there is a new logic for concentration. In Weberian terms, there is a new system of "coordination," one which focuses on the development of specific geographic control sites in the international economic order.

A second major theme of this book concerns the impact of this type of economic growth on the economic order within these cities. It is necessary to go beyond the Weberian notion of coordination and Bell's (1973) notion of the postindustrial society to understand this new urban order. Bell, like Weber, assumes that the further society evolves from nineteenth-century industrial capitalism, the more the apex of the social order is involved in pure managerial process, with the content of what is to be managed becoming of secondary importance. Global cities are, however, not only nodal points for the coordination of processes; they are also particular sites of production. They are sites for (1) the production of specialized services needed by complex organizations for running a spatially dispersed network of factories, offices, and service outlets; and (2) the production of financial innovations and the making of markets, both central to the internationalization and expansion of the financial industry. To understand the structure of a global city, we have to understand it as a place where certain kinds of work can get done, which is to say that we have to get beyond the dichotomy between manufacturing and services. The "things" a global city makes are services and financial goods.

[. . .]

A second way this analysis goes beyond the existing literature on cities concerns the financial industry. I shall explore how the character of a global city is shaped by the emerging organization of the financial industry. The accelerated production of innovations and the new importance of a large number of relatively small financial institutions led to a renewed or expanded role for the marketplace in the financial industry in the decade of the 1980s. The marketplace has assumed new strategic and routine economic functions, in comparison to the prior phase, when the large transnational banks dominated the national and international financial market. Insofar as financial "products" can be used internationally, the market has reappeared in a new form in the global economy. New York, London, and Tokyo play roles as production sites for financial innovations and centralized marketplaces for these "products."

A key dynamic running through these various activities and organizing my analysis of the place of global cities in the world economy is their capability for producing global control. By focusing on the production of services and financial innovations, I am seeking to displace the focus of attention from the familiar issues of the power of large corporations over governments and economies, or supracorporate concentration of power through interlocking directorates or organizations, such as the IMF. I want to focus on an aspect that has received less attention, which could be referred to as the *practice* of global control: the work of producing and reproducing the organization and management of a global production system and a global marketplace for finance. My focus is not on power, but on production: the production of those inputs that constitute the capability for global control and the infrastructure of jobs involved in this production.

The power of large corporations is insufficient to explain the capability for global control. Obviously, governments also face an increasingly complex environment in which highly sophisticated machineries of centralized management and control are necessary. Moreover, the high level of specialization and the growing demand for these specialized inputs have created the conditions for a freestanding industry. Now small firms can buy components of global capability, such as management consulting or international legal advice. And so can firms and governments anywhere in the world. While the large corporation is undoubtedly a key agent inducing the development of this capability and is a prime beneficiary, it is not the sole user.

[. . .]

A third major theme . . . concerns the consequences of these developments for the national urban system in each of these countries and for the relationship of the global city to its nation-state. While a few major cities are the sites of production for the new global control capability, a large number of other major cities have lost their role as leading export centers for industrial manufacturing, as a result of the decentralization of this form of production. Cities such as Detroit, Liverpool, Manchester . . . Nagoya and Osaka have been affected by the decentralization of their key industries at the domestic and international levels. According to the first hypothesis presented above, this same process has contributed to the growth of service industries that produce the specialized inputs to run global production processes and global markets for inputs and outputs. These industries – international legal and accounting services, management consulting, financial services – are heavily concentrated in cities such as New York, London, and Tokyo. We need to know how this growth alters the relations between the global cities and what were once the leading industrial centers in their nations. Does globalization bring about a triangulation so that New York, for example, now plays a role in the fortunes of Detroit that it did not play when that city was home to one of the leading industries, auto manufacturing? Or, in the case of Japan, we need to ask, for example, if there is a connection between the increasing shift of production out of Toyota City (Nagoya) to offshore locations (Thailand, South Korea, and the United States) and the development for the first time of a new headquarters for Toyota in Tokyo.

[. . .]

In all these questions, it is a matter of understanding what growth embedded in the international system of producer services and finance has entailed for different levels in the national urban hierarchy. The broader trends – decentralization of plants, offices, and service outlets, along with the expansion of central functions as a consequence of the need to manage such decentralized organization of firms – may well have created conditions contributing to the growth of regional subcenters, minor versions of what New York, London, and Tokyo do on a global and national scale. The extent to which the developments posited for New York, London, and Tokyo are also replicated, perhaps in less accentuated form, in smaller cities, at lower levels of the urban hierarchy, is an open, but important, question.

The new international forms of economic activity raise a problem about the relationship between nation-states and global cities. The relation between city and nation is a theme that keeps returning throughout this book; it is the political dimension of the economic changes I explore. I posit the possibility of a systemic discontinuity between what used to be thought of as national growth and the forms of growth evident in global cities in the 1980s. These cities constitute a system rather than merely competing with each other. What contributes to growth in the network of global cities may well not contribute to growth in nations. For instance, is there a systemic relation between, on the one

hand, the growth in global cities and, on the other hand, the deficits of national governments and the decline of major industrial centers in each of these countries?

The fourth and final theme in the book concerns the impact of these new forms of and conditions for growth on the social order of the global city. There is a vast body of literature on the impact of a dynamic, high growth manufacturing sector in the highly developed countries, which shows that it raised wages, reduced inequality, and contributed to the formation of a middle class. Much less is known about the sociology of a service economy. Daniel Bell's (1973) *The Coming of Post-industrial Society* posits that such an economy will result in growth in the number of highly educated workers and a more rational relation of workers to issues of social equity. One could argue that any city representing a postindustrial economy would surely be like the leading sectors of New York, London, and increasingly Tokyo.

I will examine to what extent the new structure of economic activity has brought about changes in the organization of work, reflected in a shift in the job supply and polarization in the income distribution and occupational distribution of workers. Major growth industries show a greater incidence of jobs at the high- and low-paying ends of the scale than do the older industries now in decline. Almost half the jobs in the producer services are lower-income jobs, and half are in the two highest earnings classes. In contrast, a large share of manufacturing workers were in the middle-earnings jobs during the postwar period of high growth in these industries in the United States and United Kingdom.

Two other developments in global cities have also contributed to economic polarization. One is the vast supply of low-wage jobs required by high-income gentrification in both its residential and commercial settings. The increase in the numbers of expensive restaurants, luxury housing, luxury hotels, gourmet shops, boutiques, French hand laundries, and special cleaners that ornament the new urban landscape illustrates this trend. Furthermore, there is a continuing need for low-wage industrial services, even in such sectors as finance and specialized services. A second development that has reached significant proportions is what I call the downgrading of the manufacturing sector, a process in which the share of unionized shops declines and wages deteriorate while sweatshops and industrial homework proliferate. This process includes the downgrading of jobs within existing industries and the job supply patterns of some of the new industries, notably electronics assembly. It is worth noting that the growth of a downgraded manufacturing sector has been strongest in cities such as New York and London.

The expansion of low-wage jobs as a function of growth trends implies a reorganization of the capital–labor relation. To see this, it is important to distinguish the characteristics of jobs from their sectoral location, since highly dynamic, technologically advanced growth sectors may well contain low-wage dead-end jobs. Furthermore, the distinction between sectoral characteristics and sectoral growth patterns is crucial: Backward sectors, such as downgraded manufacturing or law-wage service occupations, can be part of major growth trends in a highly developed economy. It is often assumed that backward sectors express declining trends. Similarly, there is a tendency to assume that advanced sectors, such as finance, have mostly good, white-collar jobs. In fact, they contain a good number of low-paying jobs, from cleaner to stock clerk. These, then, are the major themes and implications of my study.

As a further word of introduction I must sketch the reasons why producer services and finance have grown so rapidly since the 1970s and why they are so highly concentrated in cities such as New York, London, and Tokyo. The familiar explanation is that the decade of the 1980s was but a part of a larger economic trend, the shift to services. And the simple explanation of their high concentration in major cities is that this is because of the need for face-to-face communication in the services community. While correct, these clichés are incomplete.

We need to understand first how modern technology has not ended nineteenth-century forms of work; rather, technology has shifted a number of activities that were once part of manufacturing into the domain of services. The transfer of skills from workers to machines once epitomized by the assembly line has a present-day version in the transfer of a variety of activities from the shop floor into computers, with their attendant technical and professional personnel. Also, functional specialization within early factories finds a contemporary counterpart in today's pronounced fragmentation of the work process spatially and organizationally. This has been called the "global assembly line," the production and assembly of goods from factories and depots throughout the world, wherever labor costs and economies of scale make an international division of labor cost-effective. It is, however, this very "global assembly line" that creates the need for increased centralization and complexity of management, control, and planning. The development of the modern corporation and its massive participation in world markets and foreign countries has made planning, internal administration, product development, and research increasingly important and complex. Diversification of product lines, mergers, and transnationalization of economic activities all require highly specialized skills in top-level management. These have also "increased the dependence of the corporation on producer services, which in turn has fostered growth and development of higher levels of expertise among producer service firms" (Stanback and Noyelle 1982:15). What were once support resources for major corporations have become crucial inputs in corporate decision-making. A firm with a multiplicity of geographically dispersed manufacturing plants contributes to the development of new types of planning in production and distribution surrounding the firm.

[...]

The growth of advanced services for firms, here referred to as producer services, along with their particular characteristics of production, helps to explain the centralization of management and servicing functions that fueled the economic boom of the early and mid-1980s in New York, London, and Tokyo. The face-to-face explanation needs to be refined in several ways. Advanced services are mostly producer services; unlike other types of services, they are not dependent on proximity to the consumers served. Rather, such specialized firms benefit from and need to locate close to other firms who produce key inputs or whose proximity makes possible joint production of certain service offerings. The accounting firm can service its clients at a distance but the nature of its service depends on proximity to other specialists, from lawyers to programmers. Major corporate transactions today typically require simultaneous participation of several specialized firms providing legal, accounting, financial, public relations, management consulting, and other such services. Moreover, concentration arises out of the needs and expectations of the high-income workers employed in these firms. They are attracted to the amenities and lifestyles that large urban centers can offer and are likely to live in central areas rather than in suburbs.

[...]

[From Chapter 9, "Economic Restructuring as Class and Spatial Polarization"]

CASUAL AND INFORMAL LABOR MARKETS

There has been a pronounced increase in casual employment and in the informalization of work in both New York City and London. This trend is also emerging, under a different form, in Tokyo where the increase in the number of casual workers, particularly "daily laborers" and part-time workers has led the government to express alarm publicly. The increase in various types of casual work is often thought of as a function of the increased participation of women in the labor force. Indeed, part-time, temporary and seasonal jobs are more common among women than among men in all three cities. However, all the evidence points to significant increases of such jobs among men over the last decade. More generally, the industry/occupational mix prevalent among such jobs indicates that they account for a significant share of new jobs created in these economies. In addition, jobs that were once full-time ones are now being made into part-time or temporary jobs, pointing to a transformation in the employment relation. While so-called flexible work arrangements may be a development of advanced economies associated with a higher quality of life, the vast majority of casual jobs hardly fits this category. A majority are low-wage jobs, with no fringe benefits and not returns to seniority – a way of organizing work that reduces the costs for employers. However there is a new trend that I consider significant in this study: High-income professional and managerial employees in many of the new specialized service and financial firms are more vulnerable to dismissal and have fewer claims on their employers than was the case with their equivalents in the large commercial banks and insurance houses. This greater "flexibility" in

the employment relations is another way of saying that these jobs have become causalized as well. I will return to this subject, which has received no attention in the pertinent literature.

The growth of service jobs is crucial to the expansion of part-time jobs. The pressures to reduce labor costs in industries with limited profit margins, such as catering, retail, and cleaning, assume added weight when these account for a growing share of jobs. In addition, many service industries require work at night, on weekends, and on holidays, which would entail costly overtime payments for full-time workers. And since many of these jobs do not require many skills or training, they can be down-graded into part-time, more lowly paid jobs. As these services industries have grown, the gap between the work week in such industries as retail, reaching seventy hours a week, and the forty-hour full-time work week has grown in weight. Part-time jobs can recruit women more easily, create greater flexibility in filling various shifts, and reduce labor costs by avoiding various benefits and overtime payments required by full-time workers....

There is also evidence pointing to an expansion of the underground economy. Of interest to the analysis here is one particular component of the underground economy, informal work. This encompasses work that is basically licit but takes place outside the regulatory apparatus covering zoning, taxes, health and safety, minimum wage laws, and other types of standards. In other words, this is work that could be done in the formal economy, unlike the criminal activities that are also part of the underground economy. Government regulations play a particularly important role in the rise of informal production because of the costs that they impose on formal businesses through their various licensing fees, taxes and restrictions. Labor costs also have an effect on the formation and expansion of the informal economy: directly, in terms of the wage paid, and indirectly, in terms of various contributions demanded by law. One question is whether the importance of these inducements to informalization varies by industry and location.

The specification of this particular component, the informal economy, has implications for theories on the nature of advanced capitalism and the postindustrial society. While criminal activities and underreporting of income are recognized to be present in advanced industrialized economies, informal sectors are not. The literature on the informal sector has mostly focused on Third World countries and has, wittingly or not, assumed that as a social type such sectors are not to be expected in advanced industrialized countries. And the literature on industrialization has assumed that as development progresses, so will the standardization of production and generalization of the "formal" organization of work. Since much of the expansion of the informal economy in developed countries has been located in immigrant communities, this has led to an explanation of its expansion as being due to the large influx of Third World immigrants and their assumed propensities to replicate survival strategies typical of their home countries. Related to this view is the notion that backward sectors of the economy are kept backward, or even kept alive, because of the availability of a large supply of cheap immigrant workers. Both of these views posit or imply that if there is an informal sector in advanced industrialized countries, the sources are to be found in Third World immigration and in the backward sectors of the economy – a Third World import or a remnant from an earlier phase of industrialization.

A central question for theory and policy is whether the formation and expansion of informal and casual labor markets in advanced industrialized countries is the result of conditions created by advanced capitalism. Rather than assume that Third World immigration is causing informalization and the entry of mothers into the labor force is causing the casualization of work, what we need is a critical examination of the conditions that may be inducing these processes. Immigrants, insofar as they tend to form communities, may be in a favorable position to seize the opportunities represented by informalization. And women, insofar as they have children and inadequate access to child care may be interested in part-time or temporary job opportunities. But the opportunities are not necessarily created by immigrants and women. They may well be a structured outcome of current trends in the advanced industrialized economies. Similarly, what are perceived as backward sectors of the economy may not be remnants from an earlier phase of industrialization; they may well represent a downgrading of work connected to the dynamics of growth in leading sectors of the economy.

There is a strong tendency for the service sector overall to produce or make possible more part-time jobs than does manufacturing. This tendency is clearly embedded in a number of basic institutional arrangements and in specific historical conditions. The institutionalization of the family wage was closely interlinked with the rise of powerful manufacturing-based unions and a male-dominated "labor aristrocracy." The family wage is, or rather, was, the institutionalized principle that a man's wage should be high enough to support his family. Thus, it contributed to establishing the gender-based occupational work structure characteristic of industrialized economies. . . .

The shift in the economy toward a prevalence of service industries contains what one could think of as a structurally induced erosion of the – albeit limited – institutional bases for the family wage. The growth of part-time work, the growth in the numbers of female-headed households, the decline of manufacturing-based unions, and the large-scale displacement of male workers – all these conditions have contributed to an erosion of the institution of the family wage, limited as its implementation was, especially in the United States, and to an erosion of the ideology of the family wage. One important question is whether the current conditions – a sort of disarray compared with the ideal type presupposed in the family wage – represent a transition to less gender-based structures of work or are yet another step in the formation of a supply of cheap and powerless workers.

The overall effect is the causalization and informalization of the employment relation. The differences among these three cities stem partly from the distinct institutional arrangements through which work is organized. In the United Kingdom, the government has until recently played a rather fundamental role as supplier of a vast range of services and goods, from housing to health services, which in the United States are largely delivered through the private market. The net effect was to incorporate a vast number of workers who were employed by the government or public authorities. This carried with it a considerable degree of government regulation over large sectors of the labor market. The recently implemented withdrawal of the government from these various markets through the privatization of services and goods has created a situation that has facilitated the transformation of many of these jobs from regular full-time, year-round jobs with the requisite fringe benefits into various kinds of part-time temporary jobs, as well as the subcontracting of work. The historical obligation assumed by a government to enforce its own regulations covers a shrinking share of the work force in an increasingly restricted set of labor markets. . . .

In the United States, the government's role in the economy, while strong, has not been centered on the labor market or oriented toward the provision of housing and health services on a national scale. The mere absence of a national system entails a much less encompassing role for government in shaping the characteristics of jobs. Economic conditions in the postwar era – the dominance of consumer-oriented industry, large unions, the expansion of a middle class, the growth of standardized production – all promoted the expansion of a large number of jobs that respected the regulatory framework. Thus, the outcome was similar to that in Britain in the postwar era, but was effected through different channels. The key vehicle in the expansion of part-time work over the last decade was the combination of a shift to services and the dismantling of the manufacturing-based unions, which had the power to impose wage standards on broad sectors of the economy.

Beyond the trend toward an increase in part-time work arising from these various conditions, we see the expansion of informal work arrangements, at times resembling an informal sector with a fairly elaborate set of relations of production, distribution, and markets for labor and inputs – that is to say, informal arrangements that do not simply consist of a few individuals working off the books. In the case of the United States there has not been such rapid wide-scale privatization of the production and delivery of various public goods and services as in Britain, but there has been a somewhat similar development in terms of the shift of a growing number of jobs from highly regulated formal labor markets to semiregulated, unregulated, or casual labor markets. Much work in the informal economy is not casual strictly speaking in that it is part of a well-organized chain of production and involves full-time, year-round work. But much of it is casual, this flexibility being precisely the key advantage of informal work of employers or contractors.

In Japan, rapid industrialization, immensely rapid urbanization, and culture have created very specific conditions. The securing of a reliable industrial work force under conditions of extremely rapid growth and different cultural preferences and expectations from those in the West has contributed to the development of the so-called lifelong job security system. This system now accounts for only a fifth of all workers, having also suffered from the shift to and rapid growth of service industries. The growing participation of women in the labor force has assumed the form of a rapid growth in part-time and temporary jobs. However, the growth of the category of daily laborers and the rapid erosion of the institutional arrangements that are supposed to cover daily laborers, are reminiscent of the casualization and informalization of work evident in Britain and the United States.

While present in all three cities under study, these developments assume rather specific forms and operate through distinct social arrangements. The available evidence suggests that the most pronounced form over the last decade in London was the growth of part-time work; in Tokyo it was the expansion of daily labor; in New York, the growth of informal work. The following sections discuss each of these instances of the casualization of the employment relation.

INFORMALIZATION IN NEW YORK CITY

A small but growing body of evidence points to the expansion of informal work in major cities of the United States over the last decade. . . . As categories for analysis, the underground economy and the informal economy overlap only partly. Studies on the underground economy have sought to measure all income not registered in official figures, including income derived from illicit activities, such as drug dealing. Studies on the informal economy focus on the production and sale of goods and services that are licit but produced and/or sold outside the regulatory apparatus covering zoning, taxes, health and safety, minimum wage laws, and other types of standards.

[. . .]

Grover and I investigated these activities from 1986 to 1989. The concern was to specify a relationship between the growth of informal activities in New York City and overall conditions in the economy and the existing regulatory environment.

In this context, one can think of informalization as an emergent, or developing, "opportunity" structure that avoids or compensates for various types of constraints, from regulations to market prices for inputs. This type of inquiry requires an analytical differentiation of immigration, informalization – and the characteristics of the current phase of advanced industrialized economies – in order to establish the differential impact of (1) immigration and (2) conditions in the economy at large on the formation and expansion of the informal economy. The theoretical and policy implications associated with the primacy of one or the other will vary. For theory, the primacy of economic structure would point to the need for further theoretical elaboration on the current understanding of the nature of advanced capitalism. For policy; the primacy of immigration would suggest, at its crudest, that controlling immigrant activity in the informal economy would eradicate the latter; it would, then, also reinforce standard theories on advanced industrialization or the postindustrial society, which allow no room for such developments as an informal economy.

The industries covered were construction; garments; footwear; furniture; retail activity; and electronics. We did field visits in all the boroughs of New York City. . . . On the basis of our fieldwork, interviews, and secondary data analysis, we found the following profile of the informal economy in the New York City area: (1) A rather wide range of industrial sectors use informal work-apparel, general construction contractors, special trade contractors, footwear, toys and sporting goods, electronic components; and accessories. (2) Informal work is also present in lesser measure in particular kinds of activities, such as packaging notions, making lampshades, making artificial flowers, jewelry making, distribution activities, photoengraving, manufacturing of explosives, etc. (3) There is a strong tendency for informal work to be located in densely populated areas with very high shares of immigrants. (4) There is an emergent tendency for "traditional" sweatshop activity (notably in garments) to be displaced from areas undergoing partial residential and commercial gentrification; such areas engender new forms of unregistered work, catering to the new clientele.

There are several patterns in the organization of the informal economy in New York City that are of interest to an examination of its articulation with leading growth trends. One pattern is the concentration of informal activities in immigrant communities, where some activities meet a demand from the communities and others meet a demand that comes from the larger economy. A second pattern is the concentration of informal activities in areas undergoing rapid socioeconomic change, notably gentrification. A third is the concentration of informal activities in areas that emerge as a type of manufacturing and industrial servicing area in a context where both regulations and market forces do not support such activities; while these are frequently located in immigrant communities, they cater to the larger economy.

[. . .]

An examination of what engenders the demand for informal production and distribution indicates several sources: (1) One of these is competitive pressures in certain industries, notably apparel, to reduce labor costs to meet massive competition from low-wage Third World countries. Informal work in this instance represents an acute example of exploitation. (2) Another source is a rapid increase in the volume of renovations, alterations, and small-scale new construction associated with the transformation of many areas of the city from low-income, often dilapidated neighborhoods into higher-income commercial and residential areas. What in many other cities in the United States would have involved a massive program of new construction was mostly a process of rehabilitation of old structures in the case of New York City. The volume of work, its small scale, its labor-intensiveness and high skill content, and the short-term nature of each project all were conducive to a heavy proportion of informal work. (3) A third source is inadequate provision of services and goods by the formal sector. This inadequacy may consist of excessively high prices, inaccessible or difficult-to-reach location of formal provision, or actual lack of provision. It would seem that this inadequacy of formal provision involves mostly low-income individuals or areas. (4) The existence of a cluster of informal shops can eventually generate agglomeration economies that induce additional entrepreneurs to move in. (5) The existence of a rather diversified informal economy making use of a variety of labor supplies may lower entry costs for entrepreneurs and hence function as a factor inducing the expansion of the informal economy. This can be construed as a type of supply-side factor.

[. . .]

It would seem from our study that important sources of the informalization of various activities are to be found in characteristics of the larger economy of the city. Among these are the demand for products and services that lend themselves to small scales of production, or are associated with rapid transformations brought about by commercial and residential gentrification, or are not satisfactorily provided by the formal sector. This would suggest that a good share of the informal economy is not the result of immigrant survival strategies, but rather an outcome of structural patterns or transformations in the larger economy of a city such as New York. Workers and firms respond to the opportunities contained in these patterns and transformations. However, in order to respond, workers and firms need to be positioned in distinct ways. Immigrant communities represent what could be described as a "favored" structural location to seize the opportunities for entrepreneurship as well as the more and less desirable jobs being generated by informalization.

THE CASUALIZATION OF WORK IN LONDON

There has been a great increase in part-time, casual, and sweated labor in construction, clothing, catering, retailing, tourism, cleaning, and even printing in London and in the United Kingdom generally. The growth of unorganized and low-paid labor can drag down the pay and working conditions of the better-paid, organized workers. This is contributing to a further erosion of the socioeconomic conditions of low-income workers. While the vast majority of part-time workers in the United Kingdom, as well as in most developed countries, are women, the share of men has grown. . . . In the early 1970s, men accounted for a small share of part-time workers. By 1981, 19 percent of all part-time workers in the United Kingdom were men. Not only is part-time work increasing, part-time contracts are becoming shorter. . . .

Studies by the Greater London Council and the Low Pay Unit estimate that over 20 percent of workers in the hotel and catering sectors are on temporary contracts. There is a job center for casual employment, with lines forming every morning at six and waiting until the center opens at eight. The

regular use of casual labor has also increased in this industry, not simply as a response to seasonal demand in hotel and catering, but largely as a way to cut costs and avoid addressing poor working conditions and low pay. There has also been an increase of these practices in the construction industry. Every morning there is a hiring hall in certain areas and a scramble for jobs. These jobs carry no benefits, and workers are often classified as self-employed in order to exempt the employer from taxes and other responsibilities. Again we see here parallels with what is occurring in New York City.

These practices, however, are not confined to the more traditional industries. In such specialized service industries as architectural and engineering services and banking there has also been an increase of workers paid by the hours and a reduction in part-time workers with the same rights as full-time workers. These workers, mostly women, have no sick pay, no overtime pay, no holiday pay, and no job security. Yet they work as much as a regular worker. As in the garment or construction industries, these workers are also classified as self-employed by their employers. These practices contribute to the income polarization evident in the advanced services sector.

[. . .]

Homework and sweated work have also increased. The clearest case is apparel. To be able to use the cheapest workers, ethnic minorities and women, smaller textile and clothing firms now tend to be concentrated in London and other cities. . . . London has lost all its large factories, as many of the large manufacturers no longer do production: They subcontract to overseas or domestic producers. And in the 1960s the big retailers began subcontracting as well, initially to Hong Kong and Taiwan and now also to domestic firms. As a result, the number of firms and workers has increased, especially "fly-by-night" factories. The majority of the workers are from ethnic minorities and up to a third are homeworkers. Many are part-time or temporary workers, employed in the busy season. The vast majority of homeworkers and part-time workers in clothing are women. . . . As in New York and Tokyo, the whole new emphasis on fashion and luxury or simulated luxury has also led to a new need for quick turnover from design to finished product and hence for producers close by. The estimate is that 30 percent to 50 percent of all East End and northeast London garment production is through homework and that at least half of these homeworkers are Bangladeshi or Pakistani women; of the rest, a large number of women in garment work are from Cyprus. And in fieldwork in 1989, I found a significant presence of recently arrived Turkish entrepreneurs and workers. Homework is taking place not only in clothing but also in other lines of production, including the making of lampshades, electrical goods, painted toys, and zippers.

[. . .]

Reviewing the evidence for Britain, Hurstfield (1987) found that by far the greatest savings to employers derived from using part-time employees seem to be on National Insurance contributions, by keeping part-time workers earnings below the threshold. . . .

As in the United States, the government has passed legislation that further weakens the position of part-time workers, creates additional incentives for employers to use part-time workers, and legitimizes the use of part-time workers. Compared with other European countries, Britain has lost significant ground in terms of the protection of workers, though it is probably still more generous than the United States. . . .

TOKYO'S DAILY LABORERS

Daily laborers, especially in construction, construction-related industries, and longshoring – all major industries in Japan – are supposedly registered and entitled to unemployment compensation and other benefits according to their work records. They have work carnets in which are registered the days worked every week and month – necessary to establish the amount of unemployment compensation a worker is entitled to. There are specific locations where jobs are listed and allocated by government employees, who staff the various desks or counters, give workers their job slips, and write the information in the workers' carnets. These workers supposedly also can write themselves into a waiting list for housing and have access to other services. There was a time when their numbers were smaller, fairly well-paying jobs in longshoring were their main occupation, and daily laborers generally were better incorporated in mainstream society.

The recent massive expansion of this category of workers has meant that a minority of them are actually covered by these regulations. Daily labor has increasingly become a residual category, formed by those who were fired from other jobs, including white-collar workers, elderly men who no longer can work in the jobs they once held, and young men unable to get any other job. In my fieldwork in Yokohama's daily laborers' camp, I found university graduates, including one who had been a political militant at the University of Tokyo and had subsequently been blacklisted. Daily labor has also become a key employment form for new illegal immigrants from several Asian countries. In my fieldwork I found that a number of these immigrants had actually attended university in their home countries.

There are four major hiring halls for daily laborers in the country, two in the Tokyo-Yokohama area and one each in Nagoya and Osaka. The largest of these hiring halls is in the Taito ward in Tokyo. It has a reputation for being a rather dangerous place. While the Japanese version of Western-style gangsters or Mafia, the "yakusa," are known to control all four of the large hiring halls in the country, Taito's is supposed to be the worst. As I discussed above in describing Tokyo's inner city, Taito is one of the most deprived wards in the city, with growing rates of criminality, poverty, and unemployment. There is clearly a massive breakdown of the system that was supposed to protect daily laborers and low-income residents generally. The hiring halls are also frequently places for homeless men.

On my first visit to one of these halls, Kotobuki-cho in Yokohama, we walked over at five in the morning. It was still dark. There was a gray concrete structure, the equivalent of four stories, with wide-open platforms at street level and one at the equivalent of two stories up, covered by a flat, slablike roof. It was a square structure with about 50 meters per side. Both on the street level and on the second-story platform on one of the enclosed sides were what looked like train station ticket counters, with long lines of men at each one. Through the ticket window I could see lists of jobs, with wages listed. At the other end of the platform were large groups of men, lying on the ground or just rising, clearly homeless, covered with tattered clothes, lying on dirty blankets, unshaven, unhealthy. An image of absolute misery. There were also young, neat men, among them many immigrants, and many older men standing in line. Amongst this vast and varied sea of men walked about twenty flashy, flamboyant men, arrogant and aggressive looking, with dark sunglasses notwithstanding the predawn darkness. They were the yakusas. They acted in rather threatening ways toward me, circling me. But I knew I was safe for a complicated set of reasons, not the least being that murder is still extremely rare even in the absolute bottom of the Japanese social structure. At about 8 a.m., the contractors have left with their hired laborers, and the large numbers that stay behind have nothing much to look forward to. They sit at the edges of the streets and talk, play various games. There is no place beneath this place.

I visited some of the living quarters of those who had been left behind. You enter an old, minute version of a New York City tenement: a long, very dark and narrow hallway, with an extremely low ceiling. There is an endless row of roughly made wooden doors. Behind each door is a cubicle the size of a narrow double bed and, at least in the ones I saw, a small window. Some of the quarters I visited were extremely neat, the occupant clearly intent on salvaging as much as he could of his dignity. Some of the daily laborers are extremely clean and neat in appearance; they make use of the public showers, for which they are willing to pay. They have not been morally broken. At least not yet; they may still have the hope of a better job.

The meager evidence we have and the fate of the older, less employable daily laborers suggest that increasing hardship and demoralization lie ahead for the mass of daily laborers. One can still see amidst the misery and the darkness and the dankness the behavior of individuals who consider themselves integrated in a wider society; there is in fact no exit for most of them. The distance between the world of the daily laborer and the rest of society, the world of regular, full-time jobs, has grown immensely in only a few years. So has the distance between the world of the daily laborer and the society of men and women and families and children. This is a world exclusively of men. I had been warned that many of the older men had probably not seen women, live and up close, for many years, and they would come up to me and stare and try to touch me, innocently, with no mean intent. They did. This is a world far removed from the Japan we think of in the West.

[. . .]

Daily labor in Japan is, perhaps, the sharpest instance of casualization in the employment relation. Along with female part-time work, it represents a growing stratum of the labor force.

REFERENCES FROM THE READING

Bell, Daniel. 1973. *The Coming of Post-industrial Society: A Venture in Social Forecasting.* New York: Basic Books.

Hurstfield, Jennifer. 1987. *Part-timers: Under Pressure.* London: Low Pay Unit.

Sassen-Koob, Saskia and W. Grover. 1986. "Unregistered Work in the New York Metropolitan Area." Working Paper, Columbia University, Program in Urban Planning.

Stanback, Thomas N. Jr. and Theirry J. Noyelle. 1982. *Cities in Transition: Changing Job Structures in Atlanta, Denver, Buffalo, Phoenix, Columbus, Nashville, Charlotte.* Totowa, NJ: Allanheld, Osmun.

"Social Polarisation, Economic Restructuring and Welfare State Regimes"

from S. Musterd and W. Ostendorf (eds),
Urban Segregation and the Welfare State (1998)

Chris R. Hamnett

Editors' Introduction

Chris Hamnett's essay questions some of the assumptions underlying Sassen's work. Professor of geography at Kings College of the University of London, Hamnett is interested in understanding how global trends affect patterns of inequality in cities. He notes that the income distributions are becoming more polarized in all industrialized nations, which have social, political, and spatial implications for their cities. He believes, however, that polarization is a more nuanced phenomenon than many analysts, including Sassen, have allowed.

According to Hamnett, globalization scholars like Sassen think not only that a growing polarization has taken place – defined as a shift away from most people having middling incomes and occupations – but that this has been caused by the forces of economic internationalization. When Hamnett looks at the British case, however, he sees little evidence of occupational polarization and questions whether income polarization is as extreme as Sassen maintains. He suggests that Sassen and others have been overly influenced by the U.S. experience, where income and occupational distributions have grown notably more unequal. Other nations, however, have welfare and taxation policies that mitigate the impacts of global economic change. Hamnett's conclusion suggests that political differences between nation states, specifically differences in the strength of the welfare state, help us understand how global economic forces affect such different cities as, say, Los Angeles, Manchester, and Tokyo.

Chris Hamnett's research has focused on social and economic inequalities in London; he is also engaged in cross-national comparative research on this theme. He is the author of *Unequal City: London in the Global Arena*, published by Routledge in 2003, and many articles on social polarization, gentrification, and housing policies.

INTRODUCTION

[This essay] examine[s] the relationships between urban social polarisation, economic restructuring and the role of the welfare state. The existence of polarisation or dualisation – the growing division in society between the haves and the have-nots; the socially included and the excluded; and a shrinking of the size of the middle groups – has become almost a conventional wisdom regarding social change and divisions in western cities. I want to problematise the notion of polarisation, which is accepted uncritically. As Fainstein *et al.* [1992, p. 13] pointed out:

> The images of a dual or polarised city are seductive, they promise to encapsulate the outcome of a wide variety of complex processes in a single, neat and easily comprehensible phrase. Yet the hard evidence for such a sweeping and general conclusion regarding the outcome of economic restructuring and urban change is, at best, patchy and ambiguous. If the concept of 'dual' or 'polarising' city is of any real utility, it can serve only as a hypothesis, the prelude to empirical analysis, rather than as a conclusion which takes the existence of confirmatory evidence for granted.

Second, and following from this, I wish to dispute the way in which the concept of social polarisation and the associated concept of the dual city is commonly used as a general, all-purpose, signifier of growing inequality and social divisions. While I accept that social polarisation has been used in a number of different ways in Britain and North America, and that it has an important representational, ideological and rhetorical role regarding growing social divisions in cities, there is a parallel danger that, by uncritically accepting the existence of social polarisation as some sort of general, catch-all, process, we may fail to see the existence of different forms of polarisation in different cities. Social polarisation is not a single, homogeneous process, which operates in the same way in different places.

Third, and perhaps most importantly, there is the danger that, by uncritically accepting the conventional wisdom, we may fail to see that the processes driving polarisation in different cities differ/are mediated in various ways. It is thus necessary, in my view, to conceptually unpack the term 'polarisation' and to examine the extent to which different forms of polarisation are found in different contexts and to theorise the reasons for such variations. Otherwise we risk becoming slaves to unexamined, imprecise or ill-defined concepts.

I take 'polarisation' to be a term referring to a change in certain social distributions such that there is a shift away from a statistically normal or

egg-shaped distribution towards a distribution where the bottom and top ends of the distribution are growing, relatively and possibly absolutely, at the expense of the middle. This is the dominant interpretation and reflects the concerns of a number of commentators. Marcuse [1989, p. 699] put it well:

> The best image . . . is perhaps that of the egg and the hour glass: the population of the city is normally distributed like an egg, widest in the middle and tapering off at both ends; when it becomes polarised the middle is squeezed and the ends expand till it looks like an hour glass. The middle of the egg may be defined as intermediate social strata. . . . Or if the polarisation is between rich and poor, the middle of the egg refers to the middle income group. . . . The metaphor is not structural dividing lines, but of a continuum, whose distribution is becoming increasingly bi-modal.

SOCIAL POLARISATION AND WELFARE STATE REGIMES

My concern in this chapter is with the social dimensions of polarisation rather than with social segregation per se, but the two questions are clearly linked together. Finally, and following on from the concerns outlined above, I wish to challenge the dominant theory of polarisation advanced by Sassen and others which sees occupational and income polarisation as the outcome of a general shift from manufacturing to services which is particularly marked in global cities as a result of their concentration of key advanced business and financial services, gentrification and the growth of sweated manufacturing. This thesis of polarisation is essentially unicausal, focusing on change in the paid labour force. I shall argue that, whilst there is strong evidence for income polarisation in capitalist economies, there is no evidence of occupational polarisation and that the development of income polarisation may also be the result of changes in taxation, welfare benefits and unemployment rather than occupational restructuring.

The thesis I wish to put forward is that the extent and the forms of social polarisation in different countries are unlikely to be homogeneous or unidirectional; that they result from a combination of economic restructuring which is changing the structure of the labour market, the structure of occupations and incomes in the paid labour market and the division between the economically active and the inactive and unemployed. In addition, I want to argue that there may be other forces generating income polarisation such as shifts in household composition and the age structure of the population. Finally, I want to argue that the extent of polarisation in the occupational and income structure of western societies is likely to be mediated by structures of welfare provision and taxation which are instrumental in influencing both the necessity to enter the paid labour market and the incomes derived from paid employment and from welfare benefits. I further wish to argue that the extent of polarisation in many U.S. cities is a result of the specific institutional context in that country, particularly the high and growing level of immigration and its implication for labour supply, the relative paucity of welfare provisions and income support for the poor, and the absence of effective minimum wage legislation and the growth of a large, low-paid casualised service sector. The rapid growth of income polarisation in the U.K. in the 1980s and early 1990s may also reflect some similar trends, although the absence of a growing low-skilled section of the employed labour force in Britain and in other Western European countries may reflect differences in job opportunities and welfare provision which inhibit the growth of this section of the paid labour force. In recent years we have seen the emergence of two different literatures: one on polarisation and economic restructuring and another on welfare state regimes. These literatures have rarely been brought together but, as I shall try to show, the relationship between the two is crucial.

Social polarisation: Sassen's thesis

In a well-known series of works Saskia Sassen has outlined a thesis for global cities whereby social polarisation in these cities is seen as a result of a form of economic restructuring which is particularly concentrated in such cities. This restructuring has several key elements. First, there is the shift from manufacturing to services, particularly advanced business services. This shift is seen to result in a polarised occupational and income structure which

is characterised by growth both at the top and at the bottom end by virtue of the more polarised nature of the service sector, and the contraction of the manufacturing sector which contained more skilled manual, middle-income jobs.

Second, Sassen points to the growth of low-grade service jobs which are seen to be dependent on the growth at the top end of the occupational and income structure. These jobs are concentrated in the personal service sector and provide services for the wealthy. Third, Sassen points to the growth of informalisation within what remains of the 'down-graded' manufacturing sector. These low-skilled, low-paid jobs are strongly concentrated in the immigrant labour force, which is seen to be attracted to global cities by virtue of the growing labour market opportunities. This growing occupational and income polarisation is said to be linked to a growing geographical polarisation as the top and bottom of the occupational and income structure become increasingly differentiated in space.

Sassen outlined the basics of her polarisation thesis in 1984 . . . (Sassen-Koob 1984). She argued that economic restructuring has had several major implications in global cities. She suggested that new forms of economic growth no longer produce the type of jobs that "were constitutive of the massive expansion of the middle class in the post-World War II period", and she argued that there is polarisation in the occupational structure, including what she terms "a vast expansion in the supply of low-wage jobs and a shrinking supply of middle-income jobs" (1984: 139). She quoted Stanback and Noyelle's [1982, p. 133] finding that

> for the services as a whole, the important observation is that there tend to be heavy concentrations of employment in better than average and in poorer than average jobs. In contrast, in manufacturing and construction the distributions are more heavily weighted toward medium- and above-average income jobs.

In her book *The Global City* (1991), she argues that the evolving structure of economic activity in global cities has "brought about changes in the organisation of work reflected in a shift in the job supply and polarisation in income distribution and occupational distribution of workers" (1991: 9). Sassen [1991, p. 13] summarised her thesis as follows:

new conditions of growth have contributed to elements of a new class alignment in global cities. The occupational structure of major growth industries characterised by the locational concentration of major growth sectors in global cities in combination with the polarised occupational structure of these sectors has created and contributed to growth of a high-income stratum and a low-income stratum of workers.

She suggests that the process of economic restructuring and polarisation are common to all global cities, arguing that New York, London and Tokyo "have undergone massive and parallel changes in their economic base, spatial organisation and social structure" (1991: 4) and she suggests that "transformations in cities ranging from Paris to Frankfurt to Hong Kong and Sao Paulo have responded to the same dynamic." These are bold claims but not everybody accepts them. I want to argue that this thesis, while very stimulating, is simultaneously partial and over-generalised by virtue of its dual focus on (a) economic restructuring and (b) the economically active labour force. Polarisation is seen as a direct and unmediated consequence of economic restructuring in global cities.

I want to argue that, while there is undoubted evidence of growing income polarisation in some capitalist economies, and particularly in the major cities, the evidence on the occupational structure of the economically active labour force points towards growing professionalisation, rather than polarisation. This is not to say that major changes in employment structure and opportunities are not taking place, but that in many Western European countries, they are more likely to create a large and growing unemployed and economically inactive group excluded from the labour force rather than the growth of a large, low-skilled and low-paid labour force. While this may be true in the USA, with its large and growing immigrant labour force, willing to work for low wages (possibly forced to because of the limited nature of welfare provision), it is not necessarily true of all western capitalist countries.

What is missing from Sassen's treatment of polarisation is that the causes of polarisation may be multi- rather than monocausal, that polarisation may be growing in terms of income but not occupation, and that the extent of social and spatial polarisation in any country may be linked to the form of welfare state in different countries. What I want to argue is that Sassen's model of polarisation is US-based, primarily monocausal, and fails to appreciate that the economic pressures towards polarisation are mediated by different welfare state regimes. What may be happening in some cities in the U.S. is not necessarily happening in similar cities in other counties because of differences in the economic, social and institutional context.

Income polarisation: the empirical evidence in Britain

There is strong evidence that income inequality has grown considerably in Britain, the USA and several other western countries during the 1980s and early 1990s. Without going into detail, a variety of data sources shows that the distribution of household income became markedly more unequal. In Britain, the top decile's share of total income rose dramatically in the 1980s while the shares of the lowest 70 percent decreased, particularly those of the lowest 30 percent. This is also true of income from employment as well as total income.

A specific examination of income inequality in London shows that the shares of the top decile of household incomes rose from 24.8 percent of the total in 1979/80 to 28.3 percent in 1985/6 and to 33.5 percent in 1989/90. The share of the next decile rose only marginally (from 16.1 percent to 16.5 percent) whilst that of all other deciles fell. A similar, though less marked, pattern was found in all the other regions. Looking at the changing percentage of households in each region in the national top 10 percent and bottom 10 percent of gross household normal weekly income in 1979/80 and 1989/90, Stark (1992) found that the percentage of households in London in the top 10 percent of national incomes rose from 14.3 percent in 1979 to 20.1 percent in 1989 while the proportion in the bottom 10 percent also rose from 8.4 percent to 10.3 percent. This clearly indicates that households in London became more polarised in income terms relative to the national average during the 1980s. The degree of income inequality between rich and poor rose sharply and the proportions of rich and poor also increased.

Clear evidence of growing income inequality in London is also presented by Buck (1994), who shows, using FES micro data sets, that whereas the interdecile ratio between the incomes of the lowest and the highest decile in London and the UK was very similar in 1978–80 at 3.85 and 3.75 respectively, the ratios had risen to 8.17 and 5.94 in 1989–91. In London the interdecile ratio more than doubled in a decade. But the data on the socio-economic composition of the economically active workforce in London show no evidence of occupational polarisation. On the contrary, they show professional and managerial groups have grown while the manual groups have decreased.

We have, then, what in terms of Sassen's thesis is rather a paradox: growing income polarisation but no polarisation of the occupational structure. Similar evidence from the Netherlands suggests that Britain is not unique in this respect, and preliminary evidence for Paris (Preteceille 1995) agrees. Indeed, there is strong evidence that, despite the growth of part-time service jobs, the occupational structure of the United States is becoming increasingly managerial and professionalised. This interpretation is supported by Esping-Andersen (1993), who argues that occupational upgrading is inherent in the post-industrial trajectory of the United States. Kloosterman (1994: 171) adds: "During the 1970s and 1980s, a decoupling seems to have taken place between the occupational level and the wage level in the United States." ... According to Esping-Andersen, a depolarisation of the occupational structure has been accompanied in the US by a polarisation of wage structure.

To the extent that this is true, and professionalisation is not merely a statistical artifact of the devaluation of job titles which is accompanied by a downwards shift in work skill (a question Wright considers but dismisses), then the paradox of growing income polarisation without occupational polarisation may not just apply to Britain and other similar countries, but to the USA as well. How and why has this come about and what light does it shed on Sassen's thesis of economic restructuring as the major driving force of polarisation?

Economic restructuring, welfare regimes and social polarization

Sassen does touch on the issue of welfare state policy. She states in the introduction to *The Global City* that one of the three key questions it addresses is 'what happens to the relationship between state and city under conditions of a strong articulation between a city and the world economy' (1991: 14). Her answer to this question is clear. She argues that 'the nation state is becoming a less central actor in the world' (1991: 167), and that the welfare state is less important under the 'new economic regime' (1991: 338).

[...]

Sassen is correct that there has been a tendency towards the dismantling of the social welfare system in a number of capitalist countries but I would question that there has been "a generalized dismantling" or a growth of income polarisation in all countries. The significance of the welfare state has varied considerably from country to country, as does the extent and the pace of the retreat from the welfare state. In order to appreciate the causes of variations in social polarisation it is necessary to focus on differences in welfare state regimes, including the availability and level of social benefits, the extent of collective consumption such as education, health and child care, state labour market intervention and the like.

Economic restructuring is extremely important as a major force shaping the nature of western capitalist countries, but economic restructuring does not occur in a social and political vacuum. On the contrary, it everywhere and always takes place within the context of nation-states with different regulatory regimes, legal structures and welfare policies and with different national and local cultures. The outcomes of global economic restructuring are essentially variable, depending on the ways in which restructuring processes are mediated within different states. Different structures of welfare and labour market practices will cushion the impacts in different ways.

[...]

It has long been recognised in the social policy sphere that variations in state welfare structures and policies are important in shaping the patterns of social outcomes. Hilary Silver (1993) suggests that international economic restructuring has led to similar changes in employment and income structure in Britain, France and the USA. There is a shift from manufacturing to services, the absolute share of credential workers increased, more jobs became more insecure, organised labour declined in membership, female labour force participation rates rose, regional employment opportunities diverged and unemployment, income inequality and poverty all increased. But she argues that, "While all these trends move in the same direction, they also proceed at different rates in different countries, so that there has been no national convergence in social variations in socio-economic trends." She quotes Howell as saying: "it is almost a truism of comparative political economy that economic imperatives create basic constraints on states but do not determine the way in which states must deal with those constraints" (1993: 339).

Silver goes on to argue that states indirectly shape the social structures through trade and industrial policies such as labour law and industrial relations policies that influence national labour markets. She instances growth of a flexible workforce in the U.S. and U.K. compared to France, where the Labour Code retarded its development. She also argues that states directly influence social structures through welfare programmes and tax structures which redistribute wealth. Silver [1993, p. 344] concludes by pointing out that

> In recent years, the globalization of markets and rising unemployment have led many to believe in the economic impotence of nation-states. . . . [But] nation-states remain a vital mode of economic organization within the emerging international division of labour. . . . By modifying common global forces in product and labour markets and through redistributive welfare policies, nation-states continue to vary in their social structures. While undergoing similar social trends, national convergence is muted.

[. . .]

Esping-Andersen's (1990) work on welfare state regimes is well known. He argued, on the basis of labour and social security policies, that it was possible to group countries into three major types depending on the character of their welfare regimes. Esping-Andersen's more recent work on comparative changes in class structure and mobility in different welfare regimes takes this argument further, to argue that 'contemporary social stratification is heavily shaped by institutions, the welfare state in particular' (1993: 1).

The argument advanced by Esping-Andersen and his colleagues is that the welfare state revolutionised the structure of the labour market and labour market behaviour. They argue that the social wage violates the assumption that classes and life chances can be identified through common labour market conditions in that welfare states introduce the possibility of a 'welfare state client class' (1993: 19). They also argue that the sometimes massive 'expansion of welfare state employment implies not only new occupational groups, but also the emergence of a huge production and reward system isolated from the operation of market forces' (1993: 19).

They argue that this expansion manifests itself not just by education and training programmes, but by labour market measures and direct welfare state employment growth. They note that in Scandinavia, welfare state employment growth accounts for 'almost the entire net employment increase over the past decades' (1993: 19). They also note that welfare state institutions dictate the choice of non-entry via the provision of a social wage and tax and service treatment of households and their impact on child care and working women. Finally, they suggest that the welfare state also furnishes the basic means for labour market exit through the introduction of early retirement, which provides many redundant workers with the option of a social wage rather than having to move into bottom-end jobs. In sum [Esping-Anderson, 1993, p. 20]:

> the structure of the welfare state is a key feature in the contemporary process of social stratification: it creates and abolishes 'empty slots', it helps decide who fills them and how they are to be rewarded, it defines what is undertaken within them, and, finally, it shapes the pattern of mobility between them.

The principal implication of this analysis is that there are marked differences between the occupational and income structures of different western capitalist countries. They argue that, while the Scandinavian countries 'exemplify an extreme case of a gendered, welfare state service-led trajectory,' Canada and the United States are characterised by their "large low-end consumer service labor market" (1993: 4). They suggest [1993, p. 5] that

In North America, unskilled service jobs tend to be very poorly paid, predominantly filled by youth and immigrants, and function very much as first-entry, or stop-gap jobs. In both [Canada and the USA] a distinct low-end mobility circuit emerges which is unparalleled elsewhere: unskilled sales, clerical and service jobs appear to constitute a common job reservoir for people with low education.

They point out that, where trade unions are centralised, strong and bargain nationally, as in Germany and Scandinavia, the employment outcome will be very different from North America where unions tend to be weak, fragmented and localised.

[...]

They argue that Germany represents a third regime with a welfare state which is both generous and comprehensive but where social rights are strictly tied to employment record and there is a strong commitment to preserve the traditional caring functions of the family. Thus, there is an implicit discouragement of female labour market participation.

Esping-Andersen argues that one of the characteristics of welfare states is that they tend to produce a relatively large 'outsider surplus population,' consisting of people unable to enter into employment, of early retirees, long-term unemployed and others subsisting on the social wage.

In countries such as the United States, on the other hand, where the welfare state is weaker, there is a large, low-wage service proletariat. Thus, post-industrial societies can experience two alternative kinds of polarisation. In the strong welfare states the polarisation is between "a small, but highly upgraded insider structure and a large outsider surplus population. In the other case, a large service class proletariat will constitute the pivotal source of polarization" (1993: 28).

They were only able to study the occupational earnings structure for Germany, Sweden and the United States, but they state [1993, p. 50] that

the pervasive low-wage effect on job trends within the American consumer service sector stands out clearly. The American earnings distribution is almost the exact opposite of the German. Instead of an extraordinarily privileged top, the United States is characterized by its extremely underprivileged bottom. The unskilled service workers are ... a very badly paid workforce.

They conclude by arguing [1993, p. 53] that

The fordist hierarchy has everywhere experienced a marked decline of the traditional manual working class; to a degree this has been offset by a modest rise in clerical and sales occupations. Fordism is, so to speak, becoming post-industrialized. ... Despite the divergent shape of the post-industrial hierarchy, there is very little evidence to suggest strong polarization. Everywhere, the trend favours higher grade occupations such that the shape of the post-industrial occupational hierarchy is biased towards the top and the middle rather than the bottom.

Immigration, low-wage jobs and polarisation in the USA

It is clear from this analysis that the United States is distinctive in the importance of its low-skilled and low-paid consumer service jobs and Esping-Andersen notes that minorities occupy a disproportionate share of these jobs. The role of ethnic minorities in polarisation receives much attention from Sassen (1991: 299–317). She states that "It is impossible to disregard the facts of race and nationality in an examination of social and economic processes in New York. To a lesser extent this is also the case with London" (1991: 299). Sassen adds that "the large influx of immigrants into the United States from low-wage countries over the last fifteen years cannot be understood separately from this restructuring."

Sassen is quite correct in this, that the problem concerns the nature of the causal process. As I have argued elsewhere (Hamnett 1994), Sassen argues that the expansion of low-wage jobs has led to the growth of immigration. "The expansion in the supply of low-wage jobs generated by the major growth sectors is one of the key factors in the continuation of the ever-higher levels of the current immigration" (1991: 316).

In other words, economic restructuring has led to changes in labour demand and migrant flows. She is opposed to the reverse explanation, namely that the existence of large-scale immigration from low-wage countries has enabled the growth of low-wage service jobs. She argues that advanced capitalism "may promote conditions for informalization. . . . The presence of large immigrant communities . . . can be seen as mediating in the process of informalization rather than directly generating it" (1991: 282) and "it is the economy rather than immigrants which is producing low-wage jobs."

I have considerable sympathy with this argument, partly for political reasons, in that it is easy to slip into a right-wing, anti-immigrant stance, blaming immigrants for undercutting wages. I find it very difficult, however, to overlook the fact that the United States, particularly the major cities, has a level of immigration unparalleled in most other western countries, and that it is also almost the only western country to have seen the rapid growth of large numbers of low-wage service jobs.

Many of these jobs are common to all western cities, particularly global cities, but New York and Los Angeles are perhaps unique in the ease with which it is possible to hire cheap immigrant labour to undertake low-paid consumer service jobs. In part, these workers have no other option. The welfare benefits available in the Netherlands, Scandinavia, Germany and, to a lesser extent, in Britain are simply not available at the same level in the U.S. To this extent, occupational polarisation in the U.S. may be the product of both the nature of the welfare state and a high level of immigration from low-wage countries.

CONCLUSIONS

The core of my argument is that while Sassen may be right that, in New York and Los Angeles, there is occupational and income polarisation within the paid labour force, the evidence from other western cities does not support this. In most large non-American cities, income polarisation is combined with professionalisation of the paid labour force. This is not to suggest, of course, that large and growing numbers of people are not experiencing low incomes. Rather it is to argue that, in countries with

stronger welfare states, a larger proportion of the population may be able to live outside the paid labour force on state benefits. I accept, of course, that in a growing number of western countries the welfare state is under grave threat. We may be facing the sort of future currently found in New York and Los Angeles. But, this is not simply a result of the unmediated operation of market forces and privatisation. It is also an outcome of the historical legacy and also of political struggle to preserve welfare states against neo-liberal economic pressures.

REFERENCES FROM THE READING

Buck, N. (1994) "Social Divisions and Labour Market Change in London: national, urban and global factors." Unpublished paper for ESRC London Seminar.

Esping-Anderson, G. (1993) *Changing Classes: Stratification and Mobility in Post-industrial Societies.* London: Sage.

Fainstein, S., I. Gordon and M. Harloe (eds) (1992) *Divided Cities: New York and London in the Contemporary World.* Oxford: Blackwell.

Hamnett, C. (1994) "Social Polarisation in Global Cities: theory and evidence." *Urban Studies* 31: 401–24.

Kloosterman, R. (1994) "Three Worlds of Welfare Capitalism? The welfare state and the postindustrial trajectory in the Netherlands after 1980." *West European Politics* 17, 4: 166–89.

Marcuse, P. (1989) "Dual City: a muddy metaphor for a quartered city." *International Journal of Urban and Regional Research* 13: 697–708.

Sassen-Koob, S. (1984) "The New Labor Demand in Global Cities" in M.P. Smith (ed.) *Cities in Transformation.* Beverly Hills, CA: Sage.

Sassen, S. (1991) *The Global City, New York, London, Tokyo.* Princeton, NJ: Princeton University Press.

Silver, H. (1993) "National Conceptions of the New Urban Poverty: social structural change in Britain, France and the United States." *International Journal of Urban and Regional Research* 17, 3: 336–54.

Stanback, T.M. Jr. and T.J. Noyelle (1982) *Cities in Transition: Changing Job Structures in Atlanta, Denver, Buffalo, Phoenix, Columbus, Nashville.* Charlotte, NJ: Allenheld, Osmun.

PART TWO

The roots of urban politics

Plate 2 Urban infrastructure.

INTRODUCTION TO PART TWO

Any larger set of power relations within cities must rest on winning elections, particularly in holding the mayoralty. Daniel Patrick Moynihan once observed that Tammany Hall was the oldest continually existing democratic political organization anywhere in the world. Founded in 1786, the Society of St. Tammany provided crucial support under the leadership of Aaron Burr for the election of Thomas Jefferson as President in 1800. For all but ten years between 1858 and 1934, Tammany had the major hand in electing every mayor of New York City, most notably when William Marcy Tweed was its "Grand Sachem." In his era, New York City paid $13 million for about $1 million worth of materials and construction costs to erect a new courthouse, with most of the money going into the pockets of senior members of "the machine." (The building is now used as headquarters for the New York City school system.) Tammany's success allowed it to erect elegant headquarters (the Hall) first on 14th Street and then on the west side of Union Square.

Municipal corruption and the monopolization of power by those associated with the machine led to periodic attempts by middle-class notables and excluded or underrepresented ethnic groups to drive the "regular" or loyal organization Democrats out of office, reform the conduct of city business, and broaden the constituencies included in city government. While he had a number of predecessors, Republican-Fusion Mayor Fiorello LaGuardia was particularly successful in achieving these goals, serving three terms between 1934 and 1945, and setting a pattern later followed by John V. Lindsay, Rudolph Giuliani, and Michael Bloomberg. In the 1960s, reform Democrats in Manhattan also launched a campaign against Tammany, with Edward I. Koch ultimately defeating Tammany leader Carmine DeSapio.

These waves of reform pushed Tammany toward hard times. It had to sell its headquarters in 1943 and the Society of St. Tammany ultimately went bankrupt in the 1950s, after DeSapio's ouster from the New York County Democratic Party chairmanship. As candidate-centered campaigns, civic groups, community-based organizations, and municipal labor unions became the main organizational vehicles for winning urban elections, old-fashioned political clubs modeled on Tammany have become ghosts of their former selves. Meanwhile, the political outsiders running who periodically ousted entrenched party incumbents across the nation reshaped how urban governments go about their business, often placing greater emphasis on providing services through community-based organizations and being more responsive to neighborhood organizations.

Even though party organizations have atrophied, political parties remain relevant to winning and holding elective office. Many large, old Eastern cities still have strongly partisan systems, where the Democratic nominee is likely to win office. Even in formally non-partisan cities, most of those who win council and mayoral elections are Democrats, since urban voters tend to hold that partisan orientation. To the extent that the partisan orientation of a city makes its municipal elections non-competitive, party incumbents can become entrenched, even in formally non-partisan systems, producing what Raymond Wolfinger has termed machine politics without political machines.

The machine-reform dynamic has therefore been central to the analysis of urban politics. It is not the only important dynamic – subsequent sections of the Reader will examine how the interactions between the public sector and private asset holders' sectors shape urban power and at how racial succession prompted challenges from newcomers to coalitions dominated by older ethnic groups – but it continues to be fundamental. The essays in this section shed light on the tensions between "regulars" and "reformers." Part 2 opens with a defense of machine politics by Richard Croker, leader of Tammany Hall from 1886 to 1901. Born in Ireland, a pugilist, leader of the "Fourth Avenue Tunnel Gang," and minor city official, Croker rose to head the organization in the wake of the deposed and disgraced Tweed, taking command and returning it to power. This was a period of tremendous immigration to New York, rapid population growth and industrialization, and no small level of class and ethnic conflict. Croker extols the ability of the Democratic Party to tame and organize all these forces. The pattern of party dominance established in New York was echoed in many forms in other cities, perhaps none more storied than the Cook County Democratic Party under Mayor Richard J. Daley. Born in an Irish Catholic neighborhood just after the turn of the century, Daley became Mayor of Chicago in 1955 and was re-elected three more times, passing away in office in December, 1976. He was reputed to have played a decisive role in the election of John F. Kennedy in 1960. Daley governed Chicago through a tumultuous period that included the rise of urban unrest and the 1968 Democratic convention. A selection from Milton Rakove's book, *Don't Make No Waves, Don't Back No Losers*, gives a strong flavor of how this organization worked.

The consolidation of urban power in the hands of party leaders regularly provoked reactions, both within and across the cities of America. In the older, more partisan cities, reform movements drew on middle- and upper-class elites as well as ethnic groups disappointed by their slow rate of political advancement within the party organizations to mount reform challenges. Even when they succeeded in electing reform mayors, these political movements were hard to institutionalize due to the differences among their supporters. Nevertheless, they tended to modernize and reform government and forced the regular political organizations to renovate themselves in order to get back into power. In the newer cities that formed later, especially in the twentieth century, reform had taken broader hold on the countries and regular party organizations had a harder time becoming dominant. Amy Bridges and Richard Kronick examine how these factors played out in the creation of urban political systems across the country. They show the complexity of the base of support for reform.

Drawing on his fieldwork in New Haven, an old, industrial, ethnic port city in Connecticut that also happens to be home to Yale University, Raymond Wolfinger analyzes how and why party organizations can still play an important role in shaping the exercise of power even in the wake of reform movements. Looking at the same evidence as Robert Dahl (Part 3), Wolfinger draws an important distinction between centralized political machines – which have largely not survived, and may not even have been all that common in the previous century – and machine politics, the decentralized influence on party incumbents and their followings.

Finally, Michael Jones-Correa examines a contemporary rendition of a centuries-old story: how new immigrant groups face an established party organization dominated by much older ethnic groups. He tells the fascinating story of Queens, New York. Though a largely middle-class and suburban part of New York City, it also has the second highest share of foreign-born residents of any county in the United States after Miami-Dade County in Florida. Somewhat paradoxically, it is also dominated by a county Democratic Party organization led by former Congressman Thomas J. Manton, one of the last great practitioners in the art of being an Irish political boss.

"Tammany Hall and the Democracy"

from North American Review (1892)

Richard Croker

Editors' Introduction

The popular image of political machines tends to focus on their excesses. Machines are associated, as well they should be, with varieties of corruption ranging from abuse of patronage positions to outright theft of public funds. Machine politicians are known to resort to fraud and intimidation to ensure victory, bullying opponents into submission, enabling supporters to cast illegal votes, and ultimately "losing" ballots thought to favor opponents.

The career of Richard Croker, leader of New York City's Tammany Hall from 1886 to 1901, does nothing to discredit these images. The son of Irish immigrants, Croker made his mark as the leader of the "Fourth Avenue Tunnel Gang," a group of brawlers loyal to Tammany politicians; he allegedly shot and killed a man during an 1874 election day fight, though he was acquitted at trial. In reward for his youthful service, he was nominated for a variety of elective and appointive posts, from Alderman to city coroner to fire commissioner, all of which enabled him to extend his power base in the organization. Although Croker probably did not steal directly from the city treasury, as did his predecessor William Marcy Tweed, he used his connections to amass a fortune sufficient to allow him to retire comfortably to estates in Ireland and Florida.

However, his true accomplishment as leader was to institutionalize and consolidate the party's power. Under his leadership, Tammany Hall was transformed from many competing factors into the sole voice of New York City "Democracy." Croker made Tammany Hall a cohesive, hierarchical organization, rather than a loose confederation of independent ward leaders. The dispensing of patronage and contracts was more centralized; the more free-wheeling committee leaders were disciplined. Like the heads of businesses that were at just this same time becoming complex corporations, Croker was intent on forming a reliable, efficient organization.

This essay, published in 1892, demonstrates the importance he placed on disciplined party organization. To him, a political organization is like an army, fighting for victory, circumstances that make obedience to the leadership essential. Indeed, he so admires effective organization that he could not resist praising the Jacobins of the French Revolution, who initiated a violent "reign of terror," but nonetheless accomplished their aims with noteworthy organizational skill. He believes that Tammany Hall's ability to create a regimented governing hierarchy brings a host of benefits to the city's residents: New York City, he argues, is efficiently run, with good services and reasonable tax burdens. The strong electoral majorities for Tammany-backed candidates, despite Republican efforts to paint them as corrupt, is sufficient evidence of Tammany's success.

"Tammany Hall and the Democracy" provides an interesting contrast to other historical writings on the political machine familiar to students of urban politics. Works such as *Plunkitt of Tammany Hall*, which conveys the opinions of George Washington Plunkitt, a district leader under Croker, or Milton Rakove's *Don't Make No Waves, Don't Back No Losers*, a study of Richard Daley's Chicago (excerpted later in this section), show a more compassionate, humane face of the party organization. George Washington Plunkitt's district leader is forever dashing to help put out fires or hand out gifts at Jewish or Italian weddings; the committee-men described by Rakove are busy fixing parking tickets and arranging jobs for constituents. Reading these books, we learn that these tasks help keep the party organization in power, but we are still charmed by the human touch shown by the party's representatives. Croker's defense of Tammany Hall, in contrast, is unabashedly about the benefits of accumulating and using power.

The rise and consolidation of political machines is a fascinating topic that has been taken up by many political scientists and historians. Histories such as Jon C. Teaford's *The Unheralded Triumph: City Government in America* (Baltimore, MD: Johns Hopkins University Press, 1984) and Eric Monkkonen's *America Becomes Urban: The Development of US Cities and Towns, 1780–1880* (Berkeley: University of California Press, 1988) provide good overviews of the rise of machine politics in U.S. cities. An analysis of the machine in its very early stages may be found in Amy Bridges' *A City in the Republic* (Cambridge: Cambridge University Press, 1984). Martin Shefter's "The Emergence of the Political Machine: An Alternative View" (in Willis D. Hawley *et al.* (eds), *Theoretical Perspectives on Urban Politics*, Englewood Cliffs, NJ: Prentice Hall, 1976) emphasizes Richard Croker's role as the consolidator of Tammany Hall power, and stresses the coalitional nature of machine politics.

No political party can with reason expect to obtain power, or to maintain itself in power, unless it be efficiently organized. Between the aggressive forces of two similar groups of ideas, one entertained by a knot of theorists, the other enunciated by a well-compacted organization, there is such a difference as exists between a mob and a military battalion. The mob is fickle, bold, and timid by turns, and even in different portions it is at the same time swayed by conflicting emotions. In fact, it is a mere creature of emotion, while the drilled and compacted battalion is animated and supported by purpose and scientific plan. It has leaders, and these leaders are known to every man in the ranks and possess their confidence. It is thus that a single company of infantry is able to quell almost any popular outbreak in a city; and a regiment is completely master of the situation, even if it be outnumbered by the malcontents in the proportion of ten or twenty to one.

The value of organization in the case of political parties does not appear so obviously upon the surface; but in point of fact organization is one of the main factors of success, and without it there can be no enduring result. In the immense republic of the United States, which is really a congress or union of over forty separate republics, each having its interests more or less dissociated from those of the others, and yet acknowledging the bond of a common political interest, the organization of a national party must, to a large extent, be based upon a system of deferential compromise, and be an aggregation. The Democrat of New York and the Democrat of Iowa are agreed on certain fundamental doctrines, and in order to put these in action they forbear to press the acceptance of ideas as to which they are at variance. They only vote for the same candidate once in four years; at other elections they choose Governors, Representatives, etc., who are at liberty to entertain widely different views as to the extent to which certain political theories should be made to operate. Thus an ultra tariff-reformer from Nebraska and a very mild tariff-reformer from some redeemed district of Pennsylvania or Massachusetts may each be an excellent Democrat at home; and they may vote harmoniously as Congressmen on national questions; but the two are not as strong and effective as if they were both members of some political club with one watchword and one purpose.

No great army ever has the cohesive power of a regiment. The larger the mass the less perfectly do its members know the habits and purposes of its leader, having no close personal contact with him; but in the regiment, which is the unit and type of military strength, every private knows his captain and his colonel as well. In the course of service he sees all his comrades and officers in array; he sees the officers advance and salute the commander and that salute returned, and thus experiences the spirit and purpose that animate the entire body. This feeling of common purpose is the supreme aim of military organization in the direction of effectiveness; and a compacted and select political club or society is governed by the same processes.

It does not detract at all from the truth of this statement, that local political organizations composed largely of depraved men of revolutionary tendencies have often been powerful engines in government. It rather proves the essential verity of the principle; and indicates the necessity of a sound political basis. Cavalry is an important and powerful factor in war, whether it consist of a horde of Scythian robbers following some incarnate fiend of strife, or of a gallant "Six Hundred" charging down some Valley of Death in obedience to a mistaken order and led by a fearless and trained leader. When we consider the ghastly turmoil of the French Revolution, we cannot fail to admire the success, the influence, the resistless power of the Jacobin Club, not because the club was praiseworthy, since its actions were abhorrent, but because it was skillfully organized and handled. When its representatives sat in the convention, they knew their orders, and they were also conscious that it was their business to carry them out. They acted upon the principle that obedience to orders is the first duty of the soldier, and that "Politics is war." Chess is war; business is war; the rivalry of students and of athletes is war. Everything is war in which men strive for mastery and power as against other men, and this is one of the essential conditions of progress.

The city of New York today contains a political organization which, in respect of age, skillful management, unity of purpose, devotion to correct principles, public usefulness, and, finally, success, has no superior, and, in my opinion, no equal, in political affairs the world over. I mean the Tammany Democracy. I do not propose to defend the Tammany organization; neither do I propose to defend sunrise as an exhibition of celestial mechanics, nor a democratic form of government as an illustration of human liberty at its best. In the campaign of 1891 almost the only argument used by the Republicans against the Democrats was the assertion that [Tammany-picked gubernatorial candidate Roswell] Flower was the candidate of a corrupt political club, and that club was named Tammany. Tammany was accused of every vice and crime known to Republican orators: it was a

fountain-head of corruption; it was because of it that every farmer throughout the State could not at once pay off his mortgages; it took forty millions annually from the citizens of New York and gave them nothing in exchange for it. To the credit of the Democrats let us note the fact that, while this torrent of abuse was being poured upon the heads of voters, Democrats did as the inhabitants of Spain are said to do when the clouds are opened, "they let it rain." Nobody apologized for the misdeeds of the alleged malefactor; the Democrats went before the people on legitimate issues, and the result of the affair was expressed in the figures, 47,937 majority. I doubt if the Democracy would have fared anything like as well if they had defended or apologized or explained away. "He who excuses himself accuses himself" is a time-worn proverb. They let Mr. Fassett [Republican state senator and failed gubernatorial candidate in 1891] shout himself hoarse over "Tammany corruption," and they won the victory.

In fact, such a defensive attitude would have been wholly at variance with the basis on which the Tammany Democracy acts. A well-organized political club is made for the purpose of aggressive warfare. It must move, and it must always move forward against its enemies. If it makes mistakes, it leaves them behind and goes ahead. If it is encumbered by useless baggage or halfhearted or traitorous camp-followers, it cuts them off and goes ahead. While it does not claim to be exempt from error, it does claim to be always aiming at success by proper and lawful methods, and to have the good of the general community always in view as its end of effort. Such an organization has no time or place for apologies or excuses; and to indulge in them would hazard its existence and certainly destroy its usefulness.

The city and county of New York comprise a population of nearly two millions and furnish the business arena for near-by residents who represent two millions more. The political party, then, that is uppermost in New York legislates locally for the largest municipal constituency on the planet, except one. The task is clearly one of enormous magnitude, and demands a combination of skill, enterprise, knowledge, resolution, and what is known as "executive ability," which cannot be at once made to order, and cannot be furnished by any body of theorists, no matter how full may be their pockets or how righteous may be their intentions.

Since the Whig party went out of existence the Democrats have administered the affairs of New York County, rarely even losing the mayoralty except on personal grounds; always having the majority in the Board of Alderman, and as a rule the Sheriff's and County Clerk's offices. And at the same time the guiding force of the New York Democracy has proceeded from the Tammany organization.

As one of the members of this organization, I simply do what all its members are ready to do as occasion offers, and that is, to stand by its principles and affirm its record. We assert, to begin with, that its system is admirable in theory and works excellently well in practice. There are now twenty-four Assembly districts in the county, which are represented in an Executive Committee by one member from each district, whose duty it is to oversee all political movements in his district, from the sessions of the primaries down to the final counting of the ballots after the election polls are closed. This member of the Executive Committee is a citizen of repute, always a man of ability and good executive training. If he were not, he could not be permitted to take or hold the place. If he goes to sleep or commits overt acts that shock public morality, he is compelled to resign. Such casualties rarely occur, because they are not the natural growth of the system, of selection which the organization practices; but when Tammany discovers a diseased growth in her organism, it is a matter of record that she does not hesitate at its extirpation.

Coincident with the plan that all the Assembly districts shall be thoroughly looked after by experienced leaders who are in close touch with the central committees, is the development of the doctrine that the laborer is worthy of his hire; in other words, that good work is worth paying for, and in order that it may be good must be paid for. The affairs of a vast community are to be administered. Skillful men must administer them. These men must be compensated. The principle is precisely the same as that which governs the workings of a railway, or a bank, or a factory; and it is an illustration of the operation of sophistries and unsound moralities, so much in vogue among our closet reformers, that any persons who have

outgrown the kindergarten should shut their eyes to this obvious truth. Now, since there must be officials, and since these officials must be paid, and well paid, in order to insure able and constant service, why should they not be selected from the membership of the society that organizes the victories of the dominant party?

In my opinion, to ask the question is to answer it. And I add that the statement made be the enemies of Tammany that "Tammany stands by its friends," is, in fact, praise, although intended for abuse. Tammany does stand by its friends, and it always will until some such change occurs in human affairs as will make it praiseworthy and beneficial that a man or an association should stand by his or its enemies. We are willing to admit that the logical result of this principle of action would be that all the employees of the city government, from the Mayor to the porter who makes the fire in his office, should be members of the Tammany organization. This would not be to their credit.

And if any one of them commits a malfeasance, he is just as responsible to the people as though he were lined bodily out of the "Union League" or some transient "Citizen Reform Association," and he will at once find himself outside of the Tammany membership also.

Fearfully and wonderfully made are the tales that are sent out into the rural districts touching the evil effects of "Tammany rule." The trembling countryman on arriving in New York expects to fall into a quagmire of muddy streets, and while struggling through these quicksands he fears the bunco man on one side and the sandbagger on the other. Reaching some hotel, be counts on being murdered in his bed unless he double-lock his door. That his landlord should swindle him is a foregone conclusion. And when no adventure happens, and he reaches home in safety, he points to himself, among his neighbors, as a rare specimen of a survival of the dangers that accompany the sway of a Democratic majority in New York.

The facts are that New York is a center to which the criminal element of the entire country gravitates, simply because it offers at once a lucrative field for crime and a safe hiding-place. Therefore, to preserve social order and "keep the peace" in New York demands more ability and more policemen than are required in country solitudes. It is safe to say that any right-minded citizen who attends to his own affairs and keeps proper company and proper hours is as safe in New York as in any part of the globe, the most violently Republican township of St. Lawrence County not excepted. Our streets are clean and are in good order as to the paving, except where certain corporations tear them up and keep their rents gaping. . . . It is conceded that we have the best police and fire departments in the world. . . . Our tax-rate is lower than that of dozens of other American cities whose affairs are not nearly so well administered. Nor is the tax-rate low because the assessed values are high. If any real-estate owner claims that his property is overvalued, you can silence him at once by offering to buy it at the valuation. . . .

That the Tammany Hall Democracy will largely aid in organizing victory for the national ticket next November is beyond question. The national Democracy is free to choose whatever candidate it may prefer. Tammany has no desire to dictate or control the choice; its part in the conflict is to elect the candidate after he shall have been named. No matter what Republican majorities may come down to the Harlem River from the interior of the State, we propose to meet and drown them with eighty-five thousand majority from New York and Kings.

T
W
O

"The Ward Organizations"

from *Don't Make No Waves, Don't Back No Losers* (1975)

Milton Rakove

Editors' Introduction

Milton Rakove's *Don't Make No Waves, Don't Back No Losers*, from which this excerpt is drawn, provides a participant observer's look at the political organization run by Richard J. Daley, mayor of Chicago and head of the Cook County Democratic party from 1955 until his death in 1976. According to some estimates, these twin positions gave Daley control over 35,000 patronage jobs that he could use to leverage ongoing monetary and electoral support to his party. Daley's longevity and ability to maintain a disciplined, successful political machine long after the "death of the machine" had been proclaimed has made him a focus of journalistic and scholarly attention. He succeeded partly by learning to use federal programs such as Urban Renewal to his advantage, and partly by co-opting potential challengers, for example, by establishing a subordinate "sub-machine" in the African-American community.

Although much popular and scholarly attention is paid to "bosses," this reading shows the importance of the grass roots. Ward leaders are the crucial link between constituents and policy makers. They get to know the voters in their ward and mobilize them on election day. In return, the party provides them and their family members with patronage jobs, promotes them, and gives them access to opportunities for personal enrichment.

In the next reading, Raymond Wolfinger contrasts machine politicians with the issue-oriented "amateurs" who seldom succeed at the local level. The ward bosses portrayed here devote their lives to politics, yet remain strangely apolitical. Among these politicians one finds scant discussion of civil rights, urban planning, or neighborhood renewal (except if patronage jobs are at stake), and in fact Vito Marzullo, an Alderman and a party ward leader, rejects an entreaty to contribute to a local anti-abortion organization. "People are for it and people against it," he complains, recognizing that such divisive issues can lose as many votes as they win.

Milton Rakove's (1918–1983) career straddled the worlds of academic political science and real-world politics. After receiving a Ph.D. in 1957 from the University of Chicago, he joined the faculty of the University of Illinois at Chicago, where he taught until his death, but he was also actively engaged in local and national politics, working as a political consultant for Democratic candidates and office holders. (He served as precinct captain of Chicago's 49th ward and ran unsuccessfully for Cook County Commission in 1970.) These experiences gave him the insider status that allowed him to gather information for the book.

Popular accounts of Richard J. Daley and the Chicago political machine may be found in Mike Royko's *Boss* and *American Pharoah*, by Adam Cohen and Elizabeth Taylor (Little, Brown, 2000). More scholarly analyses of Chicago politics under Daley include Dick Simpson's *Rogues, Rebels and Rubber Stamps: The Politics of the Chicago City Council* (Boulder, CO: Westview Press, 2001).

The most significant of all the party's relationships with any of the constituent parts of the organization are its dealings with the ward organizations in the city. Each of Chicago's fifty wards is an entity unto itself, a fiefdom ruled in the party's name by the committeeman who is a prince of the blood, a duke, a baron, an earl, or a mere knight. The ruling coterie in the party's hierarchy is analogous to the aristocracy of a sixteenth- or seventeenth-century medieval court.

The party is not a monolithic, totalitarian dictatorship but rather a feudal structure, dominated by Chairman Richard J. Daley. But Daley is neither an absolute monarch nor a totalitarian dictator. He is, rather, like Leonid Brezhnev, in the Soviet Politburo, a prominent figure who towers above all other members of the Politburo but does not stand alone. While Daley has gathered more power into his hands than has any other politician in Chicago's history, that power rests on a foundation of a shifting coalition of power groups and powerful figures, some within the political organization and some in the community at large. Daley governs in their names only so long as he keeps a stable balance of power among the groups which make up Chicago's body politic, maintains a decent level of order in the city, and provides sufficient perquisites to pacify the aspirations and demands of the city's political and economic aristocracy.

THE ROLE OF THE COMMITTEEMAN

The committeemen are the party's representatives in the wards, the link between the wards and the party headquarters, and the conduit through which party perquisites are distributed to the wards. Once a man is elected committeeman, it is almost impossible to dislodge him from his position unless the party leadership decrees that he step down from his post. Committeemen are elected in party primaries every four years by the registered voters of the party in what are known as precinct-captain elections, elections in which the machine vote is turned out by the precinct captains, and the unreliable voters are not disturbed.

An incumbent committeeman is protected against challengers by the law requiring a candidate for ward committeeman to get a petition signed by at least 10 percent of the party's vote in the last primary election. This is five times as much as the legal requirement for getting on the ballot to run for alderman (2 percent of the vote cast in the last aldermanic election) and twenty times as much as candidates for state representative or congressmen need. Candidates for those offices are required by state law to have petitions signed by only one-half of one percent of the registered voters in the party in those legislative districts. Securing that many signatures on a petition requires an organization, precinct captains familiar with the registered Democratic voters in the ward who are willing to distribute petitions on behalf of the candidate, and a substantial number of registered Democratic voters who are willing to alienate the ward committeeman by signing the petitions of a rival. While these conditions make it possible for an opposing candidate to challenge an incumbent ward committeeman, they make it exceedingly difficult for him to meet the legal requirements, not to mention the political hazards of challenging the incumbent. Few precinct captains will distribute petitions for a challenger to an incumbent ward committeeman, and not many registered Democratic voters will alienate an incumbent committeeman by signing his rival's petitions.

Once safely ensconced in his job, the committeeman is the absolute master of the ward organization. All party electoral funds are channeled directly through him, to be distributed by him at his pleasure. All patronage positions distributed by the county central committee to the ward organizations are funneled through him. All precinct captains are appointed by him and are subject to summary dismissal without recourse to any other party authority. All favors dispensed by the party hierarchy and by city and county officials connected with the party require his approval before any party or governmental functionary will take action on such requests. No person from his ward can be hired in any patronage position on the city and county level without his sponsorship. And people who have been hired might be fired on his request from positions they hold with the city or county. In other words, he holds in his hands the destinies and livelihoods of all members of his ward organization.

The position of ward committeeman carries with it not only perquisites but also responsibilities. In return for granting him absolute authority in his ward, the party expects him to support party decisions without question; to deliver for all party candidates, regardless of his personal feelings about those candidates, and to maintain an effective ward organization which serves the interests of the party as well as the needs of the committeeman's constituents. If one is given a leadership position by the party, one is expected to lead and produce or get out. There is little room in the relationship of the ward committeemen to the county central committee for sentiment, friendship, or failure. Powerful committeemen who have been high in the party hierarchy for many years have been cast aside when they lost control of their ward organizations or failed to deliver for the party in an election. For example, in 1970, County Clerk Edward J. Barrett, who was the long-time committeeman in the 44th Ward, was called down and raked over the coals by the party's hierarchy after he had lost the aldermanic election in his ward to a liberal reform Democrat. Barrett, who was seventy years old and who had been county clerk for sixteen years, was restated for his office only on condition that he give up the committeemanship of his ward. There are numerous other examples of formerly powerful committeemen who held elective or appointive offices under the party and who lost both their committeemanships and their offices when they failed to deliver in their wards.

[. . .]

Other compensations can accrue to a committeeman in the city of Chicago and even in the suburbs. If he is an attorney, he can set up a law firm and get business for that firm from city and county agencies and from commercial firms and individuals in his ward. He can open an insurance office and sell insurance to commercial firms and major property owners in his ward. Since insurance rates are set by the state, the costs to prospective customers are no higher than other insurance brokers charge, and buying insurance from the committeeman might help to discourage building inspectors, electrical inspectors, fire inspectors, and sundry other city and county officials from disturbing the property. A diligent and enterprising ward committeeman might be able to sell performance bond insurance to contractors who are doing work for the city or county. There is the possibility of becoming a member of the board of directors of a bank or two in the ward, of having an interest in a commercial firm which sells materials or does business with the city or county agencies, or of getting tips on stock deals or beneficial business transactions from bankers or businessmen in your ward or in the Loop. He may even, if he is lucky, and close enough to the inner circle, be able to buy land which the city or county may become interested in for public purposes, and sell that land back to the city or county for a fair return on his investment. Finally, if he is exceptionally hardworking and willing to put in long hours, he may be able to hold several jobs at the same time, serving in the state legislature and holding a major appointive position with the county or city.

A committeeman's status in the pecking order of the county central committee, his influence in the selection of county or state candidates, and the allocation of patronage jobs to his ward, in terms of both number and quality of jobs, are directly related to his ability to produce votes for the party. These rules, of course, must be qualified by the preeminence of the Irish Catholic princes-of-the-blood committeemen in the power structure of the county central committee, who dominate the party organization. Even though the black wards in the city have become the bedrock strength of the Democratic party in Chicago, delivering massive percentages of Democratic votes in elections, black committeemen, as a whole, do not enjoy the power and prestige of their Irish counterparts. Since the death of the late Congressman William Dawson, who was the committeeman of the 2nd Ward, no black committeeman has achieved Dawson's status in the inner circle of the party, although several old Dawson political lieutenants have achieved substantial court status as dukes or barons. . . .

A review of the performance of every committeeman in his ward is conducted by the party leadership after every primary and every election. Every committeeman is required to give the chairman an estimate of the votes he expects to get in his ward or township before the election date, and his performance is weighed against that estimate. If he cannot deliver for the party he is in trouble. If he brings in a significantly higher vote total than the estimate he gave the county central committee, he could be in trouble, since it is then clear to the party leadership that the committeeman does not know what is going on in his ward. Losing the ward, or delivering a lower percentage of the vote than was estimated, can have serious consequences for a committeeman. He could lose his job, be "viced" and replaced by an acting ward comitteeman selected by the precinct captains of the organization at the instigation of the county chairman, or he could have his patronage cut and have important jobs that he needs to reward his precinct captains taken from him.

[. . .]

THE PATRONAGE SYSTEM

"This is my judge, this is my county commissioner, and this is my state senator," Alderman and Committeeman Vito Marzullo of the West Side's 25th Ward told me at lunch one day at his table in the Bismarck hotel. "I got an assistant state's attorney, and I got an assistant attorney general," Marzullo said to a *Chicago Daily News* reporter in 1967. "I got an electrical inspector at $10,500 a year, and street inspectors and surveyors and a $700-a-year county highway inspector. I got an administrative assistant to the zoning board, and some bailiffs and some process servers and a county building inspector at $8,400 a year . . . I got fifty-nine captains and they all have jobs. I've got ten or twelve good jobs at the state department of highways, and others with the secretary of state, county clerk and county assessor and the city department of streets." The alderman was spelling out for the *Daily News* reporter some of the perquisites that flow to his organization, one of the most effective and efficient ward organizations in the city.

Few of the committeemen are willing to publicly confront the press with the realities of ward politics in Chicago, but Alderman Marzullo was only spelling out what is a privately accepted fact of political life by Chicago Democratic politicians. It is impossible try make an accurate analysis of the number of patronage positions distributed by the Cook County Democratic Central Committee to loyal workers, precinct captains, and supporters, but a fairly educated guess would be that there may be approximately 30,000 patronage positions available to the county central committee. The jobs range from $295-a-month clerks to $25,000-a-year directors of city commissions. They are located in agencies ranging front approximately 70 jobs in the county coroner's office to approximately 6,000 jobs under the control of the county board and approximately 8,000 jobs in the city of Chicago's departments and commissions. There are also many thousands of jobs in private industry throughout the Chicago metropolitan area which require the sponsorship of Democratic ward committeemen.

Estimates of the number of jobs available to individual committeemen range from a maximum of approximately 2,000 jobs in Chairman Daley's 11th Ward to a dozen jobs in a heavily Republican township. Since there are 5,463 precincts in the county, of which 3,148 are in the 50 wards in the city, the patronage pool could average from 500 to 600 jobs per ward. If each job is worth ten votes in an election, the machine has a running start of approximately 300,000 votes derived directly from the patronage system. No person can be appointed to a patronage position at any level without the prior sponsorship of his ward committeeman. This maxim applies to all appointments, not only to local patronage positions but also to high-level federal posts such as federal judges and important positions in the federal bureaucracy. The level of appointment to which the system applies was recounted to me by a friend of mine, a young attorney in Chicago, who was working as a precinct captain in one of the Democratic ward organizations. He was being considered for an appointment to one of the major federal regulatory commissions in Washington. He was told in Washington, however, that he could not be appointed without Daley's approval. He could not get Daley's approval without the sponsorship of his ward committeeman. On a night when his ward committeeman was holding office hours at the ward headquarters, the attorney presented himself to the ward committeeman, who was eating behind his desk with a big cigar sticking out of one corner of his mouth, and two precinct captains sitting on either side of the desk. The young attorney told his ward committeeman that there was an opening on one of the federal commissions and that he was being considered for the opening but needed the sponsorship of the ward committeeman. The ward committeeman took the cigar out of his mouth, and said to the young attorney, "You want to be a United States commissioner?" "Yes," replied the young attorney. Turning to the precinct captain seated on his left, the committeeman said to the precinct captain, "Do you want to be a U.S. commissioner?" "No," replied the precinct captain. Turning to the precinct captain seated on his right, the ward committeeman repeated, "Do you want to be a U.S. commissioner?" "No," replied the second precinct captain. He turned back to the young attorney standing before him and said, "Okay, you've got it. But you have to be cleared by the mayor. Go home and wait until I call you. I'll set up an appointment with the mayor." Two nights later the attorney's telephone rang. It was the ward committeeman, telling him to meet at the mayor's office in City Hall the next day at 9 a.m. The next morning the committeeman ushered in the young attorney to see the mayor, who was seated behind his desk. The committeeman made the introductions. "I understand, young man, that you are an outstanding attorney," said the mayor. "Well, your honor," said the attorney, "I'm not an outstanding attorney, but I'm a good one." At that point all conversation ceased. The mayor sat enigmatically, looking like Buddha. The committeeman sat expectantly, waiting for some further words of wisdom to drop from the lips of the mayor. After two minutes the young attorney realized that the mayor was not about to speak. Sitting there, in an embarrassed silence, the attorney had an inspiration. Turning to the mayor, he said, "Mr. Mayor, I want you to know that if I get this appointment, I will administer my position with every due consideration for the interest of the Democratic organization of Cook County." Smiling broadly, the mayor turned to the door and beckoned to his secretary, saying, "Call in the photographers!" Patronage jobs are usually parceled out to precinct captains in a ward organization on the basis of their efficiency in their precincts, and their qualifications. Young attorneys seeking to climb the political ladder by manning a precinct in a ward organization can be appointed as an assistant state's attorney, assistant corporation counsel, or work in the state attorney general's office, if the office is held by a Democrat. The college graduate with accounting skills can work in the county comptroller's office, or in a comparable position in the city bureaucracy. A high school graduate without any particular skills or training can work on a highway crew, can be a sidewalk inspector, a house-drain inspector, a forest ranger, a tree cutter, a sewer inspector, or a street sign wiper.

A classic example of how the system works was recounted to me by a Chicago alderman. According to the alderman, Committeeman Bernard Neistein of the 29th Ward sponsored a man who was a minister in a storefront church for a position with a city department. The irate department head called Neistein to complain that the prospective job candidate could neither read nor write adequately. "Put him to work. I need him," Neistein told the bureaucrat. Turning to the alderman, Neistein showed him a picture of the minister's podium, where, facing the audience, was a sign bearing the following information for the edification of the worshipers: "____ Baptist Church. _____ Minister. Bernard Neistein, Ward Committeeman." ("Bernie Neistein," one high-level local bureaucrat told me, "is reasonable. If he sends you five guys to put to work, only two are illiterate. But Matt Bieszczat sends you five illiterates and wants you to take them all!") In most cases, a precinct captain holds his patronage position as long as he delivers his precinct or covers the precinct to the satisfaction of his ward committeeman. Like his committeeman, a precinct captain is expected to carry his share of the burden for the party in his ward, as his committeeman is expected to carry the ward's share in city or county elections. Thus, power is delegated from the county central committee to the ward committeemen, and from the ward committeemen to the precinct captains. The entire system operates on the principle of autonomy of authority at each level in the political pyramid. When a man is given a precinct, it is his to cover, and it is up to him to produce for the party. If he cannot produce for the party, he cannot expect to be rewarded by the party. "Let's put it this way," Alderman Marzullo told me. "If your boss has a salesman who can't deliver, who can't sell his product, wouldn't he put someone else in who can?" Like the ward committeeman, a precinct captain is required to estimate the results of the coming election in his precinct. And, like the ward committeeman, he is expected to predict accurately. Whereas losing a precinct may result in losing a patronage job, a captain who overestimates or underestimates the vote totals in his precinct also could be in trouble with his committeeman. Being unable to predict vote totals indicates to the ward committeeman that the precinct captain either is not working his precinct well or does not know what the voters in his precinct are thinking. Either reason is sufficient cause for disciplinary action such as the following incident recounted to me by a friend who spent election night, 1964, in the ward headquarters of former Alderman and Committeeman Mathias ("Paddy") Bauler of the 43rd Ward as the precinct captains came in to report. On the wall of Bauler's office was a tally board indicating which precincts had been carried and which had been lost in that day's election. A captain, who had lost his precinct that day, appeared before Bauler. "What kind of a job are you going to look for now?" asked Bauler of the abject precinct captain, standing before him. As the unfortunate miscreant went out the door of the office, the next precinct captain appeared before the committeeman. "Your brother didn't vote," said Bauler to the precinct captain. "My brother didn't vote!" exclaimed the captain. "I'll kill him!" "Okay," said Bauler to the captain, "bring in the evidence, and you'll keep your job."

The party policy by which each captain or committeeman is expected to carry his precinct or ward, or suffer the consequences of failure regardless of the overall result of the election, forces each individual in the party's organization to concern himself only with his part of the action, not with the total result. For, even if the party wins the election, and he does poorly in his ward or precinct, the electoral victory will have little significance for him personally. Conversely, if he does well in his ward or precinct, he will stand well with the party hierarchy, even if the party loses the election. This policy forces every individual in the organization to think locally, to concern himself almost exclusively with his bit of turf, and to leave broad policy matters and directions for the party leadership. As a consequence, the machine in Chicago is not really one citywide organization but, rather, a composite of approximately 3,148 local precinct organizations, each under the control of an individual responsible for his organization. There is no room in such a system for ideology, philosophy, or broad social concern. There is, instead, a pragmatic recognition of the need to concern oneself with one's little corner of the world, not with the interest of society as a whole, or mankind in general.

I became aware of this psychology in the election of 1968, when I worked a precinct for the Democratic organization in one of the North Side wards in Chicago. On election night at ward headquarters, I stood with my precinct co-captain before the big tally board which indicated the results in each of the ward's ninety-five precincts. Over the radio came the news of the developing results of the national and statewide elections for the Democratic party. It was clear that Vice-President Hubert Humphrey was losing the presidency to Richard Nixon, that Democratic Governor Sam Shapiro was losing the gubernatorial election to Republican Richard B. Ogilvie, and that Democratic senatorial candidate William Clark was losing the United States Senate seat from Illinois to incumbent Republican Senator Everett McKinley Dirksen. That night, even though I did not hold, and had no interest in securing a patronage job, and had taught political science for sixteen years, I found myself in the ward headquarters, my eyes intently fixed on the board carrying the tallies of my precinct, uninterested in the results in any other precinct in the ward, and almost totally unconcerned with the results of the city, county, state, and national elections for the Democratic party.

WORKING THE PRECINCT

How does a good precinct captain carry a precinct in cities like Chicago? Not by stressing ideology or party philosophy, not by stuffing mailboxes with party literature, not by debating issues with his constituents, but rather by ascertaining individual needs and by trying to serve those needs. Good precinct captains know that most elections are won or lost, not on great national, ideological issues, but rather on the basis of small, private, individual interests and concerns. If they don't know this or have forgotten it, their ward committeemen remind them in ward organization meetings. "Distinguished citizens," Alderman Marzullo addressed his precinct captains on the eve of the 1971 mayoralty election, "civic leaders, and religious leaders. This ward is not depending on the kind of publicity given to troublemakers and the nitwits and crackpots, This ward is depending on the record for sixteen years of our great mayor and the people who will put their shoulder to the wheel because they love Richard J. Daley and what he has done for the city of Chicago with the help of the Democratic party. Bring this message to your neighbors in your own language. You don't have to be an intellectual. I'm not an intellectual and I don't intend to be one."

Four months after the mayoralty election, in which Marzullo carried his ward for the mayor by a heavy majority, Marzullo leaned back in the high-back leather chair in his City Hall office and told *Sun-Times* columnist Tom Fitzpatrick, "I ain't got no axes to grind. You can take all your news media and all the do-gooders in town and move them into my 25th Ward, and do you know what would happen? On election day we'd beat you fifteen to one. The mayor don't run the 25th Ward. Neither does the news media or the do-gooders. Me, Vito Marzullo. That's who runs the 25th Ward, and on election day everybody does what Vito Marzullo tells them."

What kinds of goods and services do precinct captains in ward organizations provide? According to Marzullo, his captains work 365 days a year providing "service and communication" to his people. This includes free legal service for the destitute, repair of broken street lights, intensified police squad patrol, special anti-rodent clean-ups in the alleys, new garbage cans for tenants provided for them by their landlords, and talks with the probation officers of youngsters who are in trouble. "Anybody in the 25th needs something, needs help with his garbage, needs his street fixed, needs a lawyer for his kid who's in trouble, he goes first to the precinct captain," says Marzullo. "If the captain can't deliver, that man can come to me. My house is open every day to him."

[. . .]

Marzullo is a five-feet-six-inches seventy-seven-year-old grandfather, an Italian immigrant who came to Chicago at age twelve, who has spent fifty-five years in politics on the city's west side, who went only as far as the fourth grade in school but who lectures on politics at the University of Illinois, and who is one of the last authentic old-line ward bosses in the Democratic machine which has governed Chicago for more than forty years. On a winter night, in November 1974, I spent an evening with Marzullo in his ward headquarters on Chicago's west side. Seated behind his desk, Marzullo looked and sounded like Marlon Brando as the Godfather. "I always say, 'Vito, put yourself on the other side of the table. How would you like to be treated'" Marzullo told me. "I'm not an intellectual, but I love people. I'm not elected by the media, the intellectuals or do-gooders. I'm elected by my people. Service and communication. That's how my ward is run."

Marzullo has run for office 18 times. Only once has he had an opponent. "I beat him 15,000 to 1,000," Vito said. "I carried his precinct 3 and a half to 11. In 1940, when I first ran for state representative, I carried my precinct 525 to 14. The Republican precinct captain's mother and father voted for me."

"We got the most cosmopolitan ward in Chicago," said Marzullo. "Thirty percent black, twenty percent Polish, twelve to fifteen percent Mexican, five percent Italian, Slovenians, Lithuanians, Bohemians. We got them all."

"But I take care of all my people. Many politicians are like groundhogs. They come out once a year. On November fifth [election day], I visited every precinct polling place. On November sixth, I was in my office at City Hall at nine a.m. I'm there five days a week. On November seventh, I was here in my ward office at six-thirty p.m., ready to serve my people. My home is open twenty-four hours a day. I want people to come in. As long as I have a breathing spell, I'll go to a wake, a wedding, whatever. I never ask for anything in return. On election day, I tell my people, 'Let your conscience be your guide.'"

Marzullo's precinct captains were assembled in the rear room. After State Senator Sam Romano called the roll of the forty-eight precincts in the 25th Ward, Marzullo took the podium to remind his captains to maintain their efforts. "No man walks alone. Mingle with the people. Learn their way of life. Work and give service to your people." After the meeting with the captains, Marzullo moved to the front office to greet his constituents. A precinct captain ushered in a black husband and wife. "We got a letter here from the city," the man said. "They want to charge us twenty dollars for rodent control in our building." "Give me the letter. I'll look into it," Marzullo replied. The captain spoke up. "Your daughter didn't vote on November fifth. Look into it. The alderman is running again in February. Any help we can get, we can use." "I'm looking for a job," the woman said. "I don't have anything right now," said the alderman. The telephone rang. Marzullo listened and said, "Come to my office tomorrow morning." He hung up. "She's a widow for thirteen years. She wants to put her property in joint tenancy with her daughter. The lawyer wants a hundred dollars. I'll have to find someone to do it for nothing."

"Some of those liberal independents in the city council, they can't get a dog out of a dog pound with a ten-dollar bill," Marzullo snorted. "Who's next?"

Another captain ushered in a constituent. "Frank has a problem. Ticket for a violation of street sweeping." "Tell John to make a notation," Marzullo said. "You'll have to go to court. We'll send Freddy with you." The constituent thanked Marzullo and left. The captain said, "Alderman, how about that job in the Forest Preserves?" "You'll have to wait until after the first of the year," Marzullo responded.

The captain from the 16th precinct brought in a young black man who had just graduated from college and was looking for a job. "I just lost fifteen jobs with the city," Marzullo said. "How about private industry?" said the captain. "His family has been with us a long time." "Bring him downtown tomorrow," Marzullo instructed the captain. "I'll give him a letter to the Electric Company. They may have something." The young man left. "What about a donation to the Illinois Right to Life Committee?" asked the captain. "Nothing doing," said Marzullo. "I don't want to get into any of those controversies. People for it and people against it."

[. . .]

"On election day, every captain gets $50 to $200 for expenses in his precinct," Marzullo explained. "We buy eight tables for the $100-per-plate Cook County central committee dinner. That's $8,000. We make contributions to all of our candidates and pay assessments for people running from our ward. When the mayor runs we carry the ward for him by at least five to one. He's a great family man. A great religious person. We've been together all the way. I got six married children. He came to every one of their weddings. He invited me to the weddings of every one of his kids. You don't go back on people like that."

"The money comes from our annual ward dinner dance," Marzullo explained. "We don't charge our patronage workers any dues, or take kickbacks. We sell ads for our ad book for the dance and clear about $35,000 from the ad book." Marzullo's ad book looks like a fair-size telephone book.

A Polish truck driver came in. "I was laid off three weeks ago," he told Marzullo. "I've got six children." Marzullo countered, "I lost two truck drivers and three laborers this week. The city budget is being held down. We have to keep taxes down. But come down to my office in City Hall tomorrow morning. I'll see if there is an opening in street sweeping or snow plowing." "We got ten votes in my building," said the man. "If we get you a job, let your conscience be your guide," the alderman advised.

[. . .]

More people came in. A Mexican crane operator who was getting his hours cut back, a captain who wanted a transfer from the blacksmith shop to an easier job, a woman computer operator who was being mistreated by her supervisor. To each of them, Marzullo said, "I'll see what I can do."

The last captain came in with a sickly looking black woman. "Mamie," he said, "I want you to meet my great alderman and your great alderman. If he can help you, he'll do it."

"Alderman," Mamie said, "I need food stamps. I've been in the hospital three times this year. They're giving me pills, but I can't afford to eat."

"You have to get food stamps from your case worker," Marzullo explained. "I can't help on that." Turning to County Commissioner Charley Bonk, "Charley, give her a check for fifty dollars for food." And to Mamie, "If you need more come back again." "God bless you!" Mamie responded as she went out the door. "I guess that's it for tonight," sighed Bonk wearily. "If you meet them head-on every day, you wear them all out."

In middle-class wards on the northwest and southwest sides of the city, precinct captains can help get tax bills appealed, curbs and gutters repaired, scholarships for students to the University of Illinois, summer jobs for college students, and directions and assistance to those who need some help in finding their way through the maze of government bureaucracy with a grievance. "See the four men on this wall to the right?" Marzullo once asked Tom Fitzpatrick. "There's Mayor Daley, Congressman Frank Annunzio, County Commissioner Charles Bonk, and the other man is Vito Marzullo. We put a lot of other judges on the bench, too. And don't think that these people are ingrates. They always cooperate with the party that put them on the bench whenever they can. You see what I mean? The 25th Ward has a voice in every branch of government. That's a hobby we have in the 25th. It's our way of providing service to our people."

Most of the work of a good precinct captain in providing services to his constituents is not fixing tickets, bribing officials, or getting special favors for people, but rather ascertaining the individual needs of his people, communicating those needs through proper channels to the proper authorities, and providing help to those who are unable to find their way through the massive layers of bureaucracy in twentieth-century American government.

"The days when you went to the ward committeeman for a bucket of coal or a sweater for Johnny were on the way out when we went to the welfare programs of a more compassionate society under Roosevelt," according to the number-two man in the Cook County Democratic organization, Committeeman Tom Keane of the 31st Ward. "The political organization today [in the 1970s] is a service organization, an ombudsman and an inquiry department. I consider my ward has fifty-seven community organizations doing public service," says Keane, who calls his fifty-seven precinct captains "community representatives." Keane's analysis is echoed by county board president George Dunne, the Democratic committeeman in the 42nd Ward. "To a great extent the service we offer now is referral," says Dunne. "People ask, 'How do I get this? Where do I go to get that?' In some instances a letter from me helps and I never turn them down. Some people feel a letter always helps, but I think they would get what they need without it in most cases. . . . Even though over the years times have changed, the success of a political organization depends on giving service to the people."

[. . .]

"Writing the Rules to Win the Game: The Middle-class Regimes of Municipal Reformers"

from *Urban Affairs Review* (1999)

Amy Bridges and Richard Kronick

Editors' Introduction

Observers of politics can become so immersed in the contests unfolding before them that they lose sight of how the rules of the game shape the outcome. Imagine you were hired to coach a basketball team comprised of short players. You could try to create a style of play that would help your team offset this disadvantage, but you would fare even better if you could quietly alter the rules of the game, perhaps by increasing the value of outside shots, or (as the NBA did for a time) prohibiting dunk shots.

As in basketball, the rules of political competition and representation shape political outcomes. Amy Bridges and Richard Kronick attempt to resolve a long-standing debate by focusing our attention on the rules of the game. In the early decades of the twentieth century, party machines maintained a firm grip on the governance of some American cities, while others adopted a variety of structural reforms that weakened party influence. These reforms included innovations such as city managers, at-large councils, and nonpartisan local elections.

Why did some cities embrace these reforms while others rejected them? Political scientist Edward Banfield and historians Richard Hofstadter and Samuel P. Hays emphasized the class basis of machine and reform constituencies, suggesting that the reform movement represented the triumph of a growing middle class in some cities, whereas "blue-collar" cities remained under machine rule. Although it was true that municipal reform movements had this social base, Bridges and Kronick, among others, point out that most early twentieth-century cities had remarkably similar class structures, so a class explanation cannot account for differences among them.

Instead, using census data from the early 1900s, the authors argue that reformers in some cities were better able to push through rules restricting working-class participation at an early stage of local political development. By altering the rules about who can participate easily in the political process, reformers could restrict the electorate to one that gave them an advantage. With the voting population so restricted, there were fewer opportunities for party machines to maintain power. Further structural reforms, including city managers, at-large and nonpartisan councils, and ballot initiative provisions, were then easier to implement, institutionalizing reform influence.

The legacy of the reform movement, particularly the regional variation in the success of that movement, continues to influence American urban politics today. Reformers won changes in many city charters that required the election of council members at large, and the appointment of city managers; in many states, constitutions provided for citizen initiatives and recalls. Such arrangements are far more common in the west and southwest, where reformers had the opportunity to "write the rules of the game" before local parties were sufficiently established to thwart these efforts.

The class composition and motivation of early twentieth-century municipal reformers has long been a topic of debate among students of American political development. Whereas Richard Hofstadter (*The Age of Reform: From Bryan to FDR*, New York: Vintage, 1955) saw the reform movement as emblematic of a more assertive urban middle class, Samuel P. Hays, in his essay "The Politics of Reform in Municipal Government in the Progressive Era" (in *Social Change and Urban Politics*, ed. Daniel N. Gordon, Englewood Cliffs, NJ: Prentice Hall, 1973), claims that reformers in fact represented a very wealthy, corporate-identified elite. Martin Shefter, like Bridges and Kronick, emphasizes regional disparities between "reform" and "machine" cities in his 1983 article, "Regional Receptivity to Reform: Legacy of the Progressive Era" (in *Political Science Quarterly* 98 (3): 459–83). Other noteworthy examinations of municipal reform include: Kenneth Finegold, *Experts and Politicians: Reform Challenges to Machine Politics in New York, Cleveland and Chicago* (Princeton University Press, 1995), and Melvin G. Holli, "Urban Reform in the Progressive Era," in *The Progressive Era*, ed. Louis Gould (Syracuse University Press, 1974).

Amy Bridges, Political Science Professor at the University of California, San Diego, writes about urban political history. Her work has shed light on the formative stages of urban institutional development, most notably on political parties and the rules that shape them. Her first book, *A City in the Republic: Antebellum New York and the Origins of Machine Politics* (Cornell University Press, 1986), examines the formative years of New York's Tammany Hall. In *Morning Glories: Municipal Reform in the Southwest* (Princeton University Press, 1997), she turns her attention to the histories of seven southwestern cities, noting that their trajectories are quite different from those of the northeast. Trained as a political scientist, Richard Kronick specializes in health care policy, and is Associate Professor of Family and Preventative Medicine at the UCSD School of Medicine.

The founding of the National Municipal League in 1894 marked the organization of a nationwide municipal reform movement. The National Municipal League, the League of American Municipalities, and the Short Ballot Organization promoted the reorganization of local government. Initially, this meant nonpartisanship and citywide elections for members of the city council, who served as the city's "commission." By 1915, the city manager plan had replaced the commission as the model city charter of the National Municipal League. The league organized conferences on municipal problems, offered speakers and consultants to cities redrafting their city charters, and provided boilerplate editorials and pamphlets to town leaders. Within a generation, reformers celebrated scores of triumphs as cities across the country adopted the model city charters the league endorsed. Nevertheless, many cities resisted the appeals of municipal reform, opting for politics as usual. How can we account for the choices cities made?

The answer with longest standing is the class theory of city government. Since Hofstadter (1955) wrote *The Age of Reform*, municipal reform has been understood as a movement of the middle classes. Squeezed between the impoverished immigrant many and the few of ostentatious wealth, the mugwump organized to take control of city government. Later, Banfield and Wilson (1963) elaborated this thought in the "ethos" theory of city politics. Banfield and Wilson argued that the public-regarding values of middle-class WASPs led them to support reform regimes, but the private-regarding working-class/immigrant ethos supported the creation of political machines. Evidence for the class basis of support for different styles of local government has remained elusive. The theory suggests that cities with more middle-class residents chose reform government, whereas cities with more working-class residents retained politics as usual. Students of city politics who compared the class composition of cities and towns with their forms of government found, at best, weak evidence to support the theory, although strong relationships did appear between immigrant populations and the absence of reform. The same studies revealed resilient relationships between region and reform: In every decade in this century, northeastern cities have tenaciously maintained partisan politics, directly elected mayors, and held district elections for members of the city council. Midwestern cities likewise resisted reform. By contrast, a majority of southern and western cities and towns have chosen reform charters. If class theory provided an argument in need of evidence, discussions of region and reform have enjoyed plentiful evidence in want of a theory.

The failure to find evidence supporting the class theory was particularly frustrating because historical accounts showed for city after city that the middle classes voted for reform charters, but organized labor and less affluent neighborhoods opposed them. The most active opposition everywhere to municipal reform came from the working classes. The narrative evidence about the class basis of opposition to municipal reform is remarkably consistent. Usually, opposition centered on the antidemocratic elements in reformers' proposals. Commission government was opposed as too centralized, a criticism leveled at the small number of commissioners, citywide elections, and the concentration of responsibilities in the council. Reformers also proposed that the new commission governments have greater powers, particularly taxing powers, than the governments they were replacing, and this also provoked opposition. Commissioners were denounced as "czars" and "kaisers" (and much later, the city manager plan was similarly denounced as a "Hitler" plan because the manager was not elected, nor could he be fired, by the voters). There were famous exceptions to this pattern. Denver's businessmen loved the city's boss and abandoned the city's commission charter. There was also strong working-class support for the social reform mayors such as Hazen Pingree and Golden Rule Jones, as well as reformers with a populist cast. Structural reformers, however, found little support among the working classes.

If this is so, why does class fail to appear as a determinant of style of government in cross-sectional studies? We think there are two reasons. First, there was remarkably little variation in the class structure of cities early in the century. The Progressive nation was an industrial one, and nearly every city was overwhelmingly working class. Second, we think reformers were able to

win where they could shape the electorate by disfranchising their opponents and were most successful where their opponents were weak at the polls. In some places, this was a straightforward reflection of local society; elsewhere, it was the result of restrictions on suffrage that particularly affected working-class voters and people of color. In the Progressive Era, legal innovations restricting suffrage depressed turnout, and the victories of municipal reform in the South and West came subsequent to suffrage restriction. By 1905, turnout in the West and South was well below turnout in the Northeast and Midwest. Both what we know about targets of disfranchisement and what we know about depressed turnout generally suggest that in low participation regimes, the working classes and people of color stopped voting, creating southern and western urban electorates considerably more middle class and more WASPy than the adult population. If this is correct, then earlier studies erred by looking at local populations rather than local electorates.

Municipal reformers were successful where they could write the rules to win the game. The argument that municipal reformers succeeded by writing the rules to win the game reconciles long-standing case study evidence about collective preferences for styles of government to regional disparities in collective choices, provides suggestive evidence about the class basis of support for structural reform, and explains regional receptivity (or antipathy) to structural reform. In this article, we do not provide direct evidence of the composition of municipal electorates in the Progressive Era. We are able to show, however, that turnout is systematically related to style of government, as is the presence of immigrants.

We offer two clarifications of our argument at the outset. First, our argument concerns only the adoption of reform charters (commission or city manager government), Holli's (1969, chap. 8) "structural reform." There were many varieties of municipal reform and municipal reformers in the Progressive years – social reformers, socialist politicians, and populist reformers among them. Both structural reformers and those who opposed them were persuaded that the institutional arrangements of local government had important policy consequences for equity, efficiency, and democracy – or their absence. Second, although we argue that working-class voters tended to oppose reform and middle-class voters to support it, we reject the ethos theory that has traditionally accompanied these claims. We think that working-lass voters opposed structural reform because they saw in its proposals new institutions that would be less responsive to them. Middle-class voters found reform arguments persuasive because they saw their own interests as aligned with the civic leaders proposing new charters and continued to support reform regimes as they delivered (for a time) growth, quality services, and low taxes.

We begin by reviewing studies seeking the determinants of the adoption of reform charters. The subsequent section provides a historical sketch of efforts by municipal reformers to restrict access to the suffrage. The third section describes our own efforts to sort out these relationships in the decades just after the turn of the century. A concluding discussion reviews our argument and offers some lessons for students of city politics.

SOCIAL STRUCTURE AND CITY GOVERNMENT

A long list of authors have searched for evidence of the relationships between social structure and governmental form. . . . Knoke (1982) examined the adoption and subsequent abandonment of commission and city manager charters by the nation's 267 largest cities in the period from 1900 to 1942. Searching for demographic determinants of government style, Knoke found no significant effects for the presence of Catholics and a relationship to the presence of manufacturing in opposition to the class theory. The strongest determinant of transitions in style of government was regional adoption percentages. In other words, the more neighboring cities had reform charters, the more likely additional cities were to adopt municipal reform. Regional differences were the result of "neighborhood effects;" arising "not from social differences" but from "some type of initiation or contagion" (Knoke 1982, 1666).

A different argument about regional differences in the Progressive era was offered by Shefter (1983). Shefter observed marked differences in the adoption of nonpartisanship by western states (where nonpartisanship was favored) and by midwestern and northeastern states (where parties successfully resisted nonpartisanship). Using California as a prototype for the region, Shefter argued that the weakness of party organizations in the West in the late 19th century made the region more receptive to calls for nonpartisanship and other elements of the Progressive reform agenda.

[. . .]

Wolfinger and Field (1966) surveyed 309 cities with populations more than 50,000 in 1960 and examined the relationships between foreign stock population and various measures of class, on one hand, and at-large elections, nonpartisan elections, city manager government, and other indices of structural reform, on the other hand. They found that regional effects swamped the influence of income, class structure, and ethnicity nationwide, but the hypothesized relationships were non-existent within regions. "The salient conclusion to be drawn from these data," Wolfinger and Field wrote, "is that one can do a much better job of predicting a city's political forms by knowing what part of the country it is in than by knowing anything about the composition of its population" (p. 323). Regional differences might be explained by the attachment of those in older cities (and hence, the Midwest and Northeast) to the status quo. Charter writers in new cities in the West were likely to be moved by "contemporary political fashions" (Wolfinger and Field 1966, 324). In the South, "most municipal institutions seem to be corollaries of the region's traditional preoccupation with excluding Negroes from political power" (Wolfinger and Field 1966, 325).

[. . .]

Taken together, these efforts to find evidence of relationships between social structure and form of city government present weak and contradictory evidence about social class, stronger evidence about the presence of immigrants and resistance to reform, and resilient, persistent patterns of regional difference.

We present here a unified narrative of municipal reform. We call the narrative unified because a single pair of variables accounts for outcomes across the nation, explaining both choices for reform and resistance to it and the consequent regional pattern of these choices. We find that the presence/absence of immigrants and turnout (representing more and less participatory regimes) account both for choices to adopt or resist reform and for the regional pattern of these choices. Where there were immigrants and high participation, there was resistance to reform. Where participation was low and immigrants few, there was greater receptivity to reform.

One consequence of this argument is that, despite the disappointments of the literature, we continue to think that the Anglo middle class was at the center of municipal reform, much as we might say that the Anglo middle class has been at the center of the Republican Party. How Anglo middle-class constituents came to have that special relation to municipal reform is in good part a story of how middle-class Anglo electorates were created in the presence of considerably more diverse populations. In the first two decades of this century, that was accomplished through suffrage restriction.

MUNICIPAL REFORMERS AND SUFFRAGE RESTRICTION

The municipal reform movement was, from its beginnings in the middle of the 19th century, associated with efforts to restrict popular participation in politics. Banfield and Wilson (1963, 141) reported for Boston that "many leaders of reform were leaders of the Immigration Restriction League." In New York and elsewhere, antipathy toward immigrants took programmatic form in alliance with nativists, advocacy of immigration restriction, support for very long residency requirements for immigrant voters, voter registration, and nonpartisanship (Bridges 1986, chap. 7).

By the end of the century, these sentiments found adherents beyond the municipal reform movement. As Hofstadter (1955, 178) delicately phrased it, the mugwump "had begun to question universal suffrage out of a fear that traditional democracy might be imperiled by the decline of ethnic homogeneity." Later, nativism was given an intellectual veneer with the development of eugenics, the "science" of improving the quality of the population. In the Northeast, the targets of this concern were the most recent immigrants to the United States. "The cheap stucco manikins from Southeastern Europe," wrote Eugene Ross (father of modern sociology), "do not really take the place of the unbegotten sons

of the granite men who fell at Gettysburg and Cold Harbor" (Violas 1973, 54).

Municipal reformers and other advocates of regulating suffrage shared three motivations: opposition to corruption in politics, ethnocentrism, and antipathy to poorer voters. Vigorous efforts to regulate and restrict voting met with widespread success in state legislatures. Registration was the most frequently adopted requirement. In its least inhibiting form, registration was permanent, but in many states, the more burdensome requirement of periodic registration was enacted. Literacy tests were another popular innovation. By 1908, eight southern, four western, and three northern states had literacy tests; another three states adopted literacy tests by 1924. The secret ballot, welcomed in some places for freeing the lower classes from intimidation, served in others to disfranchise the illiterate or not-literate-enough. Finally, states had widely varying laws specifying those crimes for which citizens lost the privilege of voting. Even a glance at these laws suggests their possible abuse. For example, some states provided that anyone currently in jail (even for vagrancy or other misdemeanors) lost the right to vote.

In the southern and border states, suffrage restriction was, first, openly and emphatically in the service of white supremacy and, second directed at the potential supporters of populism and other alleged radicalisms, poorer white workers and farmers. . . . Thus, the restrictions adopted in the southern and border states disfranchised many poorer white voters as well as nearly all black voters. In Alabama, for example, property ownership delivered potential voters from the literacy test; in Georgia, "after 1898, when other devices were adopted to curtail the Negro vote, the function of the poll tax . . . was to exclude . . . whites of low economic condition" (McGovney 1949, 129, 120). Urban voters were also special targets: In Texas, secret ballot legislation affected only the state's ten largest cities. Most effective among disfranchising laws were the poll tax and the white primary (and, earlier, the grandfather clause), central to the regimes of V.O. Key's Southern Politics but not exclusive to the states of the former confederacy (New Mexico, Arizona, Rhode Island, and Maryland also had poll taxes).

Western states led the union in granting the franchise to women, but in the same states, large population groups were barred from the polls. Asians and Native Americans could not vote in any of the eleven western states, and Mormons were banned from the polls in Nevada, Idaho, and Utah. Arizona, California, Wyoming, Colorado, and Oregon adopted both registration and literacy requirements; Washington also had a literacy test. In Arizona, registration was particularly difficult. In each of these states, Mexican-Americans were the particular targets of literacy testing. Hispanics in New Mexico so strongly opposed literacy or education requirements for voting that the state constitution prohibited them. Especially relevant to municipal reformers, in California, only "freeholders" could vote on city charters.

The potential targets of disfranchisement were more vulnerable in southern, border, and western states than elsewhere because they were lacking partisan protection. By contrast, New York's Italian-Americans . . . were already employed at public works and voting for Tammany. In New Haven and Philadelphia, where Democratic machines failed to recruit Italian-Americans, Italian-Americans were voting for the Republicans. Whether their votes were cast for Republicans or Democrats, party allegiance protected immigrant voters from strict enforcement of literacy testing or other disfranchising laws. By contrast, from the municipal reformers' perspective, suffrage restriction was appealing because the most ardent opposition to structural reform came from working-class voters.

The history of municipal reform in Texas suggests the importance of suffrage restriction for the success of municipal reform. The state's voter registration law of 1891 "was a direct response to the outcome of the Dallas city election" of that year, in which the victory was heavily dependent on African-American and lower-class Anglo support. Angry Democratic Party leaders "rushed to Austin" to secure passage of the voter registration bill (Kousser 1974, 203). The most widely celebrated case of reform government, Galveston's commission plan, was installed in 1901, after a hurricane devastated the port. Galveston's government was not chosen by the city's voters but was awarded to the city by the state legislature, over substantial popular objections, as a condition of poststorm assistance. Elsewhere in Texas, although proposals for reorganizing local governments had been promoted by municipal reformers since the late 1880s,

no city adopted a reform charter until after the passage of the Terrell election law in 1903. The Terrell law and its 1905 sequel together provided for a poll tax to be paid six months before an election, as well as for the secret ballot. Terrell legislation had dramatic effects on urban electorates (white voters as well as African-Americans). In Houston, for example, almost 60 percent of registered voters disappeared from the rolls between 1900 and 1904. In Houston, Austin, and Beaumont, the prospects of reformers brightened as participation in local politics declined. These events lead us to suspect that participation and resistance to municipal reform were related, a relationship we explore in the next section.

CHOOSING LOCAL GOVERNMENT

The preceding discussion suggests that municipal reformers were attentive to the institutional arrangements of politics and consistently worked to write the rules of politics to their advantage. Like them, we think municipal reform enjoyed greater success, other things being equal, in cities in which turnout was low prior to the fight over adopting reform than in cities in which turnout was high. In this section, we examine evidence for this hypothesis in the first three decades of the century.

Structural reformers first promoted the commission form of government. Debates and referenda about the commission form were the stuff of reform politics until the introduction of the city manager plan, and between 1907 and 1915, 421 cities adopted the commission plan. After that year, however, although additional cities chose commission charters, more cities adopted the city manager plan. Between 1912 (when four cities had adopted the manager plan) and 1933, 448 cities adopted manager charters.

The largest data set reflecting these choices is found in the Municipal Year Book, 1934, which reported the form of government of the 310 cities with populations of 30,000 or more in 1930. We used the yearbook report for our investigation. We treated city manager and commission governments as reform and mayor-council governments as nonreform. Because our interest was in local choices of city charters, we eliminated the cities of Pennsylvania, where after 1912, commission government was required by state law for all but the state's three largest cities.

We examined the relationship between the 1934 city charter and three main independent variables: percentage of the population that is foreign born in the census of 1920, percentage of workers in blue-collar industries in the same year, and voting turnout in the presidential election of 1908 in the county in which the city resides (defined as the number of votes cast in the county divided by the number of males age twenty-one and older). We chose 1908 to measure turnout because it precedes the major fights over the adoption of reform in very nearly all of the yearbook cities. We would have preferred to measure the turnout of municipal electorates, but these data were unavailable. We chose 1920 because there were data for more of the cities on the 1934 list than the 1910 census and because 1920 antedates a majority of the charter decisions tolled in the list.

The 278 cities in our sample were about evenly divided between reform government (commission or city manager) and those with mayor-council government. Table 1 shows the distribution of the forms of government, turnout, and various measures of social structure by region. The strong regional cast to forms of government is apparent here: In the Northeast, two-thirds of all cities retained mayor-council government, as did more than half of midwestern cities, but two out of three southern and western cities chose reform government.

In broad outlines, the regional distribution of foreign-born residents and of 1908 turnout parallel the regional variation in forms of government. Northeastern cities, where reformers were singularly unsuccessful, had a high concentration of foreign-born residents, whereas the foreign-born presence was slight in the reform-dominated South. And turnout was high in the reform-unfriendly Midwest and again very low in the South. The West is a bit of a puzzle here: The percentage of foreign born is about average, and the 1908 turnout is only slightly below average, yet the West was quite hospitable to reform.

Blue-collar residents were distributed much more evenly. The nation's cities were overwhelmingly blue collar in every region. There were very few white-collar cities in 1920. Austin, a state capital, boasted little industry; Berkeley, already a university town, was likewise home to relatively more white-collar workers, but cities like them were few and far between. This suggests one reason scholars of the early years of this century found little role for class in their analyses of styles of government: There was precious little variation in the class structure of the nation's cities.

We investigate the relationships between form of government and turnout, percentage foreign born, and the percentage of workers in blue-collar occupations by estimating a logistic regression. The dependent variable in the analysis is a 0–1 indicator of whether the city had reform government in 1934. Results are shown in Table 2.

The first column of Table 2 simply confirms the results from Table 1. Without controlling for any other variables, region has a strong and significant effect on the form of government. The Midwest is slightly more reform than the Northeast (although not significantly so), and the West and South are much more reform.

Table 1 Form of city government in 1934, by region

	North	Midwest	South	West	Total
% of cities with reform government, 1934	33.7	43.4	68.8	65.5	48.6
% of population foreign born, 1920	26.8	15.3	5.9	18.0	17.0
Voting turnout, 1908 (%)	54.6	65.0	34.8	51.0	53.3
% of workers in blue-collar industries, 1920	71.3	68.6	68.4	63.0	68.9
Number of cities	86	99	64	29	278

Note: Form of government is from the *Municipal Year Book, 1934* for cities of 30,000 or more in 1930. Pennsylvania cities are excluded from the analysis. Foreign-born and blue-collar percentages are from the census of 1920. Voting turnout, defined as the number of votes cast in the 1908 presidential election divided by the number of males aged 21 and older, is measured in the county that includes the city.

Table 2 Logistic regression results predicting the 1934 form of government

	(1)	(2)
Intercept	0.68**	−1.85**
	(.23)	(0.25)
Midwest	−0.41	0.16
	(0.31)	(0.37)
South	−1.46**	0.11
	(0.35)	(0.59)
West	−1.32**	−0.74
	(0.45)	(0.48)
1908 turnout		1.52*
		(0.88)
1920 % foreign born		6.40**
		(1.87)

Note: Dependent variable: 0 = reform; 1 = mayor/council. Standard errors are in parentheses. N = 278 cities with a population of 30,000 or more in 1930 (excluding Pennsylvania cities). Form of government is from the *Municipal Year Book, 1934* for cities of 30,000 or more in 1930. Foreign-born percentage is from the census of 1920. Voting turnout, defined as the number of votes cast in the 1908 presidential election divided by number of males age 21 and older, is measured in the county that includes the city.
$*p < .10. **p < .01.$

Turnout has the relationship to reform government that we expected. As hypothesized, where turnout was low in 1908, reformers were more successful than in cities where turnout was high in 1908. For each ten percent increase in turnout, the probability of adopting reform declines by 3.3 percent – thus a city in a county with 70 percent turnout in 1908 would have been ten percent more likely to have nonreform government than a city with 40 percent turnout in 1908. Neither the substantive nor the statistical significance of the relationship is overwhelming; the coefficient is significant at the ten percent but not the five percent level, and the difference in reform adoption between high-turnout and low-turnout cities is substantial but not determinative.

Confirming the results of other analysts, the presence of foreign-born residents has a strong relationship to the adoption of reform. For every ten percent increase in the percentage foreign born, the likelihood of adopting reform government declines by approximately fourteen percent.

Controlling for the presence of foreign born and 1908 turnout, the independent effects of region largely disappear. The slightly greater success of reformers in the Midwest (relative to the Northeast) and the much greater success of reformers in the South can be accounted for by the weakness of their opponents in these regions, either because there were few foreign-born residents (and thus fewer opponents in the potential electorate) or because turnout was low (and thus fewer opponents in the actual electorate). The West remains slightly (although not statistically significantly) different. Reformers in the West were slightly more successful than would be predicted from either percentage foreign born or from the level of turnout. In additional regressions, we examined the effects of blue-collar industries and of city size on the form of government. Neither of these variables were significantly related to the presence of reform government (data not shown).

We can retell these relationships in narrative form, adding some observations about political parties. The most vigorous opponents of municipal reform were strong party organizations that controlled, or hoped to control, city governments – the political machines upon which reformers declared war. They resisted the abandonment of ward representation as destructive to their relations with constituents, they resisted reform assaults on corrupt practices, and they resisted assaults on their most reliable sources of votes. In the Northeast, where immigrants were many, turnout high, and parties well organized, voters and party leaders successfully resisted the seductions of the municipal reform movement. In the Midwest too, although immigrants were fewer, high participation and strong parties meant resistance to reform. Elsewhere, political parties were not so well suited to the task. In the South, where popular participation in politics was minimal and one-party politics replaced party competition, boosters of municipal reform made easy progress to their goals. In the West, there was party competition, but party organizations were very weak, making the West more receptive to reform than its immigrant populations and levels of participation alone predicted.

THE RULES OF POLITICS MATTER IN MUNICIPAL REFORM

Every student of city politics knows the class theory of city government and knows that the middle classes were the heart and soul of municipal reform. Yet aside from the suburbs, evidence for this argument has been maddeningly elusive. Historians, meanwhile, have shown for city after city that the working classes opposed municipal reform, whereas middle-class voters supported it. Historians and political scientists have noted strong and persistent regional variations in receptivity to reform but lacked explanations for them.

We have argued here that historical narratives and the class theory may be reconciled. Middle-class voters were indeed the central supporters of reform regimes, and working-class voters were its ardent opponents. Reformers succeeded by writing the rules to win, disfranchising their opponents, and so creating electorates more middle class than the adult population as a whole.

Reform succeeded in the Progressive era less from the support of most voters than from the weakness of their opponents in the West, border states, and South. Crucial to opponents' weakness was suffrage restriction, which changed the composition of local electorates, not incidentally disfranchising the strongest opponents of municipal reform. In the same places – in the South and West – the working classes lacked strong parties or political machines to defend them. That said, we think reformers succeeded by writing the rules so they could win.

Our account shows that two variables – the presence/absence of immigrants and turnout (participation) in elections – largely account for regional receptivity or antipathy to municipal reform. This account is consistent with historical narratives showing middle-class support for municipal reform and working-class and immigrant opposition to it. Where suffrage restriction was successful, electorates were likely more middle class (as well as more Anglo) than the population as a whole. Low-turnout cities were likely to adopt reform charters because the opposition – almost everywhere the organized working class – was weak at the polls.

This account has the virtue, in our view, that it explains regional variation without resorting to narratives about regional peculiarities and is independent of either ethos or "attitudinal or socioeconomic variables:" Like other authors who emphasize the importance of institutions, we recognize that preferences matter but insist that the rules of politics matter as well. Everywhere, the working classes opposed municipal reform. Where the rules of political life allowed them access to the polls and where there were political parties to defend their interests, the promoters of municipal reform were rebuffed. Where parties were absent and working-class voters few, municipal reformers were triumphant.

REFERENCES FROM THE READING

Banfield, E.C., and J.Q. Wilson. 1963. *City Politics.* New York: Vintage.

Bridges, A. 1986. *A City in the Republic: Antebellum New York and the Origins of Machine Politics.* Ithaca, NY: Cornell University Press.

Hofstadter, R. 1955. *The Age of Reform: From Bryan to FDR.* New York: Vintage.

Holli, M.G. 1969. *Reform in Detroit: Hazen S. Pingree and Urban Politics.* New York: Oxford University Press.

Knoke, D. 1982. The spread of municipal reform: Temporal, spatial, and social dynamics. *American Journal of Sociology* 87: 1314–39.

Kousser, J.M. 1974. *The Shaping of Southern Politics: Suffrage Restriction and the Establishment of the One-party South, 1880–1910.* New Haven, CT: Yale University Press.

McGovney, D.O. 1949. *The American Suffrage Medley.* Chicago: University of Chicago Press.

Shefter, M. 1983. Regional receptivity to reform: Legacy of the Progressive Era. *Political Science Quarterly* 98 (3): 459–83.

Violas, P. 1973. Progressive social philosophy: Charles Horton Cooley and Edward Alsworth Ross. In *Roots of Crisis: The American Education in the Twentieth Century,* edited by C.J. Karier, P. Violas, and J. Spring, 163–76. New York: Rand McNally.

Wolfinger, R., and J.O. Field. 1966. Political ethos and the structure of city government. *American Political Science Review* 60 (1): 306–26.

"Why Political Machines Have Not Withered Away and Other Revisionist Thoughts"

from *Journal of Politics* (1972)

Raymond E. Wolfinger

Editors' Introduction

Raymond Wolfinger uses the city of New Haven (which he knew well as a result of his research with Robert Dahl, discussed in Part 3, including a year spent working in City Hall) as a lens through which to explore the conventional wisdom about the rise and fall of urban political machines.

At the time he wrote (1972), the popular and scholarly view of urban politics was that political machines may have arisen to meet the social and political needs of the unskilled immigrants who flooded cities at the turn of the twentieth century, but the assimilation and upward mobility of these immigrants led them to new means of political participation, and the political machine "withered away."

Wolfinger challenges these contentions. First, he notes that the poor, the unskilled and the recently immigrated still concentrate in cities, and could presumably still be mobilized through the incentives offered by political machines. Second, he debunks the claim that machine politics were uniquely attractive to the poor and the immigrant, who benefited from the provision of unskilled jobs and small favors. Rather, he claims, machines doled out goodies that provided benefits across class lines, throwing business to lawyers and bankers as well as to patrolmen and street sweepers. As a result, "machine politics," which he defines as "the manipulation of certain incentives to partisan political participation: favoritism based on political criteria in personnel decisions, contracting, and administration of the laws," continued to dominate many cities in the 1970s (as it does today), even if centralized, hierarchical organizations called "political machines" had become rarer.

How has machine politics managed to survive generations of political reform and demographic change? This style of politics does best in precisely those environments that would be least hospitable to issue-oriented, ideology-driven politics. The party leaders described here (or in the Chicago described by Rakove or the Queens County studied by Jones-Correa later in this section) prosper by capturing the little-noticed, off-year elections for low-profile positions like County Clerk. Such positions are of little interest to the political "amateurs" who mobilize around charismatic candidates and high-profile issues, but they provide the life-blood (patronage jobs, contracts) of machine politics. (Max Weber's distinction between politics as an "avocation" and politics as a "vocation" comes to mind here – those involved in machine politics are those who, as Weber notes, "live off" politics.) It is thus possible for machine politics to continue on the local level "under the radar" of most of the population, which becomes engaged only periodically with higher profile state and national races.

In his conclusion, Wolfinger urges students of political development to consider regional differences when trying to account for the disparate fates of political machines and machine politics in different cities.

Scholars have indeed taken up his call; the selection by Bridges and Kronick all consider regional variations in reform successes.

Raymond Wolfinger is one of the leading political scientists of his generation, writing on urban and national politics. Along with Robert Dahl and Nelson Polsby, he used the analysis of urban politics to put forward an increasingly sophisticated pluralist theory of American politics, writing *The Politics of Progress* (Prentice-Hall, 1974). More recently he has written extensively on voter participation, including *Who Votes?* (with Steven Rosenstone, Yale University Press, 1980), *The Myth of the Independent Voter* (with five former students, University of California Press, 1992), and dozens of more recent articles. He is currently Heller Professor of Political Science at the University of California, Berkeley.

Machine politics is always said to be on the point of disappearing, but nevertheless seems to endure. Scholarly analyses of machines usually explain why they have dwindled almost to the vanishing point. Since machine politics is still alive and well in many places, this conventional wisdom starts from a false premise. More important, it has several logical and definitional confusions that impede clear understanding of American local politics. This article shows that machine politics still flourishes ... and argues that the familiar explanations both for the existence of machine politics and for its putative decline are inadequate.

THE PERSISTENCE OF MACHINE POLITICS

My first-hand experience with machine politics is limited to the city of New Haven. Both parties there had what journalists like to call "old fashioned machines," of the type whose disappearance has been heralded for most of the twentieth century. Some people in New Haven were moved to participation in local election campaigns by such civic-minded concerns as public spirit, ideological enthusiasm, or a desire to influence governmental policy on a particular issue. For hundreds of the city's residents, however, politics was not a matter of issues or civic duty, but of bread and butter. There were (and are) a variety of material rewards for political activity. Service to the party or influential connections were prerequisites to appointment to hundreds of municipal jobs, and the placement of government contracts was often affected by political considerations. Thus the stimuli for political participation in local politics were, for most activists, wholly external.

A new administration taking over New Haven's city hall had at its immediate disposal about 75 politically-appointed policy-making positions, about 300 lower-level patronage jobs, and about the same number of appointments to boards and commissions. Summer employment provided around 150 additional patronage jobs. In the winter, snow removal required the immediate attention of hundreds of men and dozens of pieces of equipment.

A hundred or more jobs in field offices of the state government were filled with the advice of the party's local leaders. The City Court, appointed by the governor with the advice of the local dispensers of his patronage, had room for two or three-dozen deserving people. The New Haven Probate Court had a considerable payroll, but its real political significance was the Judge of Probate's power to appoint appraisers and trustees of estates. Except in difficult cases, little technical knowledge was necessary for appraising, for which the fee was $1 per $1,000 of appraised worth.

A great deal of the city's business was done with men active in organization politics, particularly in such "political" businesses as printing, building and playground supplies, construction, and insurance. Competitive bidding did not seriously increase the uncertainty of the outcome if the administration wanted a certain bidder to win. As in many places, it was commonplace for city or party officials to "advise" a prime contractor about which local subcontractors, suppliers, and insurance agencies to patronize. Many government purchases were exempt from competitive bidding for one reason or another. The prices of some things, like insurance, are fixed. Thus the city's insurance business could be (and was) given to politically deserving agencies. Other kinds of services, particularly those supplied by professional men, are inherently unsuited to competitive bidding. Architects, for instance, are not chosen by cost. Indeed, some professional societies forbid price competition by their members.

The income that some party leaders received directly from the public treasury was dwarfed by trade from people who hoped to do business with the city or wanted friendly treatment at city hall, or in the courts, or at the state capitol, and thus sought to ingratiate themselves with party leaders. For example, a contractor hoping to build a school would be likely to buy his performance bonds from the bond and insurance agency headed by the Democratic National Committeeman. Similar considerations applied to "political" attorneys with part-time public jobs. Their real rewards came from clients who wanted to maximize their chances of favorable consideration in the courts or by public agencies.

Control of city and state government, then, provided either local party with a formidable array of resources that, by law, custom, and public acceptance, could be exploited for money and labor. Holders of the 75 policy-making jobs were assessed five percent of their annual salaries in municipal election years and three percent in other years. At the lower patronage levels, employees and board members gave from $25 to $100 and up. Politically-appointed employees were also expected to contribute their time during campaigns and were threatened with dismissal if they did not do enough electioneering.

Business and professional men who sold to the city, or who might want favors from it, were another important source of funds. Both sides in any public contractual relationships usually assumed that a contribution would be forthcoming, but firms doing business with the city were often approached directly and bluntly. During one mayoralty campaign a party official asked a reluctant businessman, "Look, you son of a bitch, do you want a snow-removal contract or don't you?" In the 1957 mayoralty election the biggest individual contributor, who gave $1,500 to the ruling Democratic party, was a partner in the architectural firm that designed two new high schools. A contractor closely associated with a top-ranking Democratic politician gave $1,000. A partner in the firm that built the new high schools and an apartment house in a redevelopment project gave $900. Dozens of city, court, and party officials were listed as contributors of sums ranging from $250 to $1,000.

[. . .]

Political spoils in New Haven came from several jurisdictions, chiefly the municipal government, the probate court, and the state government. The more numerous the sources of patronage, the lower the probability that all would be held by the same party, and hence the easier it was for both parties to maintain their organizations through hard times. When one party was triumphant everywhere in the state, as the Democrats were in the 1960s, there was considerable potential for intraparty disunity because the availability of more than one source of rewards for political activity made it difficult to establish wholly unified local party organizations. Inevitably state leaders would deal with one or more local figures in dispensing state patronage. This local representative need not be the same man who controlled probate or municipal patronage. Although the mayor had the power to give out city patronage, either directly or by telling his appointees what to do, he found it prudent to exercise this power in concert with those leaders who could control campaign organizations in New Haven through their access to state and probate patronage. In good measure because of the diverse sources of patronage, the loyalties of Democratic party workers went to different leaders. All this was true also of the Republican party. Thus neither local party organization was monolithic. The Republicans were badly split for much of the post-war generation. The Democrats maintained a working coalition, but not without a good deal of competition and constant vigilance on the part of the mayor and the two principal party leaders. Multiple sources of patronage are commonplace with machine politics and have important consequences, which will be explored in the next section.

A second typical feature of machine politics was that the elections most important to organization politicians were obscure primaries held on the ward level. Issue-oriented "amateurs" seldom could muster sufficient strength in these elections. The amateurs seemed to be interested chiefly in national and international affairs, and thus were most active and successful in presidential primaries and elections, where their policy concerns were salient. While the stakes in presidential contests may be global, they seldom include the topic of prime interest to machine politicians – control of patronage – and hence the regulars exert less than their maximum effort in them. Conveniently for both amateurs and regulars, the two sorts of elections are held at different times and usually in different years. When the amateurs' enthusiasm is at its peak, the professionals will be less interested; when the machine's spoils are at stake, the amateurs are less involved.

Participation in election campaigns is not the only form of political action. It is important to distinguish between electioneering and other types of political activity. In New Haven there was a major divergence between campaign and non-campaign activities. The likelihood that richer people would engage in non-campaign activity was far greater than the corresponding probability for campaigns. This divergence reflected the probability that participation in a campaign is less autonomously motivated, for in New Haven the discipline of patronage compels campaign work. There are no such external inducements for most non-campaign political action. Indeed, because such activity usually consists of trying to exert pressure on public officials, it is likely to be viewed with apprehension or disfavor by those machine politicians who dispense patronage. A sense of political efficacy, education, a white-collar job, and higher income – all are thought to be associated with those personal qualities that lead people to try to influence the outcome of government decisions. In many parts of the country, these traits are also associated with electioneering. Some people participate in New Haven elections – particularly for national office – from such motives, but most activists, including party regulars, do not. The essentially involuntary character of much political participation in cities dominated by machine politics has received scant attention from students of participation, who customarily treat the phenomenon they study as the product of solely internal stimuli.

[. . .]

MACHINE POLITICS DEFINED

The terms "machine politics" and "political machine" are commonly used so as to confuse two quite different phenomena. "Machine politics" is the manipulation of certain incentives to partisan political participation: favoritism based on political criteria in personnel decisions, contracting, and administration of the laws. A "political machine" is an organization that practices machine politics, i.e., that attracts and directs its members primarily by means of these incentives. Unfortunately, the term "machine" is also used in a quite different and less useful sense to refer to the centralization of power in a party in a major political jurisdiction: a "machine" is a united and hierarchical party organization in a state, city, or county. Now there is no necessary relation between the two dimensions of incentives and centralization: machine politics (patronage incentives) need not produce centralized organization at the city level or higher.

The availability of patronage probably makes it easier to centralize influence in a cohesive party organization, since these resources can be distributed so as to discipline and reward the organization's workers. Quite often, however, all patronage is not controlled by the same people. There may be competing organizations or factions within each party in the same area, for where patronage is plentiful, it usually is available from more than one jurisdiction. In New Haven the municipal government had no monopoly on the spoils of government, which were also dispensed by the probate court and the state government. Thus the existence of a cohesive local organization in either party did not follow from the use of patronage to motivate party workers.

This distinction between machine politics and centralized local machines is far from academic, for the former is found many places where the latter is not. Chicago presently exhibits both machine politics and a very strong Democratic machine. Forty years ago it had the former but not the latter. In Boston and New York there are the same kinds of incentives to political activity as in Chicago, but no cohesive citywide organizations. Instead, these cities have several contending party factions. In New York "the party" includes reform clubs with considerable influence as well as a variety of "regular" organizations. The frequently celebrated "decline" of Tammany Hall was not so much the subjugation of the regulars by the reformers, nor the disappearance of patronage and corruption (neither has happened yet), as the decentralization of the city's old-line Democratic organization.

[. . .]

While the distinction between incentives and centralization is useful for accurate description and definitional clarity, it also has important theoretical ramifications. Robert K. Merton's influential explanation of the persistence of machine politics (patronage) points to the presumed coordinating function of centralized political machines: "The key structural function of the Boss is to organize, centralize and maintain in good working condition the 'scattered fragments of power' which are at present dispersed through our political organization. By the centralized organization of political power, the Boss and his apparatus can satisfy the needs of diverse sub groups in a larger community which are not politically satisfied by legally devised and culturally approved social structures."

Yet machine politics exists many places where, as in New York, the party "organization" is a congeries of competing factions. In fact, cohesive organizations like Chicago's may be fairly uncommon, while pervasive favoritism and patronage – machine politics – are much less so. Hence Merton explained the persistence of the incentive system by referring to functions allegedly performed by an institution (a centralized, city wide party organization) that may or may not be found where machine politics flourishes.

The rewards that create the incentives in machine politics are not only tangible but divisible, that is, they are "allocated by dividing the benefits piecemeal and allocating various pieces to specific individuals." Moreover, they typically result from the routine operation of government, not from particular substantive policy outcomes. Any regime in a courthouse or city hall will hire roughly the same number of people, contract for roughly the same amounts of goods and services, and enforce (or fail to enforce) the same laws, irrespective of the differences in policies advocated by one party or the other. The measures adopted by an activist, enterprising administration will generate a higher level of public employment and contracting than the output of a caretaker government. Yet the differences are not enough to change the generalization that the rewards of machine politics are essentially issue-free in that they will flow regardless of what policies are followed. This excepts, of course, reform of personnel and contracting practices.

One can then distinguish two kinds of tangible incentives to political participation. The incentives that fuel machine politics are inevitable concomitants of government activity, available irrespective of the policies chosen by a particular regime. A second kind of tangible incentive results from a desire to influence the outcome of particular policy decisions. This second type includes those considerations that induce political participation by interest groups that do not want patronage, but do want the government to follow a particular line of action in a substantive policy area: lower tax rates, anti-discrimination legislation, minimum-wage laws, conservation of natural resources, and the like. A particularly pure example of a political organization animated by substantive incentives would be a taxpayers' group that acted as a political party – naming candidates, getting out the vote, etc. – in order to capture city hall for the purpose of enacting a policy of minimal expenditure. As an ideal type, such a group would not care who was hired or awarded contracts, so long as a policy of economy was followed.

[. . .]

Why Political Machines Have Not Withered Away

The conventional wisdom in American social science interprets machine politics as a product of the social needs and political techniques of a bygone era. Advocates of this position attempt to explain both the past existence of machines and their supposed current demise in terms of the functions that the machines performed. In analyzing the functions – now supposedly obsolete – that machine politics served, it is useful to consider four questions:

1. Did political machines actually perform these functions in the past?
2. Do machines still perform them?
3. Has the need for the functions diminished?
4. Is machine politics found wherever these needs exist?

It is commonly argued that various historical trends have crucially diminished the natural constituencies of machines – people who provided votes or other political support in return for the machine's services. The essential machine constituency is thought to have been the poor in general and immigrants in particular. The decline of machine politics then is due to rising prosperity and education, which have reduced the number of people to whom the rewards of machine politics are attractive or necessary. These trends have also, as Thomas R. Dye puts it, spread "middle class values about honesty, efficiency, and good government, which inhibit party organizations in purchases, contracts, and vote-buying, and other cruder forms of municipal corruption. The more successful machine [sic] today, like Daley's in Chicago, have had to reform themselves in order to maintain a good public image."

One function that machines performed was furnishing needy people with food, clothing, and other direct material assistance – those legendary Christmas turkeys, buckets of coal, summer outings, and so on. There is no way of knowing just how much of this kind of help machines gave, but it seems to have been an important means of gleaning votes. From the time of the New Deal, government has assumed the burden of providing for the minimal physical needs of the poor, thus supposedly preempting a major source of the machines' appeal. The growth of the welfare state undeniably has limited politicians' opportunities to use charity as a means of incurring obligations that could be discharged by political support. Some political clubs still carry on the old traditions, however, including the distribution of free turkeys to needy families at Christmas time.

Machines supposedly provided other tangible rewards, and the need for these has not been met by alternative institutions. The most obvious of these benefits is employment. The welfare state does not guarantee everyone a job and so the power to hire is still an important power resource. It has been argued . . . that patronage jobs, mainly at the bottom of the pay scale, are not very attractive to most people. But these positions are attractive to enough people to maintain an ample demand for them, and thus they still are a useful incentive.

A second major constituent service supplied by machine politics was helping poor and unacculturated people deal with the bureaucratic demands of urban government. Describing this function, some writers emphasized its affective dimension. Robert K. Merton put it this way: "the precinct captain is ever a friend in need. In our increasingly impersonal society, the machine, through its local agents, performs the important social function of humanizing and personalizing all manner of assistance to those in need." In Dye's view, the machine "personalized government. With keen social intuition, the machine recognized the voter as a man, generally living in a neighborhood, who had specific personal problems and wants."

[. . .]

If machine politics were a response to "our increasingly impersonal society," it would seem to follow that continuing growth in the scope, complexity, and impersonality of institutional life would produce greater need for politicians to mediate between individuals and their government. The growth of the welfare state therefore has not diminished this need but increased it and presumably offers the machine politician new opportunities for helping citizens get what they want from the government. . . . As far as local politicians are concerned, new public services may be new prizes that covetous or needy citizens can more easily obtain with political help. . . . Harold Kaplan reported that in Newark "a public housing tenant, therefore, may find it easier to secure a public housing unit, prevent eviction from a project, secure a unit in a better project, or have NHA (Newark Housing Authority) reconsider his rent, if he has the right sponsor at City Hall." There is no necessary connection, then, between expanded public services and a decline in the advantages of political help or in the number of people who want to use it. While the expansion and institutionalization of welfare may have vastly ended "the party's monopoly of welfare services," they have vastly expanded the need for information, guidance, and emotional support in relations between citizens and government officials, and thus, there is no shortage of services that machines can provide the poor and unassimilated, who are still with us.

There is no doubt that in the past 50 years income levels have risen. . . . But there are plenty of poor people in the cities, the middle classes have been moving to the suburbs for the past two generations, and the European immigrants have been succeeded by blacks, Puerto Ricans, Mexicans, and poor rural whites. . . . The argument that affluence and assimilation have choked machine politics at the roots, one familiar to scholars for decades, may now look a bit more threadbare. Yet the recent rediscovery of poverty and cultural deprivation has not had a major effect on thinking about trends in the viability of machine politics.

Along with the new interest in the urban poor has come a realization that existing institutions do not meet their needs. Among these inadequate institutions is the political machine, which, in the traditional view, should be expected to do for today's blacks, Chicanos, Puerto Ricans, and poor whites just what it is supposed to have done for yesterday's immigrants. But even in cities with flourishing machine politics there has been a tremendous development of all kinds of community action groups for advice, information exchange, and the representation of individual and neighborhood interests – precisely the functions that the machines are said to have performed. The gap between the disoriented poor and the public institutions serving them seems to be present equally in cities like Chicago, generally thought to be political anachronisms, and in places like Los Angeles that have never experienced machine politics. This leads to an important point: most American cities have had the social conditions that are said to give rise to machine politics, but many of these cities have not had machine politics for a generation or more.

This fact and the evident failure of existing machines to perform their functions cast doubt on the conventional ways of explaining both the functions of machines in their supposed heyday and the causes of their "decline." One conclusion is that the decline is real, but that the principal causes of the decline do not lie in affluence and assimilation. A second possibility is that the machines persist, but have abandoned the beneficent functions they used to perform. A third is that they are still "humanizing and personalizing all manner of assistance to those in need," but cannot cope with a massive increase in the needs of their clienteles. And a fourth alternative is that the extent to which they ever performed these functions has been exaggerated.

It does seem that a whole generation of scholarship has been adversely affected by overreaction to the older judgmental style of describing machine politics. Until a decade or two ago most work on machines was moralistic and pejorative, dwelling on the seamy side of the subject and concerning itself largely with exposure and denunciation. More contemporary social scientists have diverged from this tradition in two respects. One, apparently a reaction to the highly normative style of the old reformers, is a tendency to gloss over the very real evils they described. The other, addressed to the major problem of explaining the durability of machine politics, is the search for "functions": acculturating immigrants and giving them a channel of social mobility, providing a link between citizen and city hall, and coordinating formally fragmented government agencies. Some writers suggest that urban political organizations were a rudimentary form of the welfare state. While the tone of these later works has been realistic, some of them leaned toward idealizing their subject, perhaps in reaction to the earlier moralism or because functionalism has not been accompanied by an inclination to confront the sordid details. Thus the development of a more dispassionate social science has produced, on the descriptive level, a retreat from realism. The functionalists seem to have been somewhat overcredulous: "the precinct captain is ever a friend in need."

[. . .]

"Helping" citizens deal with government is, in this context, usually thought to be a matter of advice about where to go, whom to see, and what to say. The poor undeniably need this service more than people whose schooling and experience equip them to cope with bureaucratic institutions and procedures. But in some local political cultures advice to citizens is often accompanied by pressure on officials. The machine politician's goal is to incur the maximum obligation from his constituents, and merely providing information is not as big a favor as helping bring about the desired outcome. Thus "help" shades into "pull."

Now there is no reason why the advantages of political influence appeal only to the poor. In places where the political culture supports expectations that official discretion will be exercised in accordance with political considerations, the constituency for machine politics extends across the socio-economic spectrum. People whose interests are affected by governmental decisions can include those who want to sell to the government, as well as those whose economic or social activities may be subject to public regulation.

Favoritism animates machine politics, favoritism not just in filling pick-and-shovel jobs, but in a vast array of public decisions. The welfare state has little to do with the potential demand for favoritism, except to expand opportunities for its exercise. The

New Deal did not abolish the contractor's natural desire to minimize the risks of competitive bidding, or the landlord's equally natural desire to avoid the burdens of the housing code. It is all very well to talk about "middle-class values of efficiency and honesty," but the thousands of lawyers whose political connections enable them to benefit from the billion-dollar-a-year case load of the Manhattan Surrogates' Court are surely not members of the working class.

While "help" in dealing with the government may be primarily appealing to people baffled by the complexities of modern society and too poor to hire lawyers, "pull" is useful in proportion to the size of one's dealings with government. Certain kinds of business and professional men are more likely to have interests requiring repeated and complicated relations with public agencies, and thus are potentially a stronger constituency for machine politics than the working classes. The conventional wisdom that the middle classes are hostile to machine politics rests on several types of evidence: (1) The undeniable fact that reform candidates almost always run better in well-to-do neighborhoods. (2) The equally undeniable fact that machine politics provides, in patronage and petty favors, a kind of reward for political participation that is not available in other incentive systems. (3) The less validated proposition that middle-class people think that governments should be run with impartial, impersonal honesty in accordance with abstract principles, while the working classes are more sympathetic to favoritism and particularistic criteria. These characterizations may be true in the aggregate for two diverse such categories as "the middle class" and "the working class" (although that has not yet been established), but even if these generalizations are true, they would still leave room for the existence of a sizable subcategory of the middle class who, in some political cultures, benefit from and endorses machine politics.

[. . .]

The conventional wisdom also holds that the machines' electioneering techniques are as obsolete as the social functions they used to perform. According to this interpretation, "the old politics" based its campaigns on divisible promises and interpersonal persuasion, and these methods have been outdated by the mass media – particularly television, the growing importance of candidates'

personalities, and the electorate's craving for ideological or at least programmatic promises.

Like the other explanation of the machines' demise, this argument has serious factual and logical deficiencies. As we have seen, machine politics is an effective way of raising money for political purposes. There is no reason why the money "maced" from public employees or extracted from government contractors cannot be spent on motivational research, advertising copywriters, television spots, and all the other manifestations of mass media campaigns.

Similarly, there is no inconsistency between machine politics and outstanding candidates. Just as machine politicians can spend their money on public relations, so can they bestow their support on inspirational leaders who exude integrity and vitality. Many of the most famous "idealistic" politicians in American history owe their success to the sponsorship of machine politicians. Woodrow Wilson made his first venture into electoral politics as the gubernatorial candidate of an unsavory Democratic organization in New Jersey. (Once elected governor, Wilson promptly betrayed his sponsors.) In more recent times, such exemplars of dedicated public spirit as the elder Adlai Stevenson, Paul H. Douglas, and Chester Bowles were nominated for office as the candidates of the patronage-based party organizations in their several states.

[. . .]

CONCLUSIONS AND SUGGESTIONS

To sum up my argument: Because an increasing proportion of urban populations is poor and uneducated, it is not persuasive to argue that rising prosperity and education are diminishing the constituency for machine politics. While governments now assume responsibility for a minimal level of welfare, other contemporary trends are not so inhospitable to machine politics. Various kinds of patronage still seem to be in reasonable supply, and are as attractive as ever to those people – by no means all poor – who benefit from them. The proliferation of government programs provides more opportunities for the exercise of favoritism. The continuing bureaucratization of modern government gives more scope for the machine's putative function of serving as a link between the citizen and the state.

These trends would seem to have expanded the need for the services the machines supposedly performed for the poor. Yet surviving machines apparently are not performing these functions, and machine politics has not flourished in many cities where the alleged need for these functions is just as great.

The potential constituency for political favoritism is not limited to the poor; many kinds of business and professional men can benefit from machine politics. They do in some cities but not in others. Again, it appears that the hypothesized conditions for machine polities are found in many places where machines are enfeebled or absent.

Real and imaginary changes in campaign techniques are not inconsistent with machines' capacities. In short, machines have not withered away because the conditions that supposedly gave rise to them are still present. The problem with this answer is that the conditions are found in many places where machine politics does not exist.

Attempts to explain the growth and alleged decline of machine politics usually emphasize the importance of immigrants as a constituency for machines. Yet many cities with large immigrant populations have never been dominated by machine politics, or were freed of this dominance generations ago. Machine politics continues to flourish in some states like Indiana, where foreign stock voters are relatively scarce. In other states, like Pennsylvania and Connecticut, machines seem to have been as successful with old stock American constituents as with immigrants.

Far more interesting than differences in ethnicity or social class are regional or sub-regional variations in the practices of machine politics and in attitudes toward them. Public acceptance of patronage, for example, appears to vary a good deal from place to place in patterns that are not explained by differences in population characteristics such as education, occupation, and ethnicity. Although systematic data on this subject are not available, it does seem that voters in parts of the East, the Ohio Valley, and the South are tolerant of practices that would scandalize most people in, say, the Pacific Coast states or the Upper Midwest. The residents of Indiana, for example, seem to accept calmly the remarkable mingling of public business and party profits in that state. One researcher notes that these practices have "not been an issue in recent campaigns." Another student of midwestern politics reports that "Indiana is the only state studied where the governor and other important state officials described quite frankly and in detail the sources of the campaign funds. They were disarmingly frank because they saw nothing wrong in the techniques employed to raise funds, and neither did the opposing political party nor the press nor, presumably, the citizenry.

[...]

The reasons for these marked geographical variations in political style are not easily found, but looking for them is a more promising approach to explaining the incidence of machine politics than the search for functions supposedly rooted in the socioeconomic composition of urban populations.

REFERENCES FROM THE READING

Thomas, R. Dye, *Politics in States and Communities* (Englewood Cliffs, NJ: Prentice-Hall, Inc., 1969).

Harold Kaplan, *Urban Renewal Politics* (New York: Columbia University Press, 1963).

Robert, K. Merton, *Social Theory and Social Structure* (revised edition; Glencoe, IL: The Free Press, 1957).

"Resistance from Outside: Machine Politics and the (Non) Incorporation of Immigrants"

from *Between Two Nations* (1998)

Michael Jones-Correa

Editors' Introduction

Wolfinger and Bridges have addressed the historic links between immigrant populations and the strength of political machines. A century ago, when one-third of the population of cities like New York and Chicago was foreign born, it was hard to miss the association between immigrants and party bosses. Today, the immigrant population is once again rising in many American cities. According to the 2000 Census, 40.9 percent of the population of Los Angeles had been born in another country, as had 35.9 percent of New York's residents and 21.7 percent of those in Chicago. If new immigrants were the life-blood of the traditional political machine, should not these new waves of foreign-born be expected to reinvigorate this model of politics?

Michael Jones-Correa studies immigrants as political actors in American cities. He sets out to learn how an established political organization (the Democratic Party in Queens, one of New York City's five boroughs) mobilized, or failed to mobilize, a new immigrant group in its midst. After all, even many of those political scientists who frown on the corrupt aspects of the old machines believe that the socialization of new residents was one of its positive functions. As Croker argues, Tammany Hall helped immigrants in many ways, not only by providing specific benefits and shielding them against nativist political excesses, but by making them into a cohesive body of citizens.

Some research has cast doubt on these accomplishments. In many cities, one ethnic group dominated the political machine and offered little to others. As Steven Erie has demonstrated, even those supporting the machine often did not have much, ultimately, to show for it. When compared with other European immigrant groups, he notes, the Irish, who led many of the most dominant early twentieth-century party machines, were slow to achieve middle-class status. By contrast, Germans and Jews, among others, started businesses and gained a foothold in the professions, while the descendants of Irish immigrants remained concentrated in blue-collar, low-wage jobs in the public sector that yielded few opportunities for economic advancement.

Building on Erie's work, Jones-Correa criticizes scholars such as Dahl and Wolfinger, who assume that party machines (or political parties more generally) have an inherent interest in mobilizing new groups. They are only likely to bring in new voters, he argues, when they face competition for the chance to exercise power. (The basic point that competition stimulates electoral mobilization has been developed by democratic theorists such as Joseph Schumpeter and V.O. Key.)

In the absence of competition, a party loses its incentive to engage in the hard, costly work of bringing in new voters, and its emphasis naturally shifts to the easier job of simply discouraging challengers. Although there are exceptions, Jones-Correa notes that most urban political machines operate as near monopolies in their respective political arenas. Only rarely does a city's electoral life feature two or more truly competitive parties.

In the absence of incentives to bring in new participants, urban party organizations are unlikely to welcome newcomers. Jones-Correa does not claim that party politicians engage in large-scale, overt racial discrimination. Rather, party leaders fail to build bridges to newcomers, making it extremely unlikely that immigrants will find their way into the political process. He highlights the "very simple" strategy of the Democratic machine in Queens: "it places all the costs of mobilization and incorporation on those on the very margins of politics." In other words, the political machine's message to new immigrants is: You want to participate? We won't stop you, but we won't help you, either.

Jones-Correa's book draws on an intensively researched case study of Queens. In New York City, party politics is organized along borough (or county) lines. The Democratic Party in Queens may not be the omnipotent force it had been in the heyday of Tammany Hall, but it remains the center of gravity in local politics. In the past two decades, Queens has experienced an astounding influx of immigrants (discussed further in Part 5), making it an ideal place to observe how an established party organization confronts new immigrant groups. Since this book was written, a Latino and an Asian have been elected to New York's City Council from Queens, both with the backing of the Democratic Party, suggesting that some further incorporation of immigrants may have occurred. These shifts may have been pushed by the adoption of term limits in New York City, as council members can now serve no more than two terms. Some of the incumbency advantage described by Jones-Correa has thus been eroded.

Michael Jones-Correa is an Associate Professor of Government at Cornell University. This selection comes from *Between Two Nations: The Political Predicament of Latinos in New York City* (Ithaca, NY: Cornell University Press, 1998). Jones-Correa has also edited *Governing American Cities: Inter-ethnic Coalitions, Competition and Conflict* (New York: Russell Sage Foundation, 2001 and 2005).

Those wishing to explore the works on which Jones-Correa bases his analysis should read Robert Dahl (see Part 3); Wolfinger's article in this section, and Steven Erie, *Rainbow's End: Irish-Americans and the Dilemmas of Urban Machine Politics, 1840–1985* (Berkeley: University of California Press, 1988).

■ ■ ■ ■ ■ ■

The United States has received immigrants throughout its history, and the long experience has created its own mythology. Part of the popular imagery has it that earlier in this century the ships bringing immigrants were met at the dock by the political boss, a basket of food or clothing draped over his arm, waiting to receive the newcomers and recruit them as part of the city's political machine. . . . The urban political machine is now mostly remembered as a quintessential Americanizing institution; immigration policy allowed immigrants entry, but the machine made them voting citizens.

This memory of the political machine, which emphasizes its role as an integrating mechanism into democratic politics, is acquiring renewed relevance today as the United States again becomes the destination for millions of immigrants from around the world. Over eight million immigrants arrived in this country in the last decade, more than at any other time since the last great wave of immigration from 1900 to 1910. The United States receives twice as many immigrants as all other nations combined.

Most of these become permanent residents; many may eventually become citizens. Are they being integrated into the electoral process? How relevant is the popular historical view? What role do political parties and urban political machines currently play in the integration of immigrants into American political life?

Social scientists differ on how political machines actually work. A generation of social scientists which included Robert Dahl, Robert Merton, and Raymond Wolfinger generally portrayed political machines in a positive light. Their writing was deeply influenced by instrumentalist explanations for party politics like those of Joseph Schumpeter and Anthony Downs, which posited that cost/benefit rationales drove all political behavior. Their economic modeling of party behavior laid the groundwork for an explanation of the machine's integration of immigrants. A more recent interpretation of the urban political machine, represented by Steven Erie's work partly challenges this view. Erie makes the case that political machines only occasionally mobilized immigrant voters, in large part because machines were rarely competitive. In most cases, the machine tried to retain a monopoly on access to the political process by systematically limiting participation or demobilizing participants.

The present situation of Latin American immigrants in Queens neatly illustrates Erie's case. Democratic machine politics in Queens, though weakened, still functions. But rather than playing the integrating role in which it has been cast in the popular imagination, it aims to keep political mobilization within tightly constricted bounds, and to direct and control that mobilization for its own ends. Rather than lowering the entry costs for marginal political players, the Queens Democratic party, like the hegemonic party organizations Erie describes, raises them instead. If actors are at the margins of electoral politics, as immigrants are, then they are ignored; if political players rise to challenge the machine, they are thwarted. Only if new political actors succeed in mobilizing themselves on their own does the party organization attempt to bring them into its circle.

The strategy of the party organization in Queens is very simple: it places all the costs of mobilization and incorporation on those on the margins of politics. Successful entry into the formal political arena requires passage over a series of hurdles.

INSTRUMENTALIST VIEWS OF THE POLITICAL MACHINE

The popular view of the political machine leached into the social science literature in the 1950s and '60s, abetted and reinforced by revisionist scholars who rebutted earlier opinion that the machine was the root cause for corruption and decay in American politics, and who created a vision of the urban political machine as a democratizing influence. Drawing from the example of New Haven's integration of immigrants into the electoral process, Robert Dahl asserted, for example that, "throughout the country . . . the political stratum has seen to it that new citizens, young and old, have been properly trained in 'American' principles and beliefs. . . . The result was as astonishing an act of voluntary political and cultural assimilation as history can provide" (1961: 317–18). The key point to this reinterpretation was that the locus of the mobilizing

effort was the machine. It brought outsiders into the political system, and that process was initiated by political insiders, elites in the political parties.

The assertion that the party machinery worked to integrate immigrants drew directly from instrumentalist theories of party politics, which drew in turn on economic models of political behavior. . . . Democratic politics are inherently competitive, the theory goes, because the main interest of each player is to acquire power. Politicians, being competitive, will mobilize potential players to keep the extra edge over their opponents; otherwise outlying issues and actors will be picked up by the opposition. The driving assumption is that players in the political game will seek to maximize support. This assumption adopted in the literature of immigrant mobilization, is stated most clearly by Dahl: "In a competitive political system within a changing society, a party that neglects any important source of support decreases its chances of survival . . . as new social strata emerge, existing or aspiring party leaders will see and seize opportunities to enhance their own influence by binding these new elements to the party . . . new social elements are likely to be recruited into one or both of the parties not only as sub-leaders but ultimately as leaders" (1961: 114).

Competitive parties will choose to mobilize not only active players in the system, but potential or marginal players such as immigrant [ibid.: 34]:

Since political leaders hoped to expand their own influence with the votes of ethnic groups, they helped the immigrant overcome his initial political powerlessness by engaging him in politics. Whatever else the ethnics lacked they had numbers. Hence the politicians took the initiative; they made it easy for immigrants to become citizens, encouraged ethnics to register, put them on the party rolls, and aided them in meeting the innumerable specific problems resulting from their poverty, strangeness and lowly position.

Immigrants didn't even need to be citizens to be incorporated, because the politician guided them through the court system as fast as or faster than the law allowed, making them citizens and leading them to the polling booths. This made sense in the economy of machine politics, which traded in specific, tangible benefits. Raymond Wolfinger writes that, "votes mattered to American politicians, who solicited [immigrants] with advice, favors, petty gifts, and jobs. . . . Each nationality group in a city had leaders who bargained with politicians, trading their followers' votes for money, favors and jobs."

Social scientists such as Dahl, describing the history of immigrant political integration into American life, wrote at the juncture of common belief and economic models of politics. The economic model provided a rationalization of the popular image of political machines by providing the hypothetical figure of the boss, waiting with a basket at the pier where the immigrants' ships landed, with a rational calculus for his actions. Written in a language convincing to social scientists, Dahl's and others' essentially populist vision became the accepted account of the political machine.

Writing a generation later, Steven Erie argues that political machines never really attempted to incorporate immigrants wholesale. "Throughout most of their history," he writes, "urban machines did not incorporate immigrants other than the Irish." The argument of Dahl and others is wrong on two counts. First, it assumes that the electoral arena will remain competitive. In a competitive environment, instrumentalist logic would require that the party out of power attempt to mobilize marginal voters to build a coalition to recapture political office. Political machines, however, are not in the business of being merely competitive. Machines are not interested in maintaining a dual or multiparty system where there is free competition. The aim of the machine is to drive competitors out of business and become monopolies. Machines only mobilize potential voters until they establish control of the political system, after which their purpose is to keep core constituencies satisfied; then they can demobilize or ignore marginal or outside players. This was the case through the nineteenth century and well into the twentieth in many U.S. cities. "Having already constructed a minimal winning coalition among 'old' immigrant that is Western European-voters, the established machines had little need to naturalize, register, and vote later ethnic arrivals" from southern and eastern Europe.

Erie also believes that Dahl and others are wrong to assume that the machine had unlimited inducements to mobilize voters. Machines were constantly trying to create resources to meet ethnic demands – demands which, Erie notes, "nearly always exceeded the machine's available supply". Even if they wanted to do more, the urban machines could only reward or encourage mobilization in drips and drabs. The solution the Irish-run machines at the turn of the century found was to allocate most of the choice rewards to a core group of reliable supporters, and to apportion what was left among marginal voters. The newer immigrants were among the latter group. They were given the less valuable benefits the machine had to offer: symbolic recognition, support in getting social services provided by the state government, and lobbying for state labor and social welfare legislation.

Notwithstanding his criticisms of prior interpretations, Erie's analysis of the urban political machine is not revolutionary. He provides an important corrective to Dahl and others' analyses of the machine in particular by attaching an addendum to the earlier instrumentalist model. Dahl . . . makes generalizations about the urban machine based on a single example – in Dahl's case New Haven – and assumes that because conditions there were competitive, such conditions also existed elsewhere. Erie demonstrates that cities like New Haven are much more the exception than the rule. Political incorporation of marginal players will only take place to the extent that, first, there is competition, and, second, the supply of resources can be stretched to cover the demand. In cases where these conditions do not apply, machines will tend to act like monopolies, demobilizing and ignoring marginal players. . . . What Erie adds to the discussion of party politics is a corollary about how parties act under noncompetitive conditions.

CONTEMPORARY MACHINE POLITICS IN QUEENS

The machine politics existing today in New York City are not those of eighty or even forty years ago; they face different conditions and incentives. Tammany Hall, the notorious Manhattan-based New York City machine, was already crumbling by the early 1930s, under the pressure of reformist mayors and governors. By 1961 the atrophy of the Tammany clubhouses led to the capture of the Democratic Party by reformers and the end of a citywide machine. New York City, however, has in reality always had not one but five political systems – one in each borough of the city. Although reformers may have killed off the overarching centralized machine in Manhattan, machine politics have continued operating in the outer boroughs to the present day. Electoral politics in Queens, for instance, are dominated by a Democratic organization operating out of the political clubs and Borough Hall.

[. . .]

Democratic machine politics in Queens has two principal characteristics. First, it has inherited the remnants of a long-established machine operating in a noncompetitive environment. The Democratic presence in the area is overwhelming; the Republican Party is relegated to being a permanent minority. . . . Second, with Democratic political clubs functioning, the infrastructure of the political machine still exists in Queens. However, with the [1989] changes in the city's charter, which transferred power from the Board of Estimate to an expanded, elected City Council, the borough machinery has lost significant resources it once had at its disposal. So though the Queens machine exists, it operates in weakened form. Erie's argument that an established political machine in a noncompetitive situation will hold onto its core constituency and attempt to demobilize marginal or potential players would seem especially true during periods in which resources are constricted. If Erie is correct, the combination of these two characteristics in Queens machine politics – the absence of competition and the atrophy of the patronage system – would mean that the political machine in Queens would have little interest in bringing new players into the system. This seems indeed to be the case.

In the last thirty years the Queens immigrant population has increased enormously. The 1990 census put the borough's Latino population alone at 81,000, about 20 percent of Queens' total population – most of these being first-generation immigrants. This represents an increase of more than a hundred thousand people from the count in the 1980 census. As previously established groups leave for areas lying outside the city, the Latino presence is growing rapidly in the area. The north-central area of the borough, once composed almost entirely of middle-class Jewish, Irish, and Italian neighborhoods, is now among the most diverse in the city, with majority Asian and Latino populations. Despite the changing demographics of these neighborhoods over the last twenty-five years, however, political organization has remained overwhelmingly white. [As of the 1990s] no Latino [held] elected office in the borough, nor ha[d] the county Democratic machinery endorsed any Latino (or Asian-American) candidate – which would, in much of Queens, be tantamount to election to office. . . . [Since this was written, a Latino has been elected to City Council with the Democratic Party's endorsement. Ed.]

At the turn of the century the urban machines used sometimes brutal methods to, as Erie puts it, "deflate demand" for political resources: "As oligopolists of the working-class market, the bosses drove rival producers and product lines out of town. Both repression and corruption were used to defeat the machine's labor and Socialist Party rivals. The machine's henchmen intimidated labor party speakers and voters. The machine-controlled police force broke up Socialist meetings, revoked the business licenses of insurgent immigrant entrepreneurs, and enforced Sunday closing laws to stifle Jewish dissidents." These tactics of force and intimidation are hardly in use anymore. The machine has no need for these to demobilize its potential challengers. Instead it demobilizes passively, not by doing anything in particular, but by doing nothing. People are not turned away; they are just not invited. Those who invite themselves are let in, but not necessarily welcomed or made to feel at home. Activists are co-opted, but the population as a whole is left alone except during election time. The passivity of modern machine politics is evident in Queens in the following three areas: citizenship, voter registration, and political club membership.

Once under the purview of political parties, the acquisition of citizenship is now left entirely to immigrants themselves. The entrepreneurial activism of the machine boss who guided the immigrant through the citizenship process (however limited it may have been) withered away with the passage of the Naturalization Act of 1906 which made it more difficult for political machines to manipulate the court system, implemented literacy tests in English as a requirement for citizenship, and required proof of five years of continuous residence in the local area prior to naturalization. Changes in state electoral laws ended alien suffrage by 1910, so that even before changes in the immigration laws restricted their numbers in 1917 and again in 1929, newly arrived immigrants had lost much of their appeal to the machine as potential players. Today no mechanism currently operates in the political parties to integrate non-citizens. Whereas toward the end of the nineteenth century being a non-citizen did not necessarily mean being outside the political system, in contemporary Queens it does, and immigrants can be ignored.

For those who are citizens, voter registration is rarely encouraged by local machine politicians. Party-sponsored registration at any time other than the quadrennial presidential election year drives is practically nonexistent; one Democratic club member maintained that "the club does voter registration drives, at least once a year we do it. But as far as anyone in organized politics doing outreach getting everyone registered to vote, be serious. They don't (interview)." What little attempts there are to register Latin American immigrants are carried out by immigrant organizations themselves. These tend to be haphazard and small-scale due to the lack of resources, but have been picking up in frequency and scope in the last several years. At least three Colombian organizations as well as the Queens Hispanic Coalition completed registration drives during the 1992 campaign, for instance. . . . Even so, Latino registration in Queens is far below their proportion of the population (20 percent in 1990).

It seems curious that the same machine politicians who make almost no effort to register Latinos then use voting as a measure of political influence . . . Latinos are not perceived to have the votes. "For years," said a Democratic district leader in Jackson Heights, "I have heard talk about [Latinos] delivering votes. . . . In all my years as district leader I haven't seen anyone deliver more than a pizza (meeting notes)." Claire Shulman, the Queens Borough President, reportedly asked one Latino activist why Queens politicians should pay attention to Latinos when they don't vote. She said she would deal with Latinos when they voted, and they don't vote now. A Latina described a conversation she had with state assemblyman Ivan Lafayette, who represents the Jackson Heights area: "Lafayette said, 'you need to get your community organized.' He said, 'many years ago Latinos came to me, but they did nothing. I didn't get one vote. I've been elected all these years without the Latino vote.'"

Not only are Latinos perceived as nonvoters, but local politicians have no stake in seeing that they do vote. The Democratic Party has an organization built up in Queens with a core group of political club regulars to do the crucial petitioning and mobilizing before and during primaries and elections. Politicians of all stripes, in all situations, tend to focus their attention on those constituents who are already active, and this is particularly true in Queens. Politicians can, and do, still win elections without the Latino vote. Party outreach becomes, as one Queens insider put it, a kind of "preaching to the choir." The Democratic Party could register and mobilize Latino voters, but mobilization would only mean additional competition for scarce political resources. Queens politicians can afford to marginalize Latinos in part because the Latinos in Queens who do participate in the electoral process tend to be Democrats in any case, regardless of the way they are spurned by the local party organization. As noted previously, Queens is an overwhelmingly Democratic county. The Republican Party's organizational base hardly exists, and there is no effective competition in many areas of the borough. The Republican Party has not been successful at attracting large numbers of Latinos, in spite of their willingness to support some Hispanic candidates. A slightly higher proportion of Latinos than non-Latinos lean toward the Republican party, but whether out of ideological sympathy, or simply for the pragmatic reason that it is better to work with the party in power than with the permanent minority, most stay with the Democratic Party, in spite of their discontent and periodic threats to leave it.

Membership in the party's political clubs seems to be relatively open – it can be as simple as being a registered Democrat and paying dues every year. Applicants for membership to the clubs are rarely, if ever, turned away (although there is no attempt at recruitment either). Yet there are curious gaps and omissions in club membership. In a borough where a fifth of the population is Latino – higher in areas of north-central Queens – the composition of the clubs is almost entirely white. Maureen Allen, president of the JFK Democratic Club, said new immigrants were not interested in participating: "At least they haven't reached out to us. They may be forming their own clubs, but not in this area." In fact, Latino Democratic clubs have been formed in the area in the past, but have not survived. The first attempt began during Jimmy Carter's first presidential campaign in 1976. "This club dissolved," one Ecuadorian activist said, "[in part] because there was a lot of pressure from the Democratic Party . . . which didn't want anything to exist outside the traditional clubs. They didn't see a need for a Latino Democratic club. [They asked] why didn't Latinos work within existing clubs?" A second attempt in the early 1980s also collapsed. When Latinos do join the traditional clubs they often find the atmosphere unwelcoming: "The clubs have a 'closed' environment. . . . It's a challenge . . . [our people] feel the barrier . . . a bias which is difficult to define. Something very subtle. It's like they see one of us there and [they ask themselves] 'who is that person? Why are they here?' They don't think it's natural for us to be part of the group. There's an ambivalence. They want us there, and they don't want us there (interview)."

[. . .]

That Queens politicians seem to be making little or no effort to recruit new players into the electoral system confirms what Erie would predict. Given that the political machine in Queens is long established, and has only token competition, machine politicians have little interest in disrupting the status quo. Since electoral competition in the area dominated by an established, functioning machine is something of an oxymoron, Erie suggests that the mobilization of Italian and Jewish immigrant voters took place not because of competition but rather because of party mobilization during national campaigns. This pattern is repeated on a smaller scale in contemporary Queens politics. Registration rates of both Hispanics and non-Hispanics go up sharply in presidential election years, when the race is contested at the national level and the local party machinery is expected to contribute to the national campaign effort. For example, in the presidential election year of 1988 five and a half times more Latinos registered as new voters than were registered during 1987. In general, the number of people registering to vote in presidential election years more than doubles the levels of preceding non-election years. This pattern continued in 1992 when, for the first time since 1973, voter rolls in New York City topped 3.2 million registered voters – 613,181 voters having been added in the preceding year. On the whole,

then, machine politics in north-central Queens tends to confirm Erie's hypothesis about noncompetitive party regimes.

BARRIERS TO PARTICIPATION

It would seem that Latin American immigrants in Queens would have good reason to seek alternatives to Democratic Party hegemony – turning to another party or challenging the party themselves. But if the political machine is indifferent to the incorporation of marginal actors, it is actively hostile to the mobilization of competition. In this it is abetted by New York City's political structure, which is designed to minimize the possible threat from the competition. It is difficult in the city to run against established incumbents, and hard to build up the necessary campaign infrastructure to challenge the state's arcane electoral rules. Possible routes for challengers – the community boards and the school boards – have been co-opted by the borough's machine politicians. In short, challengers must be willing to pay a high cost to gain a foothold in the city's electoral arena.

THE POWER OF INCUMBENCY

From 1961, Democrats have continuously filled the borough presidencies in Brooklyn, Queens, and Manhattan, and lost in the Bronx only from 1961 to 1965. Democrats have filled the comptroller's post through this period, and the City Council presidency for eight of nine terms, while also holding the overwhelming majority of the seats on the Council. That Democrats have consistently held these elected positions may not, in and of itself, indicate an absence of competition within or outside the party. But less than half the Council contests between 1961 and 1989 involved a primary, and most of the general elections were not even close. In only 20 percent of elections was the margin of victory less than ten percentage points. A full 80 percent of the City Council races were not competitive. Competition for borough presidency varies across the city, reflecting the relative strength or weakness of the county party organization. In Queens, though, only one out of the last eight elections for borough president has been contested, and even that race was not close. Of all the major elected positions in the city, only the mayoralty has been consistently contested.

A good part of this record can be accounted for by the power of incumbency. No incumbent borough president from Queens, Brooklyn, or Manhattan, for instance, has ever lost his or her seat in an election. Rarely is a sitting City Council member defeated at the polls. Incumbency provides benefits that are extremely difficult for challengers to overcome. Among other things it gives candidates greater name recognition, opportunities for public exposure, and the possibility of using office to gain support. It allows officeholders control over discretionary budget items, influence over appointments to community boards and other advisory committees, and a role in running the local Democratic clubs and party machinery, which has the attendant benefit of mobilizing a cadre of trained campaign volunteers. With low voting turnout in the city as a whole, and in local elections in particular, "the ability to deliver 200 or 300 votes in a single geographic pocket can be the deciding factor," as Council-member Walter McCaffrey succinctly put it. Incumbents have enough resources at their command – through appointments, budgetary items, and so on – to make it possible, in theory, to provide the margin of votes they need to win most elections. As a result, even in a situation allowing for a larger than usual field of candidates, as happened with the expansion of the City Council in 1991, almost all those winning seats in the new Council had come up within the ranks of the Democratic party, and were running either as incumbents or with the party's blessing.

Electoral Rules

The resources at the disposal of party-backed candidates are crucial for the manipulation of the state's electoral rules. Incumbency means access to campaign workers – through the party's clubs, or through the party's appointees to the community boards, antipoverty boards, or local development corporations, who form the organizational core of a run for office. This is true almost anywhere, but it is particularly true of New York City. A cadre of volunteers gives a candidate the organizational muscle to get on the ballot. This is no easy task in

the state of New York. Just to run in the City Council elections one must collect nine hundred signatures. Once signatures are collected they must survive expensive and time-consuming legal challenges by a candidate's opponents. Under New York State's intricate elections laws, candidates in the past have been knocked off the ballot for details as petty as having their petitions fastened together with a paper clip rather than a staple. In the 1991 City Council races, for instance, three of five challengers in District 21 in Corona were knocked off the ballot by petition challenges. These challenges of often trivial details has become a full-time industry: half of all election litigation in the United States occurs in New York City. In spite of the general recognition that change is necessary, officeholders from both parties have dragged their feet on election reform, for obvious reasons: the rules heavily favor incumbents over challengers.

[...]

Latin American immigrants to Queens enter an environment that discourages participation in the formal political system. Though machine politics in Queens is in some ways unique, its passivity toward the needs of marginal political actors such as immigrants embodies the response of American political parties and the political system in general. Politics is assumed to be a matter for insiders. Not being an insider, a challenger has to pay high entrance costs – in organization, money and time – to be able to compete. It can safely be said that few challengers, and almost no new immigrants, can afford these costs.

Co-optation

The hegemonic party structure in Queens does not actively recruit new members, and does its best to discourage direct challengers. If a group or faction somehow succeeds in mobilizing itself, politicians attempt to enfold it within existing institutions. In the weak one-party system that exists in the city, they do this very selectively, maximizing the limited resources at their disposal. Incentives for inclusion in the party's coalition remain more symbolic than material.

The party has been divided as to how to deal with the new immigrants. On the one hand they are a huge presence in Queens, and demographically there is every indication that they will one day become the dominant force in the borough. On the other hand, they have not succeeded in mobilizing themselves, so the party organization has no interest in expending more resources on them than they have to. The avenue taken so far – after much debate within the party leadership about whether anything should be done at all – has been the symbolic incorporation of a small number of ethnic elites into the party. In 1993 the executive committee of the borough's Democratic party finally recognized the existence of over 600,000 new immigrants from Latin America and Asia living in Queens by approving the appointment of six minority at-large district leaders – four Latinos and two Asians. These new district leaders were chosen by the party leadership from among current minority members of the Democratic clubs, so essentially they were already insiders. . . .

WHOSE RESPONSIBILITY?

Those who study immigrants and machine politics generally view the situation from the perspective of the machine, not that of the immigrant; Dahl, Wolfinger, and Erie all emphasize the machine's response to immigrants. Their studies are about representation, not participation. The focus is on the principal decision makers in the political process, the elites who drive the machine. The concern with elites and representation can be traced back through Schumpeter, and reflects a long tradition of thought on a particular kind of liberal democratic theory. . . . The responsiveness of elites is assured by the potential for activity by citizens, not their actual activity. Most citizens are assumed to be content with being apolitical.

The theoretical assumptions that political life is episodic, and that most people stay away from politics, help explain how Queens machine politicians easily blame immigrants themselves for not becoming participants in electoral politics, and how they see their own role not as one of encouraging participation, but primarily in terms of responding to the demands and complaints of their constituents. Queens machine politicians have not only an instrumental interest in not seeing new players in the political system (an interest which Erie lays out), but an ideology that allows them to justify their inaction. Ironically enough, it is an ideology that draws from much the same tradition of democratic theory that Erie and his allies depend on. Thus, machine politicians constantly question why Latinos haven't mobilized and approached them, instead of talking about why they themselves haven't made the effort to involve Latinos. Politicians defend themselves based on the record of how they respond, as the City Council member for Jackson Heights did when defending himself to a Latino news reporter: "I always respond to invitations. If I receive an invitation I make every effort to be there. But I don't like to go where I am not invited (interview)." Given this view of their mission, Queens politicians tend to respond best either to organized groups or those individuals who contact their offices directly. In both cases, these are people who have already mobilized on their own. In neither case is the politician's goal to act as a mobilizing agent, but rather to channel or defuse existing mobilization.

[. . .]

Immigrants are in a "catch-22" situation. The only avenue for incorporation that is recognized or permitted by the Democratic Party in Queens is within the party itself. The expectation is that groups on the margins of politics will mobilize themselves; only when mobilized will they receive the attention of politicians. Once groups are mobilized they receive the bulk of the attention politicians have to bestow. Those who are already mobilized then become the targets of mobilization by politicians, while those who are not remain on the margins. This makes sense from the tactical point of view of the individual politician or party, but not from the perspective of the polity as a whole.

REFERENCE FROM THE READING

Dahl, Robert. 1961. *Who Governs?* New Haven, CT: Yale University Press.

PART THREE

Understanding urban power

Plate 3 Brandenburg Gate, Berlin, 1946 and 2006.

INTRODUCTION TO PART THREE

The readings of the previous section delved into the electoral roots of urban power, focusing on the importance of party organizations and party leaders in shaping and controlling access to elected office and the only partly successful struggles to rewrite the rules of urban politics to lessen their influence. Here, we turn to the question of how the ways that mayors and other elected officials seek to exercise their power once in office are shaped by their interactions with many different interests – including various electoral constituencies, the many private interests with important business relationships with government, and those who work in government agencies as well.

John Mollenkopf begins Part 3 with a reading that provides an overview of how theorizing about the structure of urban power has evolved in the U.S. over the last half century. Initially, sociologists had presented a model of the local power structure in which political power mirrored a highly stratified economy and society, often dominated by a few families who owned or managed the major employers. Perhaps the best-known works in this vein are the chapter on "Family X" in Robert Lynd's study of Muncie, Indiana (*Middletown in Transition*, New York: Harcourt, Brace, 1937) and Floyd Hunter's finding that a small group of top corporate executives set Atlanta's agenda (*Community Power Structure*, Chapel Hill, NC: University of North Carolina Press, 1953). Reacting against this view, Yale political scientist Robert Dahl and his graduate students, who became eminent political scientists themselves, studied New Haven, Connecticut, in the late 1950s. Their work, notably Dahl's *Who Governs?*, refuted the idea that any one group or interest systematically dominated urban politics. Their model, pluralism, argued to the contrary that varied coalitions, usually changing over time, held sway over each specific policy area, that any relatively numerous and legitimate interests could organize, gain entry to the political system, and have their concerns heard, and that these mobilized interests would usually have some impact upon policy outcomes. The selections from *Who Governs?* presented here show the great subtlety and nuance with which Dahl simultaneously recognized the existence of political, social, and economic inequality with the notion that the local political system was nonetheless characterized by fluidity and responsiveness.

The New Haven studied by Dahl and his colleagues was, despite Yale's influence, an old, industrial, white ethnic blue-collar city on the verge of a profound and traumatic economic and social transformation that they did not entirely see coming. (For a fascinating revisiting of New Haven's transformation by another Yale professor who also held a top position in city government, see Douglas Rae's *City: Urbanism and its End*, New Haven, CT: Yale University Press, 2003). For one thing, Dahl did not foresee the sharp conflicts produced by white resistance to political demands from the growing black populations of many northern cities. For another, they did not fully appreciate how city elites would use urban renewal to alter the physical fabric of most big cities, often creating high levels of neighborhood protest. As Mollenkopf points out, the urban crisis of the late 1960s and 1970s prompted a systematic rethinking of the pluralist model.

Much of this rethinking tried to set urban power relations in the context of a larger political economy. Dahl and his colleagues had essentially seen New Haven as a more or less detached microcosm of American democracy. What they found locally, they assumed, would also be a valid model for the organization of political power in American society. In subsequent decades, however, it rapidly became obvious that the political leaders of cities were not free to chart their own course, but their choices were deeply constrained by their position in a complex national system.

Many different strands of thought developed within what might loosely be called the "structuralist" alternative to the pluralist approach. One deeply influential view was offered by Paul Peterson's *City Limits*. Drawing on public choice theory and the seminal work of economist Charles Tiebout, Peterson argued that since cities compete for investment and tax-paying residents, they could not adopt policies redistributing benefits from them to relatively less well-off people without promoting the flight of asset holders. They were therefore compelled, regardless of which people held office or what constituencies elected them, to pursue policies that promoted development and investment. He believed this structural imperative drained local politics of much of its meaningful content.

A second line of argument, developed by political scientists Clarence Stone and John Mollenkopf, among others, argued that the public sector's weak ability to command the resources of the private sector meant that mayors and other leaders of city government had to forge ongoing alliances with major private asset holders to carry out the plans of local government. In a kind of hybrid of the structuralist and pluralist perspectives, this school of thought, called "regime theory" by many, argued that public and private power holders would forge long-term relationships to advance their common and reciprocal interests. These relationships, or "regimes," would tend to outlast mayoral administrations that might come and go. Mollenkopf's view (see the following reading) emphasized that private power holders needed to create "pro-growth coalitions" just as much as elected officials needed support from the private sector to pursue their goals effectively.

Elizabeth Strom concludes Part 3 by asking the theoretically important question "How well do theoretical models drawn from the American experience travel to other national settings?" Drawing on the specific case of Berlin, she notes that while the same kinds of asymmetries between the public and private sectors, or between the state and market are present in Germany, they are worked out in very different ways. The specific institutional relationships between public and private sectors and between local, state, and federal government in Berlin grant a prominence to its "expertocracy" that would never be present in the U.S.

"How to Study Urban Power"

from *A Phoenix in the Ashes* (1994)

John H. Mollenkopf

Editors' Introduction

A Phoenix in the Ashes is John Mollenkopf's reflection on a decade of New York City political history. This was a decade in which that city, under the leadership of Mayor Ed Koch, emerged from a deep fiscal crisis and experienced a surge of job creation and downtown development. Mollenkopf is interested in exploring how Koch managed to piece together a coalition that could garner continued electoral support while pursuing fiscally conservative, "pro-growth" policies.

In an effort to understand the respective roles of economic and political power in that city, Mollenkopf begins his book with this reflection on the study of the distribution and exercise of power in any city – or any polity, for that matter. His review offers a sort of "biography" of four decades of urban political economy scholarship. In his view, one can identify several major strands of theoretical analysis. The pluralists, writing in the late 1950s and early 1960s, first drew political scientists into the study of cities, conducting empirical research into the local decision-making process. Their rather sanguine view – that many interests can and do influence local policy – may have had some validity for the relative quiescent period they studied (although even that, in the view of later analysts, is suspect). But certainly after a decade of local electoral challenge, civil disobedience and urban unrest, urban scholars were pressed to look more critically at the way cities were governed.

In contrast to pluralism, with its emphasis on voluntarism, alternative theoretical views stressed the limits of political action in the face of structurally rooted constraints. For neo-Marxists writing in the structuralist tradition, it is the nature of a profit-driven economic system that circumscribes the democratic process. Because cities must provide hospitable ground for capitalist success, governments have limited choices and the participation of economically marginal groups is curbed. Public choice theorists come from a very different direction – they draw on traditional economic analysis rather than neo-Marxist theory – to arrive at similar conclusions. Mollenkopf writes that for both structuralists and public choice theorists, "Politics . . . loses its autonomy and even its explanatory relevance."

Mollenkopf, like many urban scholars writing today, seeks to acknowledge the constraints described by structuralist analysis, but he believes that local politics still deals with real and pressing issues. Political actors have space in which to form coalitions and choose from a range of policy options. In his synthesis he stresses the role of political entrepreneurs who must simultaneously consider the interests of the public sector bureaucracy, electoral constituencies, and the holders of private capital. These interests are brought together into "dominant political coalitions," for which capturing the mayoralty is a necessary but not sufficient guarantee of successful governing. This synthesis, Mollenkopf contends, improves on the pluralist formulation by noting that governing coalitions are not entirely fluid and contingent, and they must include some element of the private sector to achieve any success at all. However, in contrast to the structuralist approach, this synthesis "posits a scope for political choice and innovation."

A research agenda growing out of this theoretical synthesis will study the ways in which political entrepreneurs organize such coalitions, and which strategies enable them to stay dominant over a longer time period. Such a research agenda will also seek to compare cities and their dominant coalitions, noting that the conditions for success will vary from place to place (for example, capturing the mayoralty may be less important in "weak" mayor cities than in strong mayor cities; the public sector bureaucracy may have greater veto power in cities with strong union traditions than in cities without unions). The extensive list of references following this excerpt offers a guide to additional reading on this topic.

What is the appropriate way to conceptualize the organization of political power in New York City during the Koch era? The dialogue between the pluralist interpreters of urban power and their structuralist critics has produced a rich variety of answers to this question. In the early 1960s, pluralist political scientists launched an attack on the previously accepted view, established by sociologists, that socioeconomic elites dominated urban politics. The success of this assault enabled pluralists to establish their view as the norm in political science.

[. . .]

THE PLURALIST CONCEPTION OF THE URBAN POLITICAL ORDER

The classic pluralist studies of a generation ago, like Banfield's *Political Influence*, Dahl's *Who Governs?*, or Sayre and Kaufman's *Governing New York City*, made important theoretical and methodological advances over the so-called elitists they attacked. They did not deduce power relations from the interlocks between economic and political elites. Instead, they went into the field to examine the tangled complexity of interest alignments around actual policy decisions and disputes. Pluralist scholars showed that no model of direct control by a unified economic or status elite could easily explain what they saw.

While most pluralists did not dwell theoretically on the larger relationship between the state and the economy, they implicitly rejected the notion that some underlying structural logic subordinated local politics to the private economy. They saw politics as an autonomous realm that possessed real authority and commanded important resources.

They explicitly rejected the notion that economic or social notables controlled the state in any instrumental sense. Since they argued that every "legitimate" group commanded some important resource (if only the capacity to resist) and no one group commanded sufficient resources to control all others, pluralists argued that the bargaining among a multiplicity of groups defined the urban power structure.

In this view, coalition building was central to the definition of power. Political leaders and private interests built coalitions around specific issues, the coalitions varied from issue to issue, and they tended to be short lived. By selecting a range of different policy decisions as case studies for research, pluralists seemed to imply that urban development and social service issues had an equal importance in organizing political competition.

In the face of examples where entrenched interest groups dominated their own particular, fragmented policy areas over time to the exclusion of the public interest, the pluralist approach developed a clearly critical strand of analysis. But these scholars simply saw the dark side of the pluralist worldview without fundamentally challenging its basic assumptions or deflating the optimistic claims about system openness or responsiveness prevailing among other pluralists.

[. . .]

While the pluralist studies may have been convincing and accurate portraits of urban politics in the 1950s and early 1960s, the eruption of turmoil and political mobilization in the 1960s and the fiscal crisis of the 1970s soon revealed basic flaws in the pluralist analysis. Except for Robert Dahl's work, *Who Governs?*, these studies lacked a context in economic and political development. Despite obligatory opening chapters covering economic, social, and political trends, pluralist studies such as Sayre and Kaufman's did not treat the changing structure of urban economies or racial succession as problematic for the urban political order. It would, they thought, simply absorb and adapt to these changes. While Dahl provided a fine treatment of the transition from patrician dominance to what he argued were the dispersed inequalities of pluralist democracy in New Haven, he also failed to see that blacks might be led to challenge the system, not just participate in it as a minor interest. Neither Dahl nor his colleagues foresaw how economic transformation and racial succession might fundamentally challenge the previously observed "normal" patterns.

Dahl did seek to address the relationship between social inequality and political power. For most pluralists, however, the emphasis on analyzing overt disputes diverted them from asking whether deep and persistent economic and social inequalities might create an equivalent political inequality underneath the surface of contending interest groups.

Dahl explicitly denied that economic and social inequalities would overlap and reinforce each other in the political arena. Other pluralist scholars also did not recognize the possibility that non-elite elements of the urban population would feel systematically excluded from power and would react by pressing for greater representation and more vigorously re-distributive policies. As a result, the urban battles that erupted in the latter 1960s in New York City and elsewhere made their relatively tranquil picture of urban politics as a kind of market equilibrium-reaching mechanism seem anachronistic.

STRUCTURALIST CRITIQUES

As the pluralist political equilibrium unraveled on the ground, it came under increasing challenge from structuralist critics. The broad outlines of their progress may be traced from Peter Bachrach and Morton Baratz's classic essay on the "two faces of power" to Clarence Stone's work on "systemic power" to John Manley's "class analysis of pluralism." Bachrach and Baratz attacked pluralists for focusing on the "first face" of power, namely its

exercise, while ignoring the second, namely the way that the relationship between the state and the underlying socioeconomic system shapes the political agenda. "Power may be, and often is," they said, "exercised by confining the scope of decision-making to relatively safe issues." But while making a case for analyzing how the values embedded in institutional practices bias the rules of the game, they do not specify the mechanisms that promote some interests and issues while dampening others.

Stone advanced this line of thought by shifting the locus of analysis from decisions ("market exchange") toward the mechanisms that create systemic or strategic advantages for some interests over others ("production"). The unequal distribution of private resources, he argues, creates a differential capacity among political actors to shape the flow of benefits from the basic rules of the game, the construction of particular agendas, and the making of specific decisions. Business, in particular, derives systemic power not only from its juridical status and economic resources but from its attractiveness as an ally for those who advance any policy change and from the shared subculture from which private and public officials both emerge.

Despite the structuralist leaning in his concept of "systemic power," Stone did not break decisively with the pluralist interplay of interests around decisions. Manley's Marxist critique does make this break. He embraced the argument that the legal and structural primacy enjoyed by private ownership of capital requires the state to reinforce the systemic inequalities that result from the drive for private profit. He attacked pluralists, even the later work of Dahl and Lindblom that concedes that business enjoys a privileged position in pluralist competition, for lacking a theory of exploitation and, hence, an objective standard of a just or equal distribution. In Manley's view, the juridical protection of private property inevitably commits the state to control workers and promote capital.

While neo-Marxist work similar to Manley's stressed the systematic subordination of the state and politics to capital accumulation and the private market, a parallel and quite non-radical strand of public choice analysis reached quite similar conclusions. Focusing on the notion that cities compete to attract well-off residents and private investment, this line of analysis stretches from

Charles Tiebout's work on the quasi-market competition among local governments to Paul Peterson's "unitary theory" of urban politics [see their readings below]. Despite drastically different evaluations of the state–market relationship, this body of work is logically quite similar to some neo-Marxist critiques of pluralism.

NEO-MARXIST CRITIQUES

Structuralists have decisively transcended the pluralist vocabulary. They provided the social and economic context missing from pluralism and highlighted the ways that private property, market competition, wealth and income inequality, the corporate system, and the stage of capitalist development pervasively shape the terrain on which political competition occurs. They underscored the need to analyze how basic patterns of the economic, political, or cultural rules of the game bias the capacity of different interests to realize their ends through politics and the state.

Most importantly, neo-Marxist structuralists were able to empirically investigate these mechanisms, refuting the pluralist retort that "non-decisions" either must be studied just like decisions or else are unobservable ideological constructs. They have shown cases in which the systemic and cumulative inequality of political capacity under-girded, and indeed was ideologically reinforced by, a superficial pluralism. Structuralist studies may be flawed by economic determinism, but they are factually on target in observing and describing mechanisms that generate systemic, cumulative, political inequality, which has a more profound impact on outcomes than the coalition patterns studied by pluralists. Such critiques won relatively broad support among the younger generation of scholars, if not their elders. They may be sub-classified into theories that stress the political logic of capital accumulation, social control, or the interplay of accumulation and legitimation. Each offers a different perspective on the central mechanisms that generate cumulative political inequality.

Theorists influenced by Marx's economic works have tended to argue that the mode of production stamps its pattern more or less directly on the organization of the state and on the dynamics of political competition. Marxists as different as

David Harvey and David Gordon have both argued that the stage of capitalist development and the circuits of capital have determined urban spatial patterns, the bureaucratic state, and for Harvey even urban consciousness. While this strand of Marxist thinking made a breakthrough in orienting analysts to the importance of the process of capital accumulation, it has generally lacked a well-developed theory of the state that either identifies the instrumental mechanisms that link state actions to the power of capital or grants the relative autonomy to the state.

This literature does stress one mechanism, however: the state's dependence on private investment for public revenues. If the mobility of capital can discipline the state and constrain political competition, then competition among polities (whether cities or nations) to attract investment leads them to grant systematic benefits for capital, a dynamic that Alford and Friedland have called "power without participation." As Harvey wrote,

> The successful urban region is one that evolves the right mix of life-styles and cultural, social, and political forms to fit with the dynamics of capital accumulation. . . . Urban regions racked by class struggle or ruled by class alliances that take paths antagonistic to accumulation . . . at some point have to face the realities of competition for jobs, trade, money, investments, services, and so forth.

Sooner or later, the state and political competition will be subordinated to the needs of capital.

Several analysts, including Friedland and Palmer as well as Molotch and Logan, abstracted this mechanism from the larger Marxian vocabulary and made it central to their analysis of urban power. Friedland and Palmer argued that, while businesses do directly influence policy-making, such intervention is logically secondary. "The growth of locales depends on the fortunes of their firms," according to Friedland and Palmer, thus "dominant and mobile [corporate] actors set the boundaries within which debate over public policy takes place." As capital has become more mobile and less tied to specific locations, the need for business to intervene directly in politics has waned, while the structural subordination of local government to the general interests of business has waxed.

Molotch and Logan took a different tack on the same course. While conceding that the mobility of capital gives local government a powerful incentive to defer to capitalists, they argued that certain classes of business are not mobile: real estate developers, utilities, newspapers, and others with a fixed relationship to a place. Large sunk costs give these interests a powerful incentive to intervene in and dominate local politics in order to get local government to promote new investment. They saw this "growth machine" as a ubiquitous, inevitable, and at best weakly challenged feature of American cities."

[. . .]

While this strand of thinking argued that the multiplicity of competing local governments forces the state to reproduce and protect basic features of the advanced capitalist economy, a second, equally important school of neo-Marxist thinking stressed the way urban politics serves to dampen and regulate the conflicts inevitably generated by capitalist urbanization. Castells' work on "collective consumption" and urban social movements, Piven and Cloward's studies of urban protest, and Katznelson's studies of the absorptive capacity of local bureaucracies and the bias against class issues in urban politics represent the best of this work.

While these analysts differed over how the state coopts movements that challenge urban governments, they share the idea that this process is a central feature of urban politics in advanced capitalist societies." Not everyone, even on the left, has agreed with these contentions. Theret, Mingione, and Gottdiener have criticized the explanatory power of the notion of collective consumption, while Ceccarelli has argued that urban social movements did not turn out to be the force in West European urban politics that Castells portrayed them to be. Whatever the situation in Europe, the civil rights movement, urban unrest, and community organization clearly had a profound impact on urban politics in the United States after the 1960s, particularly in the rise of programs designed to absorb and deflect these forces.

[. . .]

A third stream of neo-Marxism, stimulated by James O'Connor's and Claus Offe's contributions to the theory of the state, attempted to develop a multivariate approach to the structure of urban

power that accorded equal place to the imperatives to promote accumulation and to achieve legitimacy. In this approach, the two imperatives are crosscutting: the state must promote accumulation but cannot be seen to be doing so without risking its legitimacy. Efforts to bolster legitimacy through expanded social spending may hinder corporate profits if they are financed through progressive taxation. The structure and political orientation of the state become a battleground where these issues are fought out.

[. . .]

Friedland, Piven, and Alford used O'Connor's distinctions to construct a compelling theory of how accumulation-oriented and legitimacy-oriented functions become segregated into different local government bureaucracies. This segregation reduces the tension between the two functions: activities that promote growth are lodged in independent authorities lacking political accountability, while those related to social spending are subject to extensive public control. . . . Piven and Friedland extended such thinking to intergovernmental relations and the fragmentation of municipal government. In this case, competition among local jurisdictions for private investment not only favors the interests of capitalists, as argued by Harvey, Molotch, and others, but reconciles the tensions local governments face between raising revenues and securing electoral majorities. "The politics of vote and revenue generation diverge," Piven and Friedland argued, "because they are acted out in different parts of the local state, with the result that voter politics and investor politics each tend to be insulated from the other.

Although he eschewed the formal language of neo-Marxism, Shefter's analysis of New York City politics expanded this theme by suggesting that local governments face not two but four imperatives: they must not only raise revenues and construct electoral majorities; they must also preserve access to the municipal bond market and regulate conflict among citizens. The core of Shefter's analysis stressed how the New York City fiscal crisis stemmed from a basic conflict between attracting electoral support and extracting revenues from private investment: the Lindsay administration expanded spending to secure the former but could finance it only with debt, not increased taxation of local investment. Fiscal crisis was the inevitable result.

PUBLIC CHOICE CRITIQUES

Neo-Marxist thinking is not the only source of structural criticism of the pluralist paradigm, however. Microeconomics, in the form of public choice theory, has contributed its own critique. Tiebout's seminal work led to Forrester's simulation of urban systems and ultimately to Paul Peterson's sophisticated "unitary" theory of urban politics. This tradition, born of the economists' distrust of state allocation of resources, has sought a functional equivalent to the marketplace in the multiplicity of local governments. They would compete, Tiebout argued, for residents of different means and desires by providing different service packages at various tax costs. An equilibrium would thus be reached in the sorting of populations across urban and suburban jurisdictions within the metropolis. This equilibrium would represent an efficient production of public goods, matching the marginal prospective resident with the jurisdiction's need to add (or subtract) residents on its own margin to provide services at the most efficient scale.

[. . .]

This analysis reached its highest form in Paul Peterson's *City Limits*. Like neo-Marxists, Peterson analyzed how external economic conditions shape and constrain the urban political arena and concluded that, "political variables no longer become relevant to the analysis." Unlike neo-Marxists, however, he posited the importance of consumer as well as investor demand and imputed a unitary interest in economic growth to all constituent urban interests. "The interests of cities," he said, lie not in an optimum size for efficient service provision nor in some pluralist bargaining among constituencies, but in "policies [that] maintain or enhance the economic position, social prestige, or political power of the city, taken as a whole."

Of these, he found economic position paramount and equated it with the health of export industries. The overwhelming importance of promoting exports means that "the issues screened out of local politics are not eliminated by local electoral devices, bureaucratic manipulations, or a one-sided press . . . [but because they] fall outside the limited sphere of local politics." In fact, local politics is so limited in Peterson's view that it cannot even generate partisan competition or serious group challenges to prevailing policy. Observed inter-group struggles are only ethnic competition over jobs and contracts. Subsequently, Peterson concluded that, owing to economic decline and racial transition, "the industrial city has become an institutional anachronism."

[. . .]

While this market-based explanation of the limits on urban politics has a markedly different and more positive evaluation of the final equilibrium than do neo-Marxist formulations, it has a similar logical structure and reaches similar conclusions. Politics – at any rate urban politics – loses its autonomy and even its explanatory relevance. Intercity competition drives redistribution off the urban political agenda and puts the promotion of economic development in top position.

STRUCTURALISM RECONSIDERED

By providing the missing economic and social-structural context, these structuralist critiques achieved a considerable advance over pluralist analysis. Cities can no longer be taken as independent entities isolated from the larger economic and social forces that operate on them. Analysts can no longer ignore the impact of global and national economic restructuring on large central cities. Since cities cannot retard these global economic trends . . . nor remake their populations at will, they clearly navigate in a sea of externally generated constraints and imperatives.

The structuralist critiques also make it clear that urban politics can no longer be considered to be unrelated to the cumulative pattern of inequality in the economy and society. They have focused attention on how the state's dependence on private investment fosters political outcomes that systematically favor business interests. Structuralists have explored specific mechanisms that produce this result, such as the invidious competition among fragmented, autonomous urban governments for investment, the segregation of local government functions into quasi-private agencies that promote investment and politically exposed agencies that absorb and deflect protest, and the organization of the channels of political representation so as to articulate interests in some ways but not others. By stressing that advanced capitalism characteristically generates urban social movements and political conflicts, some structuralists have also implied that political action can alter some of the constraints capitalism imposes on democracy.

Despite these strengths, however, structuralist perspectives also have grave flaws. The assertion that the state "must" undertake activities that favor capital tends to be functionalist. Such a standpoint begs the question of how these "imperatives" are put in place and reproduced over time, which inevitably must be through the medium of politics. As a result, structuralists may not see that political actors can fail to fulfill or to maximize their supposed imperatives. Dominant urban political coalitions have certainly done things that cost them elections and the ability to exercise power; they have persisted in increasing the tax burden on private capital and imposing exactions on private developers even after the point that they diminish further investment. Others have chosen to increase budget deficits and risk their bond ratings.

Given the right conditions, nothing is inevitable about an administration's pursuit of electoral success, private investment, well-managed social tensions, or even good bond ratings. Nothing guarantees that city government will be willing or able to fulfill the functions structuralists have assigned to it. As Piven and Friedland observed in rejecting a "smoothly functioning determinism," a structural analysis cannot be adequate until it specifies "the political processes through which . . . systematic imperatives are translated into government policies."

Structural critiques also tend not to be disconfirmable. For example, if structuralists argue that the use of legal injunctions by the conservative Republican administrations before the New Deal illustrates how the state supports capitalism, while the New Deal's recognition and promotion of trade unions also illustrate state support for capitalism, then they are explaining everything and nothing. Put another way, structural theories tend to have a hard time explaining the real and important variation over time and across places. The basic features of capitalism are common across nations and evolve slowly, while the political outcomes that capitalism is supposed to drive are highly varied and change more quickly.

[. . .]

To summarize, for all their strengths, structuralists conceptualized the political system as ultimately subordinate to economic structure. They tended to reduce urban politics to the fulfillment of economic imperatives; even social control achieved through political means serves capitalist ends. The most promising threads of structuralist thinking examined how systemic imperatives might conflict with each other or generate system-threatening conflict, thus opening the way for political indeterminacy. Here, however, they risked moving outside and beyond a structuralist paradigm. Indeed, orthodox Marxism (or for that matter orthodox neoclassical economics) simply does not provide a good basis for building a theory of politics. To the extent that structuralist theorists held true to the logic of their argument, they underplayed the importance of politics. They did not appreciate that policies that promote private investment must be constructed in a political environment that may favor but by no means guarantees this outcome. Indeed, popular, social, and communal forces pressure the state and the political process just as strongly in different, and often opposed, directions.

This tendency to trivialize politics removes a way to explain why outcomes vary even though capitalism is constant. States may be constrained, but they are also sovereign. They exercise a monopoly on the legitimate use of force, establish the juridical basis for private property, and shape economic development in myriad ways. Economies are delicate. They depend on political order and have been deformed or smashed by political disorder. State actions may be conditioned by economic structure, but they cannot be reduced to it. Many substantially different capitalisms are possible, and politics determine which ones evolve. Just as the state is dependent on the economy, economic institutions depend on and are vulnerable to the state and its changing political circumstances.

Even as dedicated a Marxist as David Harvey has conceded that the political arrangements supporting capitalism are "complex," "open to a curious mix of private and class pressures, social traditions and conventions, and political processes," "not entirely imposed from above or given from outside," and constitute a "powerful shaping force." As Schumpeter said, the state's influence on the economy and society usually "explains practically all the major features of events, in most periods it explains a great deal and there are but a few periods when it explains nothing." If politics is conceded to have causal power, then any judgment that external economic structural causes are more important than internal political ones must be substantiated, not simply assumed.

Peterson to the contrary notwithstanding, an excellent case can be made that urban politics is a high-stakes game with winners and losers, even on matters of economic development. Close to one-quarter of the Gross National Product (GNP) passes through the public sector in the United States, much of it through urban governments. Local regulation strongly shapes land values. Variations in public spending priorities can become quite significant.... Moreover, private investments in cities are sufficiently fixed in the short term – and even in the long term when a city has a strong competitive advantage – that many investors can neither escape nor easily resort to the sledge-hammer of a capital strike.

Even where capital mobility places real limits on politics, local governments still have considerable power to tax, regulate, and direct economic activity. Where cities enjoy an underlying competitive advantage, this political leverage is all the greater. While capital enjoys systemic advantages, neither capital as a system nor businesses organized as an interest group can dictate outcomes in spending, tax, and development policy. Contrary to Peterson's contention that questions of city development are essentially nonpolitical, they have been characterized by extensive conflict. The spirited political challenges to the established urban political order since the late 1960s and the depth of the private sector reaction against them suggest that participants in urban politics certainly believe they can influence important outcomes.

"Polity-centered" thinking must thus augment the "economy-centered theorizing of the structuralist critiques." This does not require an equally one-sided political determinism. Rather, it requires us to extend the lines of structuralist thinking that stress conflict among imperatives or developmental tendencies until we go beyond the limits of economic determinism.... As Manuel Castells ... reflected, "experience was right and Marxist theory was wrong" about the central theoretical importance of urban social movements and the impossibility of reducing them to a class basis.

But if we give politics an analytic weight equal to that of economic structure, how can we avoid returning to a voluntaristic pluralism? How can we develop a vocabulary for analyzing politics and state action that reconciles the political system's independent impact on social outcomes with its observed systemic bias in favor of capital? A satisfactory approach must operate at three interrelated levels: (1) how the local state's relationship to the economy and society conditions its capacity to act, (2) how the "rules of the game" of local politics shape the competition among interests and actors to construct a dominant political coalition able to exercise that capacity to act, and (3) how economic and social change and the organization of political competition shape the mobilization of these interests.

TOWARD A THEORETICAL SYNTHESIS

We can begin to build such an approach by recognizing that city government and its political leaders interact with the resident population and constituency interests in its political and electoral operating environment and with market forces and business interests in its economic operating environments. This approach emphasizes two primary interactions: first, between the leaders of city government and their political/electoral base; and second, between the leaders of city government and their economic environment. It also suggests that political entrepreneurs who seek to direct the actions of city government must contend with three distinct sets of interests: (1) public sector producer interests inside local government, (2) popular or constituency interests (which are also public sector consumer interests), especially as they are organized in the electoral system, and (3) private market interests, particularly corporations with discretion over capital investment, as they are organized in the local economy.

To be sure, these interests are highly complex ... and cannot be captured by simple dichotomies like black versus white or capitalist versus worker. ... The city's residential communities are highly heterogeneous. Terms like "minority" hide far more than they reveal; even "black" or "Latino" blur important distinctions regarding nativity and ethnicity. Business interests come in many sizes, industries, and competitive situations; even corporate elites vary greatly. Still, a focus on the relationships among state, citizenry, and marketplace provides an entry point for analyzing what determines the shape of the urban political arena.

The concept of a "dominant political coalition" gives us a focal point for this analysis. A dominant political coalition is a working alliance among different interests that can win elections for executive office and secure the cooperation it needs from other public and private power centers in order to govern. To have an opportunity to become dominant, it must first win election to the chief executive office. To remain dominant, it must use the powers of government to consolidate its electoral base, win subsequent elections, and gain support from those other wielders of public authority and private resources whose cooperation is necessary for state action to go forward. Put another way, a dominant coalition must organize working control over both its political and its private market operating environments.

This formulation improves on the pluralist approach by directing our attention toward how the relationship between politics and markets biases outcomes in favor of private market interests, as structuralist approaches have pointed out. The notion of a dominant political coalition would not sit well with pluralists, who have argued that coalitions are unstable, form or re-form according to the issue, and may be stymied by the capacity of any sizable group to resist. We posit instead that coalitions can be stable, operate across issues, and create persistent winners and losers. Challenging and supplanting such coalitions have generally been difficult, particularly for constituencies that lack resources or are particularly vulnerable to sanction. Effective challenges generally arise only at moments of crisis in periods of rapid social and economic change.

This formulation also improves on the structuralist approach by according the political/electoral arena an influence equal to that of economic forces. It also points us toward how strategies to control the direction of city government are shaped by (and in turn shape) the political environment and by the public sector producer interests that have a permanent stake in its operation. It posits a scope for political choice and innovation that is lacking in the structuralist perspective.

This approach points us toward the following central questions: how do political entrepreneurs seek to organize such coalitions, what enables them to succeed in the first instance, and how do they sustain success over time? In what ways can such coalitions be bound together? What interests do dominant coalitions include and exclude and why? How do the economic and political contexts affect these binding relationships? And what tensions or conflicts undermine dominant coalitions, opening the way for power realignment?

As structuralists have shown, one part of the answer to these questions lies in the relationship of politics to the structure of economic interests. Efforts to explore this relationship may be found in Stone's studies of Atlanta, Shefter's study of New York, and my own work on pro-growth coalitions in Boston and San Francisco. Stone distinguished three levels at which to analyze the relationship between a dominant coalition and various urban interests. The least interesting is the pluralist domain of individual decisions or "command power" in which one actor induces or coerces others to follow his or her bidding. The two others are more relevant to this analysis.

Political actors wield "coalition power" when they join together to exercise the policy powers of the state to produce a steady flow of benefits to their allies, without the need for coercing or inducing specific actions. Stone showed how the Atlanta regime used public and private subcontracts to minority business enterprises to cement its political support, but he gave relatively little attention to other aspects of how the coalition tried to dominate its political operating environment.

Instead, he emphasized the "preemptive" or "systemic power" enjoyed by private interests whose command over private resources are so great as to make their support crucial to the dominant coalition. Among the mechanisms of preemptive power in Atlanta, Stone identified the unity of a well-organized downtown business community, newspaper and television support for policies that favored downtown development, the reliance of politicians on campaign contributions from developers, business control over the equity and credit that government needed to carry out its plans, and the business community's ability to provide or deny access to upward mobility for the black middle class. These systemic powers made corporate interests ideal allies for politicians seeking to achieve and sustain political dominance.

While the arrival of a black majority in the electorate, militant new black leadership, and neighborhood mobilization eventually destabilized and modified the tradition of white dominance that prevailed in Atlanta until Maynard Jackson was elected mayor in 1973, Stone argued that they did not overturn the preemptive power of corporate interests. Jackson's successor, Andrew Young, chose to abandon the fragmented and undisciplined neighborhood movement in favor of pro-growth politics with a new face, consolidated by white business support for set-asides to minority entrepreneurs.

It is theoretically instructive to contrast New York with Atlanta, however. In New York, business interests, while large and powerful, appear to be less cohesive and less well organized than in Atlanta, while the power of city government is greater, public sector producer interests stronger, and its political constituencies better organized. Shefter's analysis of New York City politics parallels that of Stone but offers some interesting departures.

Although Shefter used the notion of "accommodation among interests" rather than "regime" or "dominant coalition" and thus veers toward pluralism, he argued that political interests reached a stable pattern of accommodation with private interests in the 1950s. As in Atlanta, these broke down under the political mobilizations of the late 1960s and early 1970s, but the fiscal crisis of 1975 set the stage for reasserting a conservative pattern of accommodation. Like Stone, he argued that the elite interests who refinanced the city's debt, particularly the large banks, were the major participants forging policy accommodations. Unlike Stone, however, Shefter argued that other parties played a major role, including the public employee unions and the fiscal monitors that were buttressed by the "reform vanguard." (The last include good-government organizations whose funding and leadership may come from corporate headquarters but whose social base lies in the city's middle-class, opinion-shaping professionals.)

Shefter argued that business elites and public sector labor leaders constructed a new, postreform dominant coalition during the fiscal crisis by shifting city spending away from social welfare and reducing the burden of government on the economy while protecting public service producer interests, particularly the municipal unions. In his view, black leaders and the dependent poor were relegated to the sidelines.

My own study of how political entrepreneurs constructed pro-growth coalitions in Boston, San Francisco, and other large cities in the late 1950s and 1960s also advanced reasons why politicians would want to forge alliances with private sector elites. Promoting private development was an obvious way to bring together such otherwise disparate elements as a Republican corporate elite, regular Democratic party organizations, and reform-oriented rising public sector and nonprofit professionals.

These perspectives on how dominant political coalitions shape development politics and budget policy to secure business support, while convincing, remain incomplete. The "preemptive power" of business interests only explains part of how political actors construct a coalition to direct city government in the exercise of its powers. As Lincoln Steffens long ago observed, "dominant coalitions needed to develop a grass-roots base of legitimacy as well as support from elite interests." However much they may need corporate support, dominant coalitions must also have support from popular constituencies organized by such organizations as political parties, labor unions, and community organizations.

Mayors can lead dominant political coalitions only when they win electoral majorities and keep potential sources of electoral challenge fragmented or demobilized. The mobilization of blacks and Latinos and the neighborhood organization that began in the mid-1960s and continue today have prompted many currently dominant political coalitions to adopt policies that do not follow from a devotion to private market interests or public sector producer interests. For example, dominant coalitions must respond to mobilizations against the negative impacts of downtown growth and inadequate public services, or demands for government programs that provide upward mobility for excluded groups.

[. . .]

The organization of political institutions and processes are especially important in determining how political leaders form coalitions. Their strategies arise not just from a pluralist "parallelogram of forces" but in response to deeply embedded political practices that govern the articulation of political interests. Questions raised in Martin Shefter's work – when and why do political parties have an incentive to mobilize some constituencies and demobilize others? – take on a central importance. The literature on the decay of political parties, the nature of one-party politics in large cities, and the evolving role of municipal labor unions and community organizations is particularly relevant. The dimensions of space, place, and community should have equal importance to the systemic power of corporate interests as building blocks of political dominance.

[. . .]

SELECT BIBLIOGRAPHY

Alford, R. and R. Friedland (1985) *Powers of Theory: Capitalism, the State, and Democracy.* New York: Cambridge University Press.

Bachrach, P. and M. Baratz (1962) "Two Faces of Power." *American Political Science Review* 56: 947–952.

Banfield, E. (1961) *Political Influence.* Glencoe, IL: Free Press.

Castells, M. (1978) *The City and the Grassroots.* Berkeley: University of California Press.

Ceccarelli, P. (1982) "Politics, Parties, and Urban Movements: Western Europe" in S. Fainstein and N. Fainstein (eds) *Urban Policy Under Capitalism.* Beverly Hills, CA: Sage.

Dahl, R. (1961) *Who Governs.* New Haven, CT: Yale University Press.

Forrester, J. (1969) *Urban Dynamics.* Cambridge, MA: MIT Press.

Friedland, R. and D. Palmer (1984) "Park Place and Main Street: Business and the Urban Power Structure." *Annual Review of Sociology* 9: 406–407.

Friedland, R., F.F. Piven and R. Alford (1977) "Political Conflict, Urban Structure, and the Fiscal Crisis." *International Journal of Urban and Regional Research* 1, 3: 447–461.

Gordon, D. (1977) "Capitalism and the Roots of Urban Fiscal Crisis" in R. Alcaly and D. Mermelstein (eds) *Fiscal Crisis of American Cities.* New York: Vintage.

Harvey, D. (1985) *The Urbanization of Capital.* Baltimore, MD: Johns Hopkins University Press.

Katznelson, I. (1981) *City Trenches: Urban Politics and the Patterning of Class in the United States.* New York: Pantheon.

Logan, J. and H. Molotch (1987) *Urban Fortunes: The Political Economy of Place.* Berkeley: University of California Press.

Manley, J. (1983) "Neo-pluralism: A Class Analysis of Pluralism I and Pluralism II." *American Political Science Review* 77, 2: 368–383.

Mingione, E. (1981) *Social Conflict and the City.* New York: St. Martin's Press.

Mollenkopf, J.M. (1983) *The Contested City.* Princeton, NJ: Princeton University Press.

O'Connor, J. (1973) *The Fiscal Crisis of the State.* New York: St. Martin's Press.

Offe, C. (1984) *Contradictions of the Welfare State.* London: Hutchinson.

Piven, F.F. and R. Cloward (1979) *Poor Peoples' Movements.* New York: Pantheon.

Shefter, M. (1985) *Political Crisis/Fiscal Crisis.* New York: Basic Books.

Stone, C. (1980) "Systemic Power in Community Decision-Making: A Restatement of Stratification Theory." *American Political Science Review* 74: 978–990.

—— (1989) *Regime Politics: Governing Atlanta.* Lawrence, KS: University Press of Kansas.

Theret, B. (1982) "Collective Means of Consumption, Capital Accumulation and the Urban Question." *International Journal of Urban and Regional Research* 6, 3: 345–371.

THREE

"Who Governs?"

from *Who Governs? Democracy and Power in an American City* (1961)

Robert Dahl

Editors' Introduction

"In a political system where nearly every adult may vote but where knowledge, wealth, social position, access to officials, and other resources are unequally distributed, who actually governs?" asks Robert Dahl. *Who Governs* aims to address that question by bringing an ambitious theoretical analysis to bear on a detailed case study of the politics of New Haven, Connecticut, where Dahl was Professor of Political Science at Yale University.

His analysis covers two centuries of the city's history, during which, he argues, the city's political system shifted from an oligarchy, in which a small group of economic elites ruled, to one described as "pluralist," meaning that a variety of different groups compete for power. Dahl's methods were central to his work. In contrast to earlier students of urban power, who had used "reputational" analysis to deduce the key community decision-makers, Dahl sought to build his argument in a more scientific way. To study power, he argued, you examine the decision-making process; influence can be deduced by seeing who participates, and whose interests are ultimately served by the outcome. Dahl chooses three policy areas on which to focus: public education, urban development and the party nominating process. Through this research, Dahl concludes that power and influence are fairly well dispersed throughout the community: there is no single elite that dominates all three policy areas, and (although he never denies the existence of inequalities) many different interests have the opportunity to participate. Leaders might exercise a certain degree of power, but just as often they seemed to be at the mercy of their constituents. Not everyone participates, and indeed most people remain thoroughly apolitical, but everyone *could* participate, and indeed, probably would, if their elected leaders pursued policies that began to displease them.

Dahl, like many of the most prominent political scientists working at that time, wrote about cities, believing that the local polity allowed for the study of larger political forces on a manageable scale. Viewing the city as a "case" of the larger question of democratic governance, American political scientists like Dahl, Raymond Wolfinger (*The Politics of Progress*, New York: Prentice-Hall, 1974), Theodore Lowi (*At the Pleasure of the Mayor: Patronage and Power in New York City, 1898–1958*, New York: Free Press, 1964), Nelson Polsby (*Community Power and Political Theory*, New Haven, CT: Yale University Press, 1963) and Edward Banfield (*City Politics*, Cambridge, MA: Harvard University Press, 1963; *The Unheavenly City*, Boston, MA: Little, Brown, 1970; *The Unheavenly City Revisited*, Boston, MA: Little, Brown, 1974) all focused on an analysis of big city political systems.

Dahl's research has generated a library of criticism, some of which is summarized in the selection by John Mollenkopf. Soon after *Who Governs* was published, Bachrach and Baratz wrote two articles challenging both the theory and method of Dahl's analysis. Of course Dahl found that all participants possessed some sorts of resources with which to influence policy outcomes, they claimed: his approach, of studying policy decision-making, rendered invisible those so lacking in resources they had no means of participating.

"Nondecisions," they argued – the issues never even taken up by the political leadership – also spoke volumes about who had and did not have power (see Peter Bachrach and Morton S. Baratz, "Decisions and Non-decisions: An Analytical Framework." *American Political Science Review* (1963) 57: 641–651; and "The Two Faces of Power." *American Political Science Review* (1962) 56: 941–952).

Critics also doubted that cities could be analyzed as mini-nation states; they operate under legal and economic constraints that make them, one could say, almost a different species of polity. As succeeding generations of political scientists have successfully argued to put these constraints at the center of urban analysis, they have set the record straight – but this new perspective may have also accelerated the marginalization of the study of the local within the larger field of political science.

Robert Dahl, the Sterling Professor Emeritus of Political Science and the Senior Research Scientist in Sociology, is one of the pre-eminent political scientists of his generation, credited with developing a sophisticated theory of pluralism, and shaping the study of American democracy. In addition to *Who Governs*, one of the true classics of the field, Dahl is the author of many noteworthy books about American politics and democratic theory, including: *A Preface to Democratic Theory* (Chicago, IL: University of Chicago Press, 1963); *A Preface to Economic Democracy* (Berkeley: University of California Press, 1986); *On Democracy* (New Haven, CT: Yale University Press 2000); *Polyarchy* (New York: Taylor & Francis 1991); and *Democracy and its Critics* (New Haven, CT: Yale University Press, 1991).

[CHAPTER 8] OVERVIEW: THE AMBIGUITY OF LEADERSHIP

One of the difficulties that confronts anyone who attempts to answer the question, "Who rules in a pluralist democracy?" is the ambiguous relationship of leaders to citizens.

Viewed from one position, leaders are enormously influential – so influential that if they are seen only in this perspective they might well be considered a kind of ruling elite. Viewed from another position, however, many influential leaders seem to be captives of their constituents. Like the blind men with the elephant, different analysts have meticulously examined different aspects of the body politic and arrived at radically different conclusions. To some, a pluralistic democracy with dispersed inequalities is all head and no body; to others it is all body and no head.

Ambiguity in the relations of leaders and constituents is generated by several closely connected obstacles both to observation and to clear conceptualization. To begin with, the American creed of democracy and equality prescribes many forms and procedures from which the actual practices of leaders diverge. Consequently, to gain legitimacy for their actions leaders frequently surround their covert behavior with democratic rituals. These rituals not only serve to disguise reality and thus to complicate the task of observation and analysis, but – more important – in complex ways the very existence of democratic rituals, norms, and requirements of legitimacy based on a widely shared creed actually influences the behavior of both leaders and constituents even when democratic norms are violated. Thus the distinction between the rituals of power and the realities of power is frequently obscure.

Two additional factors help to account for this obscurity. First, among all the persons who influence a decision, some do so more directly than others in the sense that they are closer to the stage where concrete alternatives are initiated or vetoed in an explicit and immediate way. Indirect influence might be very great but comparatively difficult to observe and weigh. Yet to ignore indirect influence in analysis of the distribution of Influence would be to exclude what might well prove to be a highly significant process of control in a pluralistic democracy.

Second, the relationship between leaders and citizens in a pluralistic democracy is frequently reciprocal: leaders influence the decisions of constituents, but the decisions of leaders are also determined in part by what they think are, will be, or have been the preferences of their constituents. Ordinarily it is much easier to observe and describe the distribution of influence in a political system where the flow of influence is strongly in one direction (an asymmetrical or unilateral system, as it is sometimes called) than in a system marked by strong reciprocal relations. In a political system with competitive elections, such as New Haven's, it is not unreasonable to expect that relationships between leaders and constituents would normally be reciprocal.

One who sets out to observe, analyze, and describe the distribution of influence in a pluralistic democracy will therefore encounter formidable problems. It will, I believe, simplify the task of understanding New Haven if I now spell out some of the theory and assumptions that guided our study of the distribution of influence.

THE POLITICAL STRATUM

In New Haven, as in other political systems, a small stratum of individuals is much more highly involved in political thought, discussion, and action than the rest of the population. These citizens constitute the political stratum.

Members of this stratum live in a political subculture that is partly but not wholly shared by the great majority of citizens. Just as artists and intellectuals are the principal bearers of the artistic, literary, and scientific skills of a society, so the members of the political stratum are the main bearers of political skills. If intellectuals were to vanish overnight, a society would be reduced to artistic, literary, and scientific poverty. If the political stratum were destroyed, the previous political institutions of the society would temporarily stop functioning. In both cases, the speed with which the loss could be overcome would depend on the extent to which the elementary knowledge and basic attitudes of the elite had been diffused. In an open society with widespread education and training in civic attitudes, many citizens hitherto in the apolitical strata could doubtless step into roles that had been filled by members of the political stratum. However, sharp discontinuities and important changes in the operation of the political system almost certainly would occur.

In New Haven, as in the United States, and indeed perhaps in all pluralistic democracies, differences in the subcultures of the political and the apolitical strata are marked, particularly at the extremes. In the political stratum, politics is highly salient; among the apolitical strata, it is remote. In the political stratum, individuals tend to be rather calculating in their choice of strategies; members of the political stratum are, in a sense, relatively rational political beings. In the apolitical strata, people are notably less calculating; their political choices are more strongly influenced by inertia, habit, unexamined loyalties, personal attachments, emotions, transient impulses. In the political stratum, an individual's political beliefs tend to fall into patterns that have a relatively high degree of coherence and internal consistency; in the apolitical strata, political orientations are disorganized, disconnected, and unideological. In the political stratum, information about politics and the issues of the day is extensive; the apolitical strata are poorly informed. Individuals in the political stratum tend to participate rather actively in politics; in the apolitical strata citizens rarely go beyond voting and many do not even vote. Individuals in the political stratum exert a good deal of steady, direct, and active influence on government policy; in fact some individuals have a quite extraordinary amount of influence. Individuals in the apolitical strata, on the other hand, have much less direct or active influence on policies.

Communication within the political stratum tends to be rapid and extensive. Members of the stratum read many of the same newspapers and magazines; in New Haven, for example, they are likely to read the *New York Times* or the *Herald Tribune*, and *Time* or *Newsweek*. Much information also passes by word of mouth. The political strata of different communities and regions are linked in a national network of communications. Even in small towns, one or two members of the local political stratum usually are in touch with members of a state organization, and certain members of the political stratum of a state or any large city maintain relations with members of organizations in other states and cities, or with national figures. Moreover, many channels of communication not designed specifically for political purposes – trade associations, professional associations, and labor organizations, for example – serve as part of the network of the political stratum.

In many pluralistic systems, however, the political stratum is far from being a closed or static group. In the United States the political stratum does not constitute a homogeneous class with well-defined class interests. In New Haven, in fact, the political stratum is easily penetrated by anyone whose interests and concerns attract him to the distinctive political culture of the stratum. It is easily penetrated because (among other reasons) elections and competitive parties give politicians a powerful motive for expanding their coalitions and increasing their electoral followings.

[. . .]

Not only is the political stratum in New Haven not a closed group, but its "members" are far from united in their orientations and strategies. There are many lines of cleavage. The most apparent and probably the most durable are symbolized by affiliations with different political parties. Political parties are rival coalitions of leaders and subleaders drawn from the members of the political stratum. Leaders in a party coalition seek to win elections, capture the chief elective offices of government, and insure that government officials will legalize and enforce policies on which the coalition leaders can agree.

In any given period of time, various issues are salient within the political stratum. Indeed, a political issue can hardly be said to exist unless and until it commands the attention of a significant segment of the political stratum. Out of all the manifold possibilities, members of the political stratum seize upon some issues as important or profitable; these then become the subject of attention within the political stratum. To be sure, all the members of the political stratum may not initially agree that a particular issue is worthy of attention. But whenever a sizable minority of the legitimate elements in the political stratum is determined to bring some question to the fore, the chances are high that the rest of the political stratum will soon begin to pay attention.

Although political issues are sometimes generated by individuals in the apolitical strata who begin to articulate demands for government action, this occurs only rarely. Citizens in the apolitical strata are usually aware of problems or difficulties in their own circle; through word of mouth or the mass media they may become aware of problems faced by people in other circles. But to be aware of a problem is by no means equivalent to perceiving a political solution or even formulating a political demand. These acts are ordinarily performed only by members of the political stratum. Within the

political stratum, issues and alternatives are often formulated by intellectuals, experts, and reformers, whose views then attract the support of professionals. This is how questions as abstract and difficult as the proper rate of growth in the Gross National Product are injected into national politics; and, as we shall see, this is roughly the route by which urban redevelopment came into the politics of New Haven.

However, in gaining attention for issues, members of the political stratum operate under constraints set by party politicians with an eye on the next election. Despite the stereotype, party politicians are not necessarily concerned only with winning elections, for the man who is a party politician in one role may, in another, be a member of a particular interest group, social stratum, neighborhood, race, ethnic group, occupation or profession. In this role he may himself help to generate issues. However, simply qua party politician, he not only has a powerful incentive to search for politically profitable issues, but he has an equally strong motive for staying clear of issues he thinks will not produce a net gain in votes in the next election.

Because of the ease with which the political stratum can be penetrated, whenever dissatisfaction builds up in some segment of the electorate party politicians will probably learn of the discontent and calculate whether it might be converted into a political issue with an electoral payoff. If a party politician sees no payoff, his interest is likely to be small; if he foresees an adverse effect, he will avoid the issue if he can. As a result, there is usually some conflict in the political stratum between intellectuals, experts, and others who formulate issues, and the party politicians themselves, for the first group often demands attention to issues in which the politicians see no profit and possibly even electoral damage.

The independence, penetrability, and heterogeneity of the various elements of the political stratum all but guarantee that any dissatisfied segments will find spokesmen in the political stratum, but to have a spokesman does not insure that the group's problems will be solved by political action. Politicians may not see how they can gain by taking a position on an issue; action by government may seem to be wholly inappropriate; policies intended to cope with dissatisfaction may be blocked; solutions may be improperly designed;

indeed, politicians may even find it politically profitable to maintain a shaky coalition by keeping tension and discontent alive and deflecting attention to irrelevant "solutions" or alternative issues. In his search for profitable issues, the party politician needs to estimate the probable effects various actions he might take will have on the future votes of his constituents. Although he is generally unaware of it, he necessarily operates with theory, a set of hypotheses as to the factors that influence the decisions of various categories of voters and the rough weights to assign to these factors.

The subculture of the political stratum provides him with the relevant categories – businessmen, Italians, wage earners, and the like. It also furnishes him with information as to the voting tendencies of these groups, e.g., their predisposition to vote Democratic or Republican. Given a category and its voting tendency, the party politician typically operates on the simple but sound assumption that human responses can be influenced by rewards and deprivations, both past and prospective. His task then is to choose a course of action that will either reinforce the voting tendency of categories predisposed in favor of him or his party, or weaken the voting tendency of categories predisposed to vote against him or his party. This he does by actions that provide individuals in these categories with rewards or the expectation of rewards.

SOME POLITICAL AXIOMS

Most of the people in the political stratum at any given moment take for granted a number of assumptions so commonplace in the political culture of the time and so little subject to dispute that they function as "self-evident" axioms. The axioms include both factual and normative postulates. In New Haven, the most relevant current axioms among the political stratum would appear to be the following:

▪ To build an effective political coalition, rewards must be conferred on (or at least promised to) individuals, groups, and various categories of citizens.
▪ In devising strategies for building coalition and allocating rewards, one must take into account a large number of different categories of citizens.

▪ Although a variety of attributes are relevant to political strategy, many different attributes can either be subsumed under or are sometimes overridden by ethnic, racial, and religious affiliations.
▪ In allocating rewards to individuals and groups, the existing socio-economic structure must be taken as given, except for minor details.

Although a certain amount of legal chicanery is tolerable, legality and constitutionality are highly prized. The pursuit of illegal practice on sizable scale is difficult to conceal; illegal actions by public officials ordinarily lead, when known, to loss of public office; unconstitutional action is almost certain to become entangled in a complex network of judicial processes. The use of violence as a political weapon must be avoided; if it were used it would probably arouse widespread alarm and humility.

The American creed of democracy and equality must always be given vigorous and vociferous support. No one who denies the validity of this creed has much chance of winning political office or otherwise gaining influence on the local scene. Among other things, the creed assumes that democracy is the best form of government, public officials must be chosen by majority vote, and people in the minority must have the right to seek majority support for their beliefs.

In practice, of course, universalistic propositions in the American creed need to be qualified. Adherence to the creed as a general goal . . . for a good government and a good society does not mean that the creed is, or as a practical matter can be, fully applied in practice. Some elements in the political stratum are deeply disturbed by the gap between ideal and reality. Most people in the political stratum, however, are probably either unaware of any sharp conflict between ideal and reality, or are indifferent to it, or take the gap for granted in much the same spirit that they accept the fact that religious behavior falls short of religious belief.

[. . .]

DEMOCRACY, LEADERSHIP, AND MINORITY CONTROL

It is easy to see why observers have often pessimistically concluded that the internal dynamics of political associations create forces alien to popular control and hence to democratic institutions. Yet the characteristics I have described are not necessarily dysfunctional to a pluralistic democracy in which there exists a considerable measure of popular control over the policies of leaders, for minority control by leaders within associations is not necessarily inconsistent with popular control over leaders through electoral processes. For example, suppose that (1) a leader of a political association feels a strong incentive for winning an election; (2) his constituents comprise most of the adult population of the community; (3) nearly all of his constituents are expected to vote; (4) voters cast their ballot without receiving covert rewards or punishments as a direct consequence of the way, they vote; (5) voters give heavy weight to the overt policies of a candidate in making their decision as to how they will vote; (6) there are rival candidates offering alternative policies; and (7) voters have a good deal of information about the policies of the candidates. In these circumstances, it is almost certain that leaders of political associations would tend to choose overt policies they believed most likely to win the support of a majority of adults in the community. Even if the policies of political associations were usually controlled by a tiny minority of leaders to each association, the policies of the leaders who won elections to the chief elective offices in local government would tend to reflect the preferences of the populace. I do not mean to suggest that any political system actually fulfills all these conditions, but to the extent that it does the leaders who directly control the decisions of political associations are themselves influenced in their own choices of policies by their assumptions as to what the voting populace wants.

Although this is an elementary point, it is critical to an understanding of the chapters that follow. We shall discover that in each of a number of key sectors of public policy, a few persons have great direct influence on the choices that are made; most citizens, by contrast, seem to have rather little direct influence. Yet it would be unwise to underestimate the extent to which voters may exert indirect influence on the decisions of leaders by means of elections.

In a political system where key offices are won by elections, where legality and constitutionality are highly valued in the political culture, and where nearly everyone in the political stratum publicly adheres to a doctrine of democracy, it is likely that the political culture, the prevailing attitudes of the political stratum, and the operation of the political system itself will be shaped by the role of elections. Leaders who in one context are enormously influential and even rather free from demands by their constituents may reveal themselves in another context to be involved in tireless efforts to adapt their policies to what they think their constituents want.

To be sure, in a pluralistic system with dispersed inequalities, the direct influence of leaders on policies extends well beyond the norms implied in the classical models of democracy developed by political philosophers.

But if the leaders lead, they are also led. Thus the relations between leaders, subleaders, and constituents produce in the distribution of influence a stubborn and pervasive ambiguity that permeates the entire political system.

SOME HYPOTHESES

Given these assumptions, one might reasonably expect to find in the political system of New Haven that the distribution of influence over important decisions requiring the formal assent of local governmental officials is consistent with the following hypotheses:

- First, only a small proportion of the citizens will have much direct influence on decisions in the sense of directly initiating proposals for policies subsequently adopted or successfully vetoing the proposals of others.
- Second, the leaders – i.e., citizens with relatively great direct influence – will have a corps of auxiliaries or subleaders to help them with their tasks.
- Third, because a democratic creed is widely subscribed to throughout the political stratum, and indeed throughout the population, the public or overt relationships of influence between leaders and subleaders will often be clothed in the rituals and ceremonies of "democratic" control, according to which the leaders are only the spokesmen or agents of the subleaders, who are "representatives" of a broader constituency.

Fourth, because of the need to win elections in order to hold key elective offices, leaders will attempt to develop followings of loyal supporters among their constituents.

Fifth, because the loyalty and support of subleaders, followings, and other constituents are maintained by memories of past rewards or the expectation of future rewards, leaders will shape their policies in an attempt to insure a flow of rewards to all those elements whose support is needed. Consequently, in some circumstances, subleaders, followings, and other constituents will have significant indirect influence on the decisions of leaders. The existence of this indirect influence is an important source of ambiguity in understanding and interpreting the actions of leaders in a pluralistic system.

Finally, conflicts will probably occur from time to time between leaders' overt policies, which are designed to win support from constituents, and their covert policies, which are shaped to win the support of subleaders or other leaders. The keener the political competition, the more likely it is that leaders will resolve these conflicts in favor of their overt commitments.

To determine whether these propositions actually fit the political system of New Haven, I now propose to turn to three "issue-areas" where possible to examine decisions to see what processes of influence are at decisions in two of these areas, public education and urban development, require the formal assent of local government officials at many points. The third, the process of making nominations in the two major parties for local elective offices, is only quasi-governmental, but I have chosen it on the assumption that whoever controls nominations might be presumed to occupy a critical role in any effort to gain the assent of local officials.

[CHAPTER 27] STABILITY, CHANGE, AND THE PROFESSIONALS

New Haven, like most pluralistic democracies, has three characteristics of great importance to the operation of its political system: there are normally "slack" resources; a small core of professional politicians exert great influence over decisions; and

the system has a built-in, self-operating limitation on the influence of all participants, including the professionals.

SLACK IN THE SYSTEM

Most of the time, as we have already seen, most citizens use their resources for purposes other than gaining influence over government decisions. There is a great gap between their actual influence and their potential influence. Their political resources are, so to speak, slack in the system. In some circumstances these resources might be converted from nonpolitical to political purposes; if so, the gap between the actual influence of the average citizen and his potential influence would narrow. The existence of a great deal of political slack seems to be a characteristic of pluralistic political systems and the liberal societies in which these systems operate. In liberal societies, politics is a sideshow in the great circus of life. Even when citizens use their resources to gain influence, ordinarily they do not seek to influence officials or politicians and family members, friends, associates, employees, customers, business firms, and other persons engaged in nongovernmental activities. A complete study of the ways in which people use their resources to influence ethers would require a total examination of social life. Government, in the sense used here, is only a fragment of social life.

THE PROFESSIONALS

The political system of New Haven is characterized by the presence of two sharply contrasting groups of citizens. The great body of citizens use their political resources at a low level; a tiny body of professionals within the political stratum use their political resources at a high level. Most citizens acquire little skill in politics; professionals acquire a great deal. Most citizens exert little direct and immediate influence on the decisions of public officials; professionals exert much more. Most citizens have political resources they do not employ in order to gain influence over the decisions of public officials; consequently there is a great gap between their actual and potential influence. The professionals alone narrow the gap; they do so by using their political resources to the full, and by using them with a high degree of efficiency.

The existence of a small band of professionals within the political stratum is a characteristic of virtually all pluralistic systems and liberal societies. The professionals may enjoy much prestige or little; they may be rigidly honest or corrupt; they may come from aristocracies, the middle strata, or working classes. But in every liberal society they are easily distinguished by the rate and skill with which they use their resources and the resulting degree of direct influence they exert on government decisions.

Probably the most important resource of the professional is his available labor time. Other citizens usually have occupations that demand a large part of their labor time; they also feel a need for recreation. Measured by the alternatives he has to forego, the average citizen finds it costly to sacrifice at most more than a few hours a week to political activities.

The professional, by contrast, organizes his life around his political activities. He usually has an occupation that leaves him freer than more citizens to engage in politics; if he does not, he is likely to change jobs until he finds one that fits easily into political routines. Celentano was an undertaker, Lee a public relations man for Yale, DiCenzo a lawyer, Golden an insurance broker – all occupations that permit innumerable opportunities for political work. [These are all politically active New Haveners profiled elsewhere in the book – Ed.] As a public official, of course, the politician can work virtually full-time at the tasks of politics.

Most citizens treat politics as an avocation. To the professional, politics is a vocation, a calling. Just as the artist remains an artist even as he walks down a city street, and the scientist often consciously or unconsciously remains in his laboratory when he rides home in the evening, or the businessman on the golf course may be working out solutions to his business problems, so the successful politician is a full-time politician. The dedicated artist does not regard it as a sacrifice of precious time and leisure to paint, the dedicated scientist to work in his laboratory, nor the dedicated businessman to work at his business. On the contrary, each is likely to look for ways of avoiding all other heavy claims on his time. So, too, the dedicated politician does not consider it a sacrifice to work at politics.

He is at it, awake and asleep, talking, negotiating, planning, considering strategies, building alliances, making friends, creating contacts – and increasing his influence.

It is hardly to be wondered at that the professional has much more influence on decisions than the average citizen. The professional not only has more resources at the outset than the average citizen, but he also tends to use his resources more efficiently. That is to say, he is more *skillful*.

SKILL

Skill in politics is the ability to gain more influence than others, using the same resources. Why some people are more skillful than others in politics is a matter of great speculation and little knowledge. Because skill in politics is hard to measure, I shall simply assume here that professionals are in fact more skillful. However, two hypotheses help to account for the superior skill of the politician.

First, the stronger one's motivation to learn, the more one is likely to learn. Just why the professional is motivated to succeed in politics is as obscure as the motives of the artist, the scientist, or the businessman. But the whole pattern of his calling hardly leaves it open to doubt that the professional is more strongly motivated to acquire political skills than the average citizen.

Second, the more time one spends in learning, the more one is likely to learn. Here the professional has an obvious advantage, as we have just seen: he organizes his life, in effect, to give him time to learn the art of politics.

I have just said the art of politics. Although politicians make use of information about the world around them, and hence depend on "scientific" or empirical elements, the actual practice of politics by a skilled professional is scarcely equivalent to the activities of an experimental physicist or biologist in a laboratory.

Even the professional cannot escape a high degree of uncertainty in his calculations. If the professional had perfect knowledge of his own goals, the objective situation, and the consequences of alternative strategies, then his choice of strategy would be a relatively simple and indeed a "scientific" matter. But in fact his knowledge is highly imperfect. He cannot be sure at what point rival professionals will begin to mobilize new resources against his policies. When new opposition flares up, he cannot be sure how much further the battle may spread or what forces lie in reserve. He cannot even be certain what will happen to his own resources if he pursues his policies. He may lose some of his popularity; campaign contributions may fall off in the future; the opposition may come up with a legal block, an ethnic angle, a scandal.

Because of the uncertainty surrounding his decisions, the politician, like the military leader, rarely confronts a situation in which his choice of strategies follows clearly and logically from all the information at his disposal, even when he happens to be well-informed as to his own goals. Surrounded by uncertainty, the politician himself necessarily imputes a structure and meaning to the situation that goes beyond empirical evidence and scientific modes of analysis. What the politician imputes to the situation depends, in sum, not only on the information at his disposal but also on his own inner predispositions. His strategy therefore reflects his predispositions for caution or bold-ness, impulsiveness or calculation, negotiation or toughness, stubbornness or resilience, optimism or pessimism, cynicism or faith in others. The strategies of professionals may vary depending on the forces that generate needs for approval, popularity, domina-tion, manipulation, deception, candor, and so on. The effect of inner dispositions on a professional's strategies is by no meant clear or direct. But as one works back from a given situation with all its uncertainties to the professional's interpretation of the situation and his choice of strategies, usually some element in the interpretation or choice is difficult to account for except as a product of his own special dispositions imposing themselves on his selection of strategies.

[. . .]

Just as individuals vary, so professionals vary in the extent to which they use all the resources at their disposal. Some professionals seem driven not only to use all the resources they have but to create new resources and thus to pyramid their influence. They are a kind of political entrepreneur. In an authoritarian milieu perhaps the political entrepreneur might even be driven to dictatorship. But in a pluralistic political system, powerful self-limiting tendencies help to maintain the stability of the system.

THE ART OF PYRAMIDING

We have seen that in the pluralistic political system of New Haven, the political order that existed before 1953 – the pattern of petty sovereignties – was gradually transformed into an executive-centered order. How could this change take place? There were few formal changes in the structure of government and politics. The city charter not only remained unaltered, but as we have seen a proposed charter that in effect would have conferred full legality and legitimacy on the executive, centered order was turned down decisively in the same election in which the chief of the new order was re-elected by one of the greatest popular majorities on record.

The transformation of petty sovereignties into an executive-centered order was possible only because there were slack resources available to the mayor which, used skillfully and to the full, were sufficient to shift the initiative on most questions to the chief executive. Initially the new mayor had access to no greater resources than his predecessor, but with superb skill he exploited them to the limit. In this way, he managed to accumulate new resources; he rose to new heights of popularity, for example, and found it increasingly easy to tap the business community and their campaign con-tributions. His new resources in turn made it easier for him to secure the compliance of officials in city agencies, enlarge his staff, appoint to office the kinds of people he wanted, obtain the coopera-tion of the Boards of Finance and Aldermen, and gain widespread support for his policies. Thus the resources available to the mayor grew by com-parison with those available to other officials. He could now increase his influence over the various officials of local government by using these new resources fully and skillfully. An executive-centered order gradually emerged.

This transformation had two necessary conditions. First, when the new mayor came into office he had to have access either to resources not available to his predecessor or to slack resources his predecessor had not used. In this instance, the new mayor initially relied on a fuller and more efficient use of substantially the same resources available to his predecessor. By using slack resources with higher efficiency the new mayor moved his actual influ-ence closer to his potential influence. Then because of his greater influence he was able to improve his access to resources. In this fashion he pyramided both his resources and his influence. He was, in short, a highly successful political entrepreneur.

There is, however, a second necessary condition for success. The policies of the political entrepreneur must not provoke so strong a counter-mobilization that he exhausts his resources with no substantial increase in his influence.

What then stops the political entrepreneur short of dictatorship? Why doesn't the political entrepreneur in a pluralistic system go on pyramiding his resources until he overturns the system itself? The answer lies in the very same conditions that are necessary to his success. If slack resources provide the political entrepreneur with his dazzling opportunity, they are also the source of his greatest danger. For nearly every citizen in the community has access to unused political resources; it is precisely because of this that even a minor blunder can be fatal to the political entrepreneur if it provokes a sizable minority in the community into raising its political resources at a markedly higher rate in opposition to his policies. . . . Yet almost every policy involves losses for some citizens and gains for others. Whenever the prospect of loss becomes high enough, threatened citizens begin to take up some of the slack in order to remove the threat. The more a favorable deal increases in importance to the opposition, the more resources they have to withdraw from other uses and pour into the political struggle; the more resources the opposition employs, the greater the cost to the politician entrepreneur if he insists on his policy. At some point, the cost became so high that the policy is no longer worth it.

This point is almost to be reached whenever the opposition includes a majority of the electorate, even if no election takes place. Normally, however, far before this extreme situation is approached the expected costs will already have become so excessive that an experienced politician will capitulate or, more likely, search for a compromise that gives him some of what he wants at lower cost.

Three aspects of Mayor Lee's situation made it possible for him to avoid costly opposition. These were: the wide degree of latent support for redevelopment that already existed in New Haven and needed ordinarily to be awakened; the evident need for a high degree of coordination among city agencies if redevelopment were to be carried out; and the Mayor's unusual skill at negotiating agreement and damping down potential disagreements before they flared into opposition. These aspects of Lee's situation are not prevalent in New Haven all the time, nor, certainly, do they necessarily exist in other cities. In the absence of any one of them, opposition might have developed, and the attempt to transform the independent sovereignties into an executive-centered order might have become altogether too costly.

Thus the distribution of resources and the ways in which they are or are not used in a pluralistic political system like New Haven's constitute an important source of both political change and political stability. If the distribution and use of resources gives aspiring leaders great opportunities for gaining influence, these very features also provide a built-in throttle that makes it difficult for any leader, no matter how skillful, to run away with the system.

These features are not, however, the only source of stability. Widespread consensus on the American creed of democracy and equality . . . is also a stabilizing factor. The analysis in the preceding pages surely points, however, to the conclusion that the effectiveness of the creed as a constraint on political leaders depends not only on the nature of the political consensus as it exists among ordinary citizens but also as it exists among members of their political stratum, particularly the professionals themselves.

"The Interests of the Limited City"

from *City Limits* (1981)

Paul E. Peterson

Editors' Introduction

Paul Peterson, Henry Lee Shattuck Professor of Government at Harvard University, created quite a stir among urban political scientists when his *City Limits* was published in 1981. His analysis of urban policy making in the United States contends that "local politics is not like national politics." Drawing on the traditions of mainstream public choice theories that stress the ability of people to choose their place of residence, and the neo-Marxist teaching that "the social and economic context within which the city is embedded limits choice," Peterson sets out to show that city politics and policy making take place within important institutional constraints. Unlike national governments, cities do not control their borders, allowing people and capital to move freely. As a result, cities must pursue policies that will attract residents and investments.

What sorts of policies are these likely to be? He labels policies that improve the economic well-being of the entire city as development policies. Everyone in the city benefits from the infrastructure improvements and urban renewal schemes that lead to the expansion of the local economy. Even those who suffer individual detriment because, perhaps, they have lost their home to a new industrial plant, also benefit as a part of the population that will enjoy the gains of a more robust tax base. On the other hand, cities should not engage in redistributive policies – that is, those policies that tax one group (usually the affluent) to provide services to another (usually the poor). Given the permeability of local borders, Peterson claims, such policies will convince the tax-paying population to move elsewhere, leading to the overall decline of the city's economic status. Peterson is not against welfare programs; rather, he suggests such programs should be nationally controlled.

In Peterson's view, these principles are imbedded in the nature of the U.S. economic and political systems, and are not subject to change. City politics, then, is pretty marginal. The only policies open to real contestation are what he calls "allocational" policies – decisions about which groups get the public sector jobs and contracts, or whose street gets paved first. This is a far cry from the sorts of issues decided at the national level.

Many urban scholars reacted harshly to these claims, in part because the suggestion that urban politics is marginal leads to the conclusion that those who study this topic are marginal as well. For some ten years after *City Limits* was published, nearly every new book or article on urban governance began with a critique of some part of Peterson's argument. First, even if local government's goal is to sustain the local economy, are there not numerous ways to achieve this goal? A city can invest in job training and small business incubators, or can give tax abatements to builders of office towers and luxury housing. Even within the arena of development policy, there is much room for political contestation. Second, although Peterson describes the politics of allocation as "low stakes," principles such as affirmative action in public sector hiring have a wider significance.

Finally, Peterson does not make it clear whether his analysis is meant to apply beyond the United States. Many of the limiting boundary conditions he describes would not prevail in countries where national governments offer more substantial intergovernmental subsidies and more centralized regulatory structures. Much of the literature challenging Peterson sought to explore the great variety of urban policies responses found cross-nationally (see in particular John Logan and Todd Swanstrom, (eds), *Beyond the City Limits*, Philadelphia, PA: Temple University Press, 1990).

Most of Peterson's work does not focus on urban politics. His primary fields of scholarship are U.S. public policy and federalism – indeed, *City Limits* may perhaps be understood as rooted in the questions about the appropriate role of different levels of government in the U.S. federal system, as discussed in such books as *When Federalism Works*, and *The Price of Federalism* (Washington, DC: Brookings 1995). His most recent work has taken up the problems of education policy and school reform. Recent titles include: *No Child Left Behind? The Politics and Practice of School Accountability* (Brookings, 2003); *The Future of School Choice* (Stanford, CA: Hoover Institution, 2003); *Our Schools and our Future. . . . Are We Still At Risk?* (Hoover, 2003); *The Education Gap: Vouchers and Urban Schools* (Brookings, 2002); *Charters, Vouchers, and Public Education* (Brookings, 2001); *Earning and Learning: How Schools Matter* (Brookings, 1999); *Learning From School Choice* (Brookings, 1998); and *The Politics of School Reform: 1870–1940* (University of Chicago Press, 1985).

Like all social structures, cities have interests. Just as we can speak of union interests, judicial interests, and the interests of politicians, so we can speak of the interests of that structured system of social interactions we call a city. Citizens, politicians, and academics are all quite correct in speaking freely of the interests of cities.

DEFINING THE CITY INTEREST

By a city's interest, I do not mean the sum total of the interests of those individuals living in the city. For one thing, these are seldom, if ever, known. The wants, needs, and preferences of residents continually change, and few surveys of public opinion in particular cities have ever been taken. Moreover, the residents of a city often have discordant interests. Some want more parkland and better schools; others want better police protection and lower taxes. Some want an elaborated highway system; others wish to keep cars out of their neighborhood. Some want more inexpensive, publicly subsidized housing; others wish to remove the public housing that exists. Some citizens want improved welfare assistance for the unemployed and dependent; others wish to cut drastically all such programs of public aid. Some citizens want rough-tongued ethnic politicians in public office;

others wish that municipal administration were a gentleman's calling. Especially in large cities, the cacophony of competing claims by diverse class, race, ethnic, and occupational groups makes impossible the determination of any overall city interest – any public interest, if you like – by compiling all the demands and desires of individual city residents.

Some political scientists have attempted to discover the overall urban public interest by summing up the wide variety of individual interests. The earlier work of Edward Banfield, still worth examination, is perhaps the most persuasive effort of this kind. He argued that urban political processes – or at least those in Chicago – allowed for the expression of nearly all the particular interests within the city. Every significant interest was represented by some economic firm or voluntary association, which had a stake in trying to influence those public policies that touched its vested interests. After these various groups and firms had debated and contended, the political leader searched for a compromise that took into account the vital interests of each, and worked out a solution all could accept with some satisfaction. The leader's own interest in sustaining his political power dictated such a strategy.

Banfield's argument is intriguing, but few people would identify public policies as being in the interest of the city simply because they have been formulated according to certain procedures. The political leader might err in his judgment; the interests of important but politically impotent groups might never get expressed; or the consequences of a policy might in the long run be disastrous for the city. Moreover, most urban policies are not hammered out after great controversy, but are the quiet product of routine decision-making. How does one evaluate which of these are in the public interest? Above all, this mechanism for determining the city's interest provides no standpoint for evaluating the substantive worth of urban policies. Within Banfield's framework, whatever urban governments do is said to be in the interest of their communities. But the concept of city interest is used most persuasively when there are calls for reform or innovation. It is a term used to evaluate existing programs and to discriminate between promising and undesirable new ones. To equate the interests of cities with what cities are doing is to so impoverish the term as to make it quite worthless.

The economist Charles Tiebout employs a second approach to the identification of city interests. Unlike Banfield, he does not see the city's interests as a mere summation of individual interests but as something which can be ascribed to the entity, taken as a whole. As an economist, Tiebout is hardly embarrassed by such an enterprise, because in ascribing interests to cities his work parallels both those orthodox economists who state that firms have an interest in maximizing profits and those welfare economists who claim that politicians have an interest in maximizing votes. Of course, they state only that their model will assume that firms and politicians behave in such a way, but insofar as they believe their model has empirical validity, they in fact assert that those constrained by the businessman's or politician's role must pursue certain interests. And so does Tiebout when he says that communities seek to attain the optimum size for the efficient delivery of the bundle of services the local government produces. In his words, "Communities below the optimum size seek to attract new residents to lower average costs. Those above optimum size do just the opposite. Those at an optimum try to keep their populations constant."

Tiebout's approach is in many ways very attractive. By asserting a strategic objective that the city

is trying to maximize – optimum size – Tiebout identifies an overriding interest, which can account for specific policies the city adopts. He provides a simple analytical tool that will account for the choices cities make, without requiring complex investigations into citizen preferences and political mechanisms for identifying and amalgamating the same. Moreover, he provides a criterion for determining whether a specific policy is in the interest of the city – does it help achieve optimum size? Will it help the too small city grow? Will it help the too big city contract? Will it keep the optimally sized city in equilibrium? Even though the exact determination of the optimum size cannot presently be scientifically determined in all cases, the criterion does provide a most useful guide for prudential decision-making.

The difficulty with Tiebout's assumption is that he does not give very good reasons for its having any plausibility. When most economists posit a certain form of maximizing behavior, there is usually a good commonsense reason for believing the person in that role will have an interest in pursuing this strategic objective. When orthodox economists say that businessmen maximize profits, it squares with our understanding in everyday life that people engage in commercial enterprises for monetary gain. The more they make, the better they like it. The same can be said of those welfare economists who say politicians maximize votes. The assumption, though cynical, is in accord with popular belief – and therefore once again has a certain plausibility.

[...]

The one reason Tiebout gives for expecting cities to pursue optimum size is to lower the average cost of public goods. If public goods can be delivered most efficiently at some optimum size, then migration of residents will occur until that size has been reached. In one respect Tiebout is quite correct: local governments must concern themselves with operating local services as efficiently as possible in order to protect the city's economic interests. But there is little evidence that there is an optimum size at which services can be delivered with greatest efficiency. And even if such an optimum did exist, it could be realized only if migration occurred among residents who paid equal amounts in local taxes. In the more likely situation, residents pay variable prices for public services (for example, the amount paid in local property taxes varies by the value of the property). Under these circumstances, increasing size to the optimum does not reduce costs to residents unless newcomers pay at least as much in taxes as the marginal increase in costs their arrival imposes on city governments. Conversely, if a city needs to lose population to reach the optimum, costs to residents will not decline unless the exiting population paid less in taxes than was the marginal cost of providing them government services. In most big cities losing population, exactly the opposite is occurring. Those who pay more in taxes than they receive in services are the emigrants. Tiebout's identification of city interests with optimum size, while suggestive, fails to take into account the quality as well as the quantity of the local population.

The interests of cities are neither a summation of individual interests nor the pursuit of optimum size. Instead, policies and programs can be said to be in the interest of cities whenever the policies maintain or enhance the economic position, social prestige, or political power of the city, taken as a whole.

Cities have these interests because cities consist of a set of social interactions structured by their location in a particular territorial space. Any time that social interactions come to be structured into recurring patterns, the structure thus formed develops an interest in its own maintenance and enhancement. It is in that sense that we speak of the interests of an organization, the interests of the system, and the like. To be sure, within cities, as within any other structure, one can find diverse social roles, each with its own set of interests. But these varying role interests, as divergent and competing as they may be, do not distract us from speaking of the overall interests of the larger structural entity.

The point can be made less abstractly. A school system is a structured form of social action, and therefore it has an interest in maintaining and improving its material resources, its prestige, and its political power. Those policies or events, which have such positive effects are said to be in the interest of the school system. An increase in state financial aid or the winning of the basketball tournament are events that, respectively, enhance the material well-being and the prestige of a school system and are therefore in its interest. In ordinary speech this is taken for granted, even when we also recognize that teachers, pupils, principals, and board members may have contrasting interests as members of differing role-groups within the school.

Although social roles performed within cities are numerous and conflicting, all are structured by the fact that they take specific spatial location that falls within the jurisdiction of some local government. All members of the city thus come to share an interest in policies that affect the well-being of that territory. Policies which enhance the desirability or attractiveness of the territory are in the city's interest because they benefit all residents – in their role as residents of the community. Of course, in any of their other social roles, residents of the city may be adversely affected by the policy. The Los Angeles dope peddler – in his role as peddler – hardly benefits from a successful drive to remove hard drugs from the city. On the other hand, as a resident of the city, he benefits from a policy that enhances the attractiveness of the city as a locale in which to live and work. In determining whether a policy is in the interest of a city, therefore, one does not consider whether it has a positive or negative effect on the total range of social interactions of each and every individual. That is an impossible task. To know whether a policy is in a city's interest, one has to consider only the impact on social relationships insofar as they are structured by their taking place within the city's boundaries.

[. . .]

To say that people understand what, generally, is in the interest of cities does not eliminate debate over policy alternatives in specific instances. The notion of city interest can be extremely useful, even though its precise application in specific contexts may be quite problematic. In any policy context one cannot easily assert that one "knows" what is in the interest of cities, whether or not the residents of the city agree. But city residents do know the kind of evidence that must be advanced and the kinds of reasons that must be adduced in order to build a persuasive case that a policy is in the interest of cities. And so do community leaders, mayors, and administrative elites.

ECONOMIC INTERESTS

Cities, like all structured social systems, seek to improve their position in all three of the systems of stratification – economic, social, and political – characteristic of industrial societies. In most cases, improved standing in any one of these systems helps enhance a city's position in the other two. In the short run, to be sure, cities may have to choose among economic gains, social prestige, and political weight. And because different cities may choose alternative objectives, one cannot state any one overarching objective – such as improved property values – that is always the paramount interest of the city. But inasmuch as improved economic or market standing seems to be an objective of great importance to most cities, I shall concentrate on this interest and only discuss in passing the significance of social status and political power.

Cities constantly seek to upgrade their economic standing. Following [German sociologist Max] Weber, I mean by this that cities seek to improve their market position, their attractiveness as a locale for economic activity. In the market economy that characterizes Western society, an advantageous economic position means a competitive edge in the production and distribution of desired commodities relative to other localities. When this is present, cities can export goods and/or services to those outside the boundaries of the community.

Some regional economists have gone so far as to suggest that the welfare of a city is identical to the welfare of its export industry. As exporters expand, the city grows. As they contract, the city declines and decays. The economic reasoning supporting such a conclusion is quite straightforward. When cities produce a good that can be sold in an external market, labor and capital flow into the city to help increase the production of that good. They continue to do so until the external market is saturated – that is, until the marginal cost of production within the city exceeds the marginal value of the good external to the city. Those engaged in the production of the exported good will themselves consume a variety of other goods and services, which other businesses will provide. In addition, subsidiary industries locate in the city either because they help supply the exporting industry, because they can utilize some of its by-products, or because they benefit by some economies of scale provided by its presence. Already, the familiar multiplier is at work. With every increase in the sale of exported commodities, there may be as much as a four- or fivefold increase in local economic activity.

The impact of Boeing Aircraft's market prospects on the economy of the Seattle metropolitan area illustrates the importance of export to regional economies. In the late sixties defense and commercial aircraft contracts declined, Boeing laid off thousands of workmen, the economy of the Pacific Northwest slumped, the unemployed moved elsewhere, and Seattle land values dropped sharply. More recently, Boeing has more than recovered its former position. With rapidly expanding production at Boeing, the metropolitan area is enjoying low unemployment, rapid growth, and dramatically increasing land values.

The same multiplier effect is not at work in the case of goods and services produced for domestic consumption within the territory. What is gained by a producer within the community is expended by other community residents. Residents, in effect, are simply taking in one another's laundry. Unless productivity increases, there is no capacity for expansion.

If this economic analysis is correct, it is only a modest oversimplification to equate the interests of cities with the interest of their export industries. Whatever helps them prosper redounds to the benefit of the community as a whole – perhaps four and five times over. And it is just such an economic analysis that has influenced many local government policies. Especially the smaller towns and cities may provide free land, tax concessions, and favorable utility rates to incoming industries.

The smaller the territory and the more primitive its level of economic development, the more persuasive is this simple export thesis. But other economists have elaborated an alternative growth thesis that is in many ways more persuasive, especially as it relates to larger urban areas. In their view a sophisticated local network of public and private services is the key to long-range economic growth. Since the world economy is constantly changing, the economic viability of any particular export industry is highly variable. As a result, a community dependent on any particular set of export industries will have only an episodic economic future. But with a well-developed infrastructure of services the city becomes an attractive locale for a wide variety of export industries. As older exporters fade, new exporters take their place and the community continues to prosper. It is in the city's interest, therefore, to help sustain a high-quality local infrastructure generally attractive to all commerce and industry.

I have no way of evaluating the merits of these contrasting economic arguments. What is important in this context is that both see exports as being of great importance to the well-being of a city. One view suggests a need for direct support of the export industry; the other suggests a need only for maintaining a service infrastructure, allowing the market to determine which particular export industry locates in the community. Either one could be the more correct diagnosis for a particular community, at least in the short run. Yet both recognize that the future of the city depends upon exporting local products. When a city is able to export its products, service industries prosper, labor is in greater demand, wages increase, promotional opportunities widen, land values rise, tax revenues increase, city services can be improved, donations to charitable organizations become more generous, and the social and cultural life of the city is enhanced.

To export successfully, cities must make efficient use of the three main factors of production: land, labor, and capital.

Land

Land is the factor of production that cities control. Yet land is the factor to which cities are bound. It is the fact that cities are spatially defined units whose boundaries seldom change that gives permanence to their interests. City residents come and go, are born and die, and change their tastes and preferences. But the city remains wedded to the land area with which it is blessed (or cursed). And unless it can alter that land area, through annexation or consolidation, it is the long-range value of that land which the city must secure – and which gives a good approximation of how well it is achieving its interests.

Land is an economic resource. Production cannot occur except within some spatial location. And because land varies in its economic potential, so do the economic futures of cities. Historically, the most important variable affecting urban growth has been an area's relationship to land and water routes.

On the eastern coast of the United States, all the great cities had natural harbors that facilitated commercial relations with Europe and other coastal communities. Inland, the great industrial cities all were located on either the Great Lakes or the Ohio River–Mississippi River system. The cities of the West . . . prospered according to their proximity to East–West trade flows. Denver became the predominant city of the mountain states because it sat at the crossroads of land routes through the Rocky Mountains. Duluth, Minnesota, had only limited potential, even with its Great Lakes location, because it lay north of all major routes to the West.

Access to waterways and other trade routes is not the only way a city's life is structured by its location. Its climate determines the cost and desirability of habitation; its soil affects food production in the surrounding area; its terrain affects drainage, rates of air pollution, and scenic beauty. Of course, the qualities of landscape do not permanently fix a city's fate – it is the intersection of that land and location with the larger national and world economy that is critical. For example, cities controlling access to waterways by straddling natural harbors at one time monopolized the most valuable land in the region, and from that position they dominated their hinterland. But since land and air transport have begun to supplant, not just supplement, water transport, the dominance of these once favored cities has rapidly diminished.

Although the economic future of a city is very much influenced by external forces affecting the value of its land, the fact that a city has control over the use of its land gives it some capacity for influencing that future. Although there are constitutional limits to its authority, the discretion available to a local government in determining land use remains the greatest arena for the exercise of local autonomy. Cities can plan the use of local space; cities have the power of eminent domain; through zoning laws cities can restrict all sorts of land uses; and cities can regulate the size, content, and purpose of buildings constructed within their boundaries. Moreover, cities can provide public services in such a way as to encourage certain kinds of land use. Sewers, gas lines, roads, bridges, tunnels, playgrounds, schools, and parks all impinge on the use of land in the surrounding area. Urban politics is above all the politics of land use, and it is easy to see why. Land is the factor of production over which cities exercise the greatest control.

Labor

To its land area the city must attract not only capital but productive labor. Yet local governments in the United States are very limited in their capacities to control the flow of these factors. Lacking the more direct controls of nation-states, they are all the more constrained to pursue their economic interests in those areas where they do exercise authority.

Labor is an obvious case in point. Since nation-states control migration across their boundaries, the industrially more advanced have formerly legislated that only limited numbers of outsiders – for example, relatives of citizens or those with skills needed by the host country – can enter. In a world where it is economically feasible for great masses of the population to migrate long distances, this kind of restrictive legislation seems essential for keeping the nation's social and economic integrity intact. Certainly, the wage levels and welfare assistance programs characteristic of advanced industrial societies could not be sustained were transnational migration unencumbered.

Unlike nation-states, cities cannot control movement across their boundaries. They no longer have walls, guarded and defended by their inhabitants. And as Weber correctly noted, without walls cities no longer have the independence to make significant choices in the way medieval cities once did. It is true that local governments often try to keep vagrants, bums, paupers, and racial minorities out of their territory. They are harassed, arrested, thrown out of town, and generally discriminated against. But in most of these cases local governments act unconstitutionally, and even this illegal use of the police power does not control migration very efficiently.

Although limited in its powers, the city seeks to obtain an appropriately skilled labor force at wages lower than its competitors so that it can profitably export commodities. In larger cities a diverse work force is desirable. The service industry, which provides the infrastructure for exporters, recruits large numbers of unskilled workers, and many manufacturing industries need only semiskilled workers. When shortages in these skill levels appear, cities may assist industry in advertising the work and living opportunities of the region. In the nineteenth century when unskilled labor was in short supply, frontier cities made extravagant claims to gain a competitive edge in the supply of ordinary labor.

Certain sparsely populated areas, such as Alaska, occasionally advertise for unskilled labor even today. However, competition among most cities is now for highly skilled workers and especially for professional and managerial talent. In a less than full-employment economy, most communities have a surplus of semiskilled and unskilled labor. Increases in the supply of unskilled workers increase the cost of the community's social services. Since national wage laws preclude a decline in wages below a certain minimum, there may be a shortage of highly skilled technicians and various types of white collar workers. Where shortages develop, the prices these workers can command in the labor market may climb to a level where local exports are no longer competitive with goods produced elsewhere. The economic health of a community is therefore importantly affected by the availability of professional and managerial talent and of highly skilled technicians.

When successfully pursuing their economic interests, cities develop a set of policies that will attract the more skilled and white collar workers without at the same time attracting unemployables. Of course, there are limits on the number of things cities can do. In contrast to nation-states they cannot simply forbid entry to all but the highly talented whose skills they desire. But through zoning laws they can ensure that adequate land is available for middle-class residences. They can provide parks, recreation areas, and good quality schools in areas where the economically most productive live. They can keep the cost of social services, little utilized by the middle class, to a minimum, thereby keeping local taxes relatively low. In general, they can try to ensure that the benefits of public service outweigh their costs to those highly skilled workers, managers, and professionals who are vital for sustaining the community's economic growth.

Capital

Capital is the second factor of production that must be attracted to an economically productive territory. Accordingly, nation-states place powerful controls on the flow of capital across their boundaries. Many nations strictly regulate the amount of national currency that can be taken out of the country. They place quotas and tariffs on imported goods. They regulate the rate at which national currency can be exchanged with foreign currency. They regulate the money supply, increasing interest rates when growth is too rapid, lowering interest rates when growth slows down. Debt financing also allows a nation-state to undertake capital expenditures and to encourage growth in the private market. At present the powers of nation-states to control capital flow are being used more sparingly and new supranational institutions are developing in their place. Market forces now seem more powerful than official policies in establishing rates of currency exchange among major industrial societies. Tariffs and other restrictions on trade are subject to retaliation by other countries, and so they must be used sparingly. The economies of industrialized nations are becoming so interdependent that significant changes in the international political economy seem imminent, signaled by numerous international conferences to determine worldwide growth rates, rates of inflation, and levels of unemployment. If these trends continue, nation-states may come to look increasingly like local governments.

But these developments at the national level have only begun to emerge. At the local level in the United States, cities are much less able to control capital flows. In the first place, the Constitution has been interpreted to mean that states cannot hinder the free flow of goods and monies across their boundaries. And what is true of states is true of their subsidiary jurisdictions as well. In the second place, states and localities cannot regulate the money supply. If unemployment is low, they cannot stimulate the economy by increasing the monetary flow. If inflationary pressures adversely affect their competitive edge in the export market, localities can neither restrict the money supply nor directly control prices and wages. All of these powers are reserved for national governments. In the third place, local governments cannot spend more than they receive in tax revenues without damaging their credit or even running the risk of bankruptcy. Pump priming, sometimes a national disease, is certainly a national prerogative.

Local governments are left with a number of devices for enticing capital into the area. They can minimize their tax on capital and on profits from capital investment. They can reduce the costs of capital investment by providing low-cost public utilities, such as roads, sewers, lights, and police and fire protection. They can even offer public land free of charge or at greatly reduced prices to those investors they are particularly anxious to attract. They can provide a context for business operations free of undue harassment or regulation. For example, they can ignore various external costs of production, such as air pollution, water pollution, and the despoliation of trees, grass, and other features of the landscape. Finally, they can discourage labor from unionizing so as to keep industrial labor costs competitive.

This does not mean it behooves cities to allow any and all profit-maximizing action on the part of an industrial plant. Insofar as the city desires diversified economic growth, no single company can be allowed to pursue policies that seriously detract from the area's overall attractiveness to capital or productive labor. Taxes cannot be so low that government fails to supply residents with as attractive a package of services as can be found in competitive jurisdictions. Regulation of any particular industry cannot fall so far below nationwide standards that other industries must bear external costs not encountered in other places. The city's interest in attracting capital does not mean utter subservience to any particular corporation, but a sensitivity to the need for establishing an overall favorable climate.

In sum, cities, like private firms, compete with one another so as to maximize their economic position. To achieve this objective, the city must use the resources its land area provides by attracting as much capital and as high a quality labor force as is possible. Like a private firm, the city must entice labor and capital resources by offering appropriate inducements. Unlike the nation-state, the American city does not have regulatory powers to control labor and capital flows. The lack thereof sharply limits what cities can do to control their economic development, but at the same time the attempt by cities to maximize their interests within these limits shapes policy choice.

LOCAL GOVERNMENT AND THE INTERESTS OF CITIES

Local government leaders are likely to be sensitive to the economic interests of their communities. First, economic prosperity is necessary for protecting the fiscal base of a local government. In the United States, taxes on local sources and charges for local services remain important components of local government revenues. Although transfers of revenue to local units from the federal and state governments increase throughout the postwar period, as late as 1975–76 local governments still were raising almost 59 percent of their own revenue. Raising revenue from one's own economic resources requires continuing MINE economic prosperity. Second, good-government is good-politics. By pursuing policies, which contribute to the economic prosperity of the local community, the local politician selects policies that redound to his own political advantage. Local politicians, eager for relief from the cross-pressures of local politics, assiduously promote goals that have widespread benefits. And few policies are more popular than economic growth and prosperity. Third, and most important, local officials usually have a sense of community responsibility. They know that, unless the economic well-being of the community can be maintained, local business will suffer, workers will lose employment opportunities, cultural life will decline, and city land values will fall. To avoid such a dismal future, public officials try to develop policies that assist the prosperity of their community – or, at the very least, that do not seriously detract from it. Quite apart from any effects of economic prosperity on government revenues or local voting behavior, it is quite reasonable to posit that local governments are primarily interested in maintaining the economic vitality of the area for which they are responsible.

Accordingly, governments can be expected to attempt to maximize this particular goal – within the numerous environmental constraints with which they must contend. As policy alternatives are proposed, each is evaluated according to how well it will help to achieve this objective. Although information is imperfect and local governments cannot be expected to select the one best alternative on every occasion, policy choices over time will be limited to those few which can plausibly be shown to be conducive to the community's economic prosperity. Internal disputes and disagreements may affect policy on the margins, but the major contours of local revenue policy will be determined by this strategic objective.

[. . .]

CONCLUSIONS: EFFICIENCY VERSUS EQUALITY

Efficiency in local government promotes city interests. By efficiency I am referring to a state in which no person can be made better off without some other person being made worse off. In Tiebout's world local governments operate with perfect efficiency such that everyone receives the services for which he has an economic demand and no one receives services unless he has such a demand. At the same time the price paid equals the lowest average cost of producing the service. No one can be made better off without someone else bearing the cost.

Although this utopia is quite beyond the capacity of local governments, the closer any locality moves toward this ideal match between taxes and services, the more attractive a setting it is for residents, and the more valuable its land becomes. It is thus in the interest of local governments to operate as efficiently – in this Pareto-optimal sense of the word – as possible.

Operating efficiently hardly means operating so as to enhance quality. As many critics of Pareto's definition of efficiency have pointed out, an efficient distribution of resources, as Pareto defines it, is incompatible with gross inequalities. On cannot redistribute wealth without making some worse off at the same time others are made better off. If a society has great inequalities in the beginning, it does not reduce these inequalities merely by increasing its efficiency. Consequently, the pursuit of a city's economic interests, which requires an efficient provision of local services, makes no allowance for the care of the needy and unfortunate members of the society. Indeed, the competition among local communities all but precludes a concern for redistribution.

Recall the finding that the benefit/tax ratio for the average taxpayer is always less than 1.0. The person who pays the mean dollar in taxes always receives less in benefits than he pays in taxes (while at the same time having unsatisfied demands for services). Then consider the fact that this benefit/tax ratio declines as the amount of redistributive activity by local governments increases. Since the person or entity that pays the mean dollar in taxes is likely to be better off than the low-income residents of the community, increased redistribution from the richer to the poorer implies a reduction in the services the person paying the mean tax dollar receives as a proportion of the amount he pays in taxes. From the point of view of this average taxpayer, the local government service-delivery system appears highly inefficient, however ably the redistributive service-delivery system is run. If other communities provide fewer services to the needy, he will see an economic advantage in migrating. Over time, this adversely affects the economic well-being of the community with a mind to redistribute.

"Looking Back to Look Forward: Reflections on Urban Regime Analysis"

from *Urban Affairs Review* (2005)

Clarence N. Stone

Editors' Introduction

Clarence Stone's analysis of postwar politics in Atlanta and his subsequent writing on urban power changed the vocabulary of urban political science. Prior to the publication of Stone's book, the study of power in cities had been through a number of phases. The pluralist model had added needed nuance to earlier sociological studies of urban elites, but its insistence that local political systems were open, and that once in power local political leaders could actually accomplish important policy goals, seems, by the 1970s, overly sanguine.

As Mollenkopf describes in the first reading in Part 3, North American and European urbanists began to apply the work of neo-Marxist analysts of the capitalist state to the study of local politics. The local political arena was not a place of free exchange, open to anyone armed merely with a vote and a voice. Rather, the nature of capitalism served to privilege those with wealth, who more easily sought and gained power. Moreover, regardless of who sat in City Hall, the nature of capitalism required local governments to pursue policies that allowed investors to prosper – so it almost did not matter who got elected; local leaders were all ultimately going to adopt similar, business-enhancing policies.

While the privileged position of capital is clear to any student of local politics, most notably in the U.S., it remained nonetheless incomplete and unsatisfying to fully adopt theories that erased the importance of agency. After all, even if business interests tended to predominate, different cities, led by different sorts of leaders, clearly pursued different policies. And within many cities, especially the larger and more complex ones, "capitalists" were themselves a disparate bunch, with different sorts of businesses preferring different sorts of policies. Would urban political analysts really have to choose between pluralism's indifference to inequality, and structuralism's one-size-fits-all determinism?

Regime analysis was an attempt to navigate between these poles. Yes, argued Stone, the pluralists were right to maintain that coalitions of elites and masses will form around different issues, depending on whose interests are most affected. Politics matters; leaders can bring interests together in different ways. However, politics is constrained by structural factors – not all citizens, or even all interests, are as likely to be represented, to achieve political voice, to be considered crucial enough to be invited to the political table, or to have the power to shape policies once represented in office. As Stone suggests in this reading, those hoping to exercise political influence must meet "threshold tests," usually the possession of the sorts of capital (money, education, social status) that are in short supply among what Stone calls "the lower strata of social stratification."

We can synthesize structure and agency, he argues, by examining urban "regimes," which he defines in *Regime Politics: Governing Atlanta* (Lawrence, KS: University Press of Kansas, 1989) as "the informal arrange-ments by which public bodies and private interests function together in order to be able to make and carry out governing decisions" (p. 6). The composition of a governing regime varies from city to city, but in all cases must include those able to help generate electoral majorities, and those able to command economic resources. Thus, not all members of the community are equally likely to be part of the governing regime. Those holding resources to facilitate the governing process make the most attractive partners, and given the importance of private capital to all aspects of governance, the business sector has real advantages. Unlike the structuralists, Stone doesn't believe that business dictates policy, or that one can predict that specific political or policy outcomes will ensue just because businesses prefer them. In his Atlanta case, a successful grassroots campaign led to the election of Maynard Jackson in 1973, the city's first African-American mayor, despite the opposition of most business leaders. Yet, whereas other interests can be safely ignored, business interests must generally be accommodated somehow – as Jackson learned when he tried to adopt an economic development agenda in the face of business resistance.

In the reading excerpted below, Stone revisits the question of regime analysis, restating his core argument while revising parts of it in response to subsequent scholarship. Some have argued that his choice of city (Atlanta, with its history of business involvement in politics) and of issue (downtown development) produced an analysis that overstated the centrality of business leadership. Here, he acknowledges that the leading role of business is common but not inevitable, and varies according to location (different cities have different interest constellations) and issue (education policy operates at a greater distance from business influence than does development policy). Stone also uses this opportunity to revisit the importance of what he calls "larger purposes" in motivating an urban agenda. When he attempted to explain what held disparate interests together in a governing regime in his *Regime Politics*, Stone had emphasized what he called "selective incentives" and "small opportunities." Coalitions can hang together, in this view, without larger consensus if its members all get something of value from their participation – in the case of Atlanta, for instance, minority business participation programs encouraged African-American leaders to support a pro-growth development agenda that suited the needs of corporate executives. In this reading, while maintaining that selective incentives effectively bind coalition partners, he acknowledges that larger, perhaps ideologically based goals can also serve to motivate collective activity.

Since the publication of *Regime Politics*, countless other scholars have adopted, challenged, and adapted the "regime" concept to explain the politics of cities in the U.S. and elsewhere. Mickey Lauria's edited volume *Reconstructing Regime Theory* (Thousand Oaks, CA: Sage, 1997) includes theoretical critiques, as well as case studies using the regime approach. David Imbroscio's *Reconstructing City Politics: Alternative Economic Development and Urban Regimes* (London: Sage, 1997) offers another perspective on regime analysis.

Below I address four topics:

I.   How urban regime analysis differs from pluralism.
II. The inadequacies of elite engagement as a focus for research.
III. The role of selective incentives in relation to ideas and especially to purpose.
IV. The place of urban regime analysis in a broader landscape of theory.

I. PLURALISM

Fragmentation is a common complaint about the contemporary world. Functional specialization has proceeded far, and we live in times in which no ruling group runs the show. Does this mean that in some sense we are all pluralists? I would say that the answer is yes *only* if we define pluralism loosely as the absence of rule by a cohesive elite. However, something more stringent than that is needed if the term is to be useful in political analysis.

Here I propose to equate pluralism with the classic understanding put forward by the Yale school, and its strong kinship with the work of David Truman. The central idea, and the one I wish to challenge explicitly, is that universal suffrage makes politics into an open and penetrable process, yielding to those who become active around particular interests salient to them.

Dahl saw people as largely concerned with things that are typically apolitical. Yet and still, he argues, where there are basic civil liberties and reasonable transparency in the conduct of public affairs, popular control is a reality. The ballot box countervails wealth and social status, and the fragmentation of a highly differentiated society further assures that political inequalities are dispersed. In Dahl's eyes, a great leveler is that people are concerned mainly with matters close at hand. People guard their immediate interests, and the centrifugal force of this pattern assures that no group is in a position to accumulate wide control over others. Coalitions are unstable and realign as issues shift with changing times and conditions. Furthermore, no group is confined to the position of permanent political exclusion.

Although pluralism captures aspects of political reality, it is fundamentally flawed. That assessment separates urban regime analysis from classic pluralism, but, given that several scholars have equated regime analysis with pluralism, I want to be explicit about the differences between the two schools of thought. Whereas *Who Governs?*, for example, explains what would make politics open and penetrable, urban regime analysis looks in a different direction to explain why politics is mainly accessible to those who can meet substantial threshold tests.

The axial issue between classic pluralism and urban regime analysis is about elections and whether control of elected office is the center point around which politics rotates. We need not revisit here all of the reasons why the vote is a limited instrument of popular control, but I want to underscore the point that, even if ideal conditions existed so that citizens were able to exercise rigorous oversight of all top officials through an ample array of ballot choices, the authority of government has only a limited writ. Standing alone, government is by itself an inadequate problem-solver. Public policy impacts (and therefore public policies) depend on complementary actions from non-governmental sources. At best, electoral accountability reaches only part of the process of shaping public policy, and for that part electoral accountability is in reality very far from being a robust process.

A second divide separates urban regime analysis from pluralism. In classic pluralism, all issues take shape on the same plane. Immediacy of concern is a pervasive condition, with centrifugal forces therefore dominant. Big issues require a wide, but

inevitably unstable, coalition. In narrow issue arenas, David is fully capable of slaying Goliath.

From Bachrach and Baratz's "Two Faces of Power" (1962) onward, critics have challenged pluralism with a counter-argument that politics is waged on multiple levels. In the Preface to *Regime Politics*, I quoted "Big Jim" Folsom, governor of Alabama in a past era: "Nothing just happens. Everything is *arranged*." Politics, I went on to argue, is "the art of arranging" (1989, p. xii), and at its heart is the capacity to structure the relationships through which a community or society is governed. That capacity is not widely dispersed, and it is different from a short-term and narrowly focused effort to influence a particular policy decision. In an earlier article [Stone 1982, p. 276], I argued:

> Just as the economy involves more than buying and selling consumer goods, so the polity involves more than the pressures and counter-pressures on discrete policy choices. Decisional politics is, in a manner of speaking, the politics of the consumer market, not the politics of investment and production. And, to continue the analogy, expending time and resources on particular substantive decisions is often a low-return form of political activity. The return is higher if one can invest in creating or maintaining arrangements.

The capacity to build, modify, or reinforce governing arrangements requires resources and skills that are not widely available. Inequalities in that capacity are substantial, systemic, and persistent – qualities that run counter to classic pluralism's idea of an open and penetrable system. Instead of the kind of fluid politics Dahl ascribed to New Haven, I find relationships to be structured. Established relationships are not readily changed; it is easier to maintain relationships than to build new ones. For that reason, those in an established position are not easily pressured into adopting a new line of action.

[. . .]

As viewed through regime analysis, the political world is complex, but, despite universal suffrage, open and penetrable it is not. In order to understand why, it is necessary to see politics as a process of shaping arrangements. That understanding separates urban regime analysis sharply from pluralism. Whereas pluralism makes group analysis central and largely dismisses class, urban regime analysis pays heed to the system of social stratification as a source of social and economic inequalities and how they work against an open and penetrable form of politics. Political activity often takes shape in a group form, but class (i.e., the system of social stratification) provides the context within which group action takes shape. For groups with a history of political, social, and economic marginality, having a political impact calls for much more than simply becoming active around a few issues of immediate concern. It calls for breaking into the "politics of investment" and becoming part of a locality's governing arrangements. Reaching such a position rests on several interrelated factors, and at the heart of them are the abilities to contribute significantly to a widely desired outcome and to enlist allies. Although politics is not a process irrevocably closed to any group, meaningful political influence rests on an ability to meet important threshold tests. For those in the lower strata of the system of social stratification, meeting those tests involves a long and difficult journey. Understanding that journey is the task of urban regime analysis.

II. THE INADEQUACY OF ELITE ENGAGEMENT

One of the criticisms of urban regime analysis has been that it deals only with elites and their relations to one another, not to the larger context of elite–mass relations. This is an important and telling criticism. Is urban regime analysis, then, a form of elitism? The answer is no, but urban regime analysis does not accord to mass opinion in and of itself an ability to direct the course of events. As observed by Gerry Stoker, "for actors to be effective regime partners two characteristics seem especially appropriate: first possession of strategic knowledge of social transactions and a capacity to act on that knowledge; and second, control of resources that make one an attractive coalition partner" (1995, p. 60). Thus a governing coalition consists typically of members based in a locality's major institutions.

It is a mistake, however, to think of urban regimes as composed of a fixed body of actors, taking on an ever-changing agenda. Instead the question is about who needs to be mobilized in order to take on a given problem effectively. One of the fundamental tenets of urban regime analysis, what Stoker once labeled its "iron law," is that the governing coalition must be able to draw together the resources commensurate with its policy agenda (Stoker 1995, p. 61). As a heuristic device, urban regime analysis holds that the issue addressed determines whose participation is needed. *Neither business nor any other group is necessarily a required member of the governing coalition.* The question about who needs to play an active part centers on resources, skills in action, effort applied to task, and thus is much more than participation in a token or pro forma manner. In the U.S., for many issues, business would be a needed member of the governing coalition, but the need for business participation derives partly from the nature of the agenda. Hence, the urgency of business participation varies with the situation, both as to the character of the problem addressed and the availability of alternative sources of needed resources.

As one moves away from economic development as an issue and away from the special place that business occupies in local civic life in the U.S., business participation may become less essential. ... For many issues – e.g., improved academic achievement in schools, workforce development, counteracting youth violence, reducing crime, the conservation and upkeep of neighborhoods or social housing estates, and many more – grass roots engagement is necessary but often hard to achieve. Histories of past neglect and frustration, alienation, and lack of confidence that conditions can be improved and opportunities realized may stand as huge barriers not likely to be overcome by a modest initiative unaccompanied by some concerted and grounded effort to enlist the hearts and minds of the target population.

Economic development and showcase projects such as hosting the Olympics often hold a position of high priority, and for such projects business involvement may be quite important. But we need to bear in mind that the high visibility of the business sector in these issues is not a sign that it holds an equally crucial position on other issues. Yet we should also not overlook the point that business is frequently well organized and in larger cities is almost sure to contain many enterprises that have deep pockets from which to contribute to the initiatives on which they look with favor. And American business has a long history of multifaceted participation in the civic and philanthropic life of local communities. In the U.S. especially, business enjoys ready-made advantages as a willing and able participant in priority agendas that it helps to set.

III. SELECTIVE INCENTIVES AND THE ROLE OF PURPOSE

Looking back on *Regime Politics*, I am least satisfied with my treatment of selective material incentives. Although they have played a major role in Atlanta, I gave the impression in some passages that they were the overriding source of biracial cooperation. They were exceedingly important, but in retrospect I see that my analysis understated the importance of purpose – specifically large purposes. Let me elaborate.

First I want to reiterate what *Regime Politics* does claim. In describing the mix of incentives at work in Atlanta, I said [Stone 1989, pp. 186–7]:

Because not everyone is narrowly opportunistic, selective incentives are not the whole story of collective action. Efforts on behalf of a group purpose may be intrinsically satisfying. The political movements and other activities heavily dependent on volunteer efforts rely on emotional commitment to motivate adherents. Volunteer activities may afford opportunities for sociability or an identity with a larger group or purpose, which are not dependent on an external system of rewards and punishments. And if adherence to group obligations is widespread enough, norms or conventions of cooperation may prevail over individual opportunism.

I also made the argument that there is a "small-opportunities" phenomenon, that is, "most people most of the time are guided, not by a grand vision of how the world might be reformed, but by the pursuit of particular opportunities" (p. 193). . . . People are motivated to pursue purposes, but few are positioned to frame a large purpose *and* bring together the resources needed to pursue such a purpose. They are positioned to see attainable and more immediate purposes, and they do put significant energy and effort into their pursuit. But where there is a larger set of arrangements that can further or hinder such pursuits, people generally

"go along," accommodating to those established relationships for their immediate purposes rather than expend the large effort to create alternative new relationships.

The key point behind the small-opportunities phenomenon is that motives are more complex than selective *material* incentives. In Atlanta, the business elite has appreciated the advantages of side payments and small opportunities. Moreover, they pursued a conscious strategy of placing the capacity to offer rewards, purposive and material, as far away from the electoral process as possible. With selective incentives as the shorthand for a complex of particular moves, we can see why control of the process was an important matter. In this broad sense, selective incentives were central in the functioning of Atlanta's regime. Two things need to be pointed out, however. One is that struggle over the control of selective incentives had much to do with relations *within* the governing coalition. Business sought to keep an upper hand in this process and to keep it as distinct as possible from the electoral process. Hence, during the time center-city land use was being reshaped, the redevelopment agency and the main planning process were never directly under the control of the mayor's office. Moreover, significant entities such as the Metropolitan Community Foundation were outside the government arena completely and responsive mainly to its business backers. In looking back at my study of Atlanta, I wish to say that, without denying that selective incentives often play an important part in bolstering and giving shape to political arrangements, I failed to convey the importance of big purposes. They are not the stuff of retail politics, but for substantial institutions with ample resources and extensive capacities to plan and assess, large purposes are vital building blocks. And broad civic purposes can enlist important individual efforts.

The point is not that masses of individual actors spontaneously gather under the banner of social causes. The process is complicated. But what *Regime Politics* underplayed was the importance of purpose in the form of a large change agenda, captured in the slogan "the city too busy to hate." The complex reality behind that slogan made a dual contribution to the city's governing regime. First of all, the combined agenda of business-oriented redevelopment and racial change provided a framework within which bargains could be negotiated, rewards handed out, and civic pride boosted. It was a means through which multiple forms of motivation could be tapped, and thereby durable governing arrangements put into place.

Selective incentives played a part, but they were in the service of some large objectives. The framework of a durable purpose means that governing arrangements do not have to be reinvented issue by issue. In economic terms, a big-purpose agenda can minimize transaction costs by providing established and familiar ways of getting things done. A big-purpose agenda thus can provide a useful framework, and it does so on a basis that gives structure to interactions over time.

But the second point is that purpose is not just an instrumental device for organizing an otherwise chaotic world. Purpose can itself be a motivator, and occasionally a very powerful one. Pursuit of a broad social or civic purpose involves sacrifices of time, energy, and resources. Sometimes individuals and the organizations they guide make huge contributions and do so without receiving immediate and particular material gains. Therefore we should not understate the role of large, socially worthy purposes. The "politics of investment," i.e., of establishing governing arrangements, requires bringing together substantial resources, both tangible and intangible, from a variety of players. Broad purposes deemed to be socially worthy play a vital part.

Competing purposes

Acknowledging the role of purpose does not eliminate the collective-action problem. Purposes compete with one another, and individuals face an abundance of worthy claims, some of which concern matters immediate in their everyday lives. Indeed, under the constraints of bounded rationality, human beings are focused on what is immediate. As suggested by the small-opportunities phenomenon, people are drawn more easily into limited aims than broad social purposes. It is easier to imagine one's role and potential impact in a small purpose than a large one.

Even so, some broad purposes do gain agenda status and others falter in the process. From early on, regime analysis differentiated the successful from the unsuccessful on the basis of the ability to garner enough resources to be feasible. So achievability is one screen for separating successful from

unsuccessful purposes. Most people agree with the policy activist who said, "if it's not going anywhere, it's not something I want to spend any time on" (quoted in Kingdon 1995, p. 171). Put another way, "What can be accomplished not infrequently affects what we want to accomplish" (Jones 2001, p. 147). Feasibility, we should bear in mind, is not an objective, self-defining condition. It is very much a matter of shared perception. . . . The real test, then, is that of *perceived* feasibility.

[. . .]

Summary

In *Regime Politics*, I highlighted selective incentives, including "small opportunities." In that book, the large purposes embodied in the slogan "the city too busy to hate" came in for less analytical consideration than was warranted. Taking stock many years later, I see a need for a thorough examination of the role of purposes. People are meaning-seeking creatures, as recognized in passing in *Regime Politics*, and not rational egoists preoccupied with satisfying immediate material interests.

Purpose has motivational force, but it is not easy to pin that force down. The appeal of broad purposes is particularly elusive and often mercurial. Such purposes are highly susceptible to attention shift. Yet within a supportive interpersonal and interorganizational context, broad purposes can inform an agenda, generate a significant appeal, and, of special importance, provide a framework within which many lesser aims and multiple forms of motivation operate. Just because big purposes do not emerge spontaneously from a sea of popular sentiment, we should not conclude that they are insignificant forces. It does mean that, given the bounded rationality of human beings, sustained purposes depend on the reinforcing support of networks. Selective incentives may be an important part of the picture, but it is purpose that provides the framework within which they operate. Narrowly understood, the free-rider problem does not capture very well the challenge of setting and maintaining an agenda. This is not to dismiss the importance of side payments, especially as compensation for losses that may result from the pursuit of a far-reaching agenda. But it is to maintain that a focus on purpose invites examination of the questions of what gives a large purpose perceived feasibility, how the problem of attention shift is overcome, and how does the interplay between purpose and supporting network alter an agenda over time. The framing of an agenda can provide a focus for creating or reshaping networks, but the networks may also help alter the agenda as new circumstances emerge. Purposes, after all, are not static. There is a reciprocal dynamic between how a purpose is framed and the character of the supporting networks.

IV. THEORY

[...]

Ideas, power and purpose

Much current discussion concerns the role of ideas. John Kingdon, for example, asserts that the "context of ideas themselves far from being mere smokescreens or rationalizations, are integral parts of decision making in and around government" (1995, p. 125). He contrasts the role of ideas with that of power, influence, and pressure. Mark Moore, on the other hand, cautions against focusing mainly on the intellectual properties of ideas; he gives more attention to their contextual properties (1988, pp. 78–80). In explaining which ideas matter, Moore offers: "If powerful people are made heroes and weaker ones villains, and if work is allocated to people who want it and away from people who do not, an idea has a greater chance of becoming powerful" (p. 80).

Instead of positing dual political processes, one of ideas and another of power and influence, Moore points us toward an alignment. Regime analysis sides with Moore, and purpose provides the link between ideas and power. Within the framework of urban regime analysis, abstract ideas gain legs only to the extent that they are embodied in concrete purposes. For creatures of bounded rationality, concreteness is a necessary connection to social meaning. Furthermore, as the case of Atlanta illustrates, purposes can be blended into an agenda despite ideological differences within the supporting coalition. The concreteness of purpose can also serve to connect immediate action with long-term and indirect benefits.

[...]

Urban regime analysis posits relationships that are necessarily neither antagonistic nor indifferent. Attracting and attaching are also a possibility. Whether a relationship is predominantly one of bonding together turns on purpose, how it is framed and understood, and the terms on which partners are linked to one another. Experience and interaction determine whether a positive relationship is simply a short-term and narrow arrangement or something broader and more long-lasting. Therefore urban regime analysis treats the matter as contingent.

To view relationships as contingent is not, however, to assume that all relationships are purely happenstance, with all having equal likelihood. Relationships take place in a structured context. Purpose enters the picture, but context, including past experience, mixes with purpose in such a way as to make some resources easier to bundle than others. If small bits of resources are widely dispersed, they may matter little in the process of forming arrangements to pursue a priority agenda. For that reason, inequalities in economic situation, social status, and organizational position carry great weight in building relationships for governing (Stone 1980). The path of least resistance is often to bring together arrangements that reinforce inequality.

Finding an alternative but viable path is the political challenge – the normatively based quest – that urban regime analysis faces. Socio-economic inequalities shape and spill over into the political arena, creating a strong potential for reinforcing rather than dispersing inequalities. The central political question is about the potential of purpose as an instrument of political agency to build arrangements that can, to a significant degree, counteract society's inequalities.

The heart of the beast

The political condition that urban regime analysis probes is weakness in the foundation for democratic politics. Why is one person/one vote an insufficient foundation?

In U.S. localities, for many structural reasons, business typically has a heavy presence in local civic life. The character of land ownership and of land-use planning, the nature of the system of taxation and revenue distribution, the pattern of city–suburb relations, and the importance of private credit to public borrowing are among the contributors to this pattern, and they differentiate the situation in the U.S. from that in Europe. This is not to suggest that private business is an insignificant force in European local politics, nor to deny that the role of business may be on the rise. But the U.S. has a long history of a Main Street/City Hall alliance around issues of development. That history need not be reviewed here, but it bears keeping in mind that, particularly in large U.S. cities with their highly organized business sectors, a changing economy has been fertile ground for government and business leaders to make common cause in devising responses. In America, with its weak party system and anemic labor movement, local business has been in a position to become part of the fabric of governing. Typically business did not so much pressure government as become part of the governing process around a priority agenda of adapting the city to a changing economic role.

Mobilizing to become part of a concerted effort to develop and pursue a priority agenda of development, in some sense, came easy, but nevertheless required substantial effort and detailed planning. . . . In middle-class communities, the meshing of governmental and non-governmental efforts in such areas as public safety and the schooling of young children often required no special mobilization. An advantaged position in the system of social stratification enabled the secure and affluent middle class, especially, to blend smoothly into day-to-day governance in education and public safety. If demographic change undermined an easy melding of governmental and household efforts, joining an exodus to the suburbs was an alternative. Exit rather than voice and mobilization has thus sometimes been a response to heightened concern about public-service performance.

Faced with political, social, and economic obstacles to exit, disadvantaged populations have had to cope with a history of marginality and confinement to aging urban areas. They are weakly connected to the political and civic life of cities, and have no ready means of mobilization – along with little confidence that it would make a difference even if attempted. Often alienated from both schools and police and subject to damaging stereotypes, disadvantaged populations are weakly positioned to become part of the fabric of governing. Meshing household and public efforts has happened only in limited and sporadic ways, but enough constructive activity has occurred to show that change is possible, but only if a strategy of inclusion and integration can achieve priority status.

Social problems are deeply embedded in a system of stratified inequality that impedes the ability of lower status groups to contribute to the amelioration of these problems. The dilemma is that these groups are weakly positioned to contribute to governing, but, without becoming an integral part of local governance, they can do little to further an agenda of needed change. Because pressure on government and other established institutions has little lasting effect, the question is whether a set of orchestrated moves can gradually alter the position of lower-status groups in the desired direction.

Thus urban regime analysis is about more than why economic development so often occupies a priority position in agenda-setting; it is also about what it would take to build and maintain a different priority agenda, one that is aimed at ameliorating social problems.

The analytics of local governance

As this discussion has indicated, the political challenge probed by urban regime analysis has far-reaching elements. Yet, at its core, urban regime analysis centers on the question of how local communities are governed. How do they establish and pursue problem-solving priorities? Structural forces form the context, but, because the formal authority of government is but one component in addressing problems of community-wide import, nongovernmental (and sometimes informal) links are an integral part. Let me now set forth the core elements of a model of local governance.

Proceeding from an assumption that winning an election is no guarantee of being able to assemble needed resources, I posit that governing arrangements stem from a strategic set of connections. Four elements are key:

- an agenda to address a distinct set of problems;
- a governing coalition formed around the agenda, typically including both governmental and non-governmental members;
- resources for the pursuit of the agenda, brought to bear by members of the governing coalition; and,

given the absence of a system of command, a scheme of cooperation through which the members of the governing coalition align their contribution to the task of governing.

In this model of governing, an ability to gain elected office may be an important factor but not necessarily the central one. Moreover, continuity of arrangements is not a matter of sustained, favorable ratings in opinion polls, nor is it a matter of coming out on top in all controversial issues. Instead, continuity rests on recognition by a set of resource-commanding actors that they need to act together in order to solve problems and pursue goals. The importance of winning or losing elections turns on a broader issue about the adequacy of resources. The key question is about what resources are needed and who will be motivated to provide them. The term "resources" includes not just material matters, but also such things as skills, expertise, organizational connections, informal contacts, and level and scope of contributing effort by participants.

With resources at center stage, we can see why governing *tends* to reflect the inequalities of society's system of social stratification. The "hidden hand" of narrow and immediacy-minded political agency, celebrated in classic pluralism, serves only to perpetuate and even accentuate inequality. The question is whether, in the service of a heightened sense of interdependence and an enlarged awareness of collective interest, political agency can ameliorate inequalities sufficiently to pursue aims of social equity and inclusion.

Merely imagining a social-reform agenda guarantees nothing. There is in place no effective capacity to govern into which the latest policy concern can be inserted. For any policy issue, governing arrangements have to be fashioned. Thus, for regime analysis, the presence of a model does not assume that every locality has a strong and effective set of governing arrangements. Instead, the model provides a way of looking at an array of arrangements and determining if, in fact, strength and stability are related to a congenial fit among agenda, coalition, resources, and scheme of cooperation. The model is about how parts fit together. It has an analytical capacity applied to current arrangements, and it also predicts what needs to be in place for a social-reform initiative to take hold and succeed.

In and of itself the model does not explain several things about the particulars of governance:

- From what specific concerns the agenda emanates (this is a matter of human agency but with the changing economy expected to be a wellspring of issues);
- What, in particular, motivates actors to play a part in the governing coalition (though motivation would presumably be closely connected to the agenda's content, along with past experience and resource capability);
- What resources are relevant (this would depend on the particular problems the agenda addresses);
- The origin of the practices that go into a scheme of cooperation (but with a blend of shared purpose, selective incentives, and networks expected to figure prominently).

The combined strength of the elements in the model account theoretically for capacity to govern and level of stability. The particular locality provides the contingent content. The model itself is not an explanation of why various cities have the actors that they do, but rather a guide to help identify which elements are key, how they are related, and how changes in those elements can account for continuity and change in capacity to govern. In short, the elements of the model itself are generic and therefore conceptually distinct from the various contexts in which they operate, but the context provides specifics that fit into the model. Hence localities can differ according to the role of the business sector in relation to a priority agenda – united and engaged, factionalized, or largely disengaged. Similarly, voting blocs can be reliable and stably engaged or volatile and subject to change. Schemes of cooperation can be weak and largely ad hoc or they can be embedded in tradition and reliably brought into play over a range of issues. The composition of the players themselves may vary. In some places the newspaper or a charitable foundation could play a significant role, but in other places be missing as an actor. Political party leaders play a part in some places but not others. Governmental officials, elected and appointed, may play wide and leading roles, but sometimes may not. Intergovernmental channels of communication may be a significant part of the arrangement or not, depending on the issues and the place. Once particulars are fit into the model and their contribution assessed, then the model can be used to predict the capacity of the arrangements to sustain their governing effort. The hard part is assessing the particulars that make up the model.

Overall, urban regime analysis has two sides. One concerns how particular regimes come into being. It is about *significant historical details* – how an agenda came to be framed in a particular way, what brought coalition partners together (or, after a period of time, what caused a break), why coalition partners devised the scheme of cooperation they did, etc. On this side, the concept of an urban regime serves to identify important foci of historical research. It does not in itself predict the course of history.

The other, more abstract side of urban regime analysis centers on a *model* of how governing arrangements operate. The model focuses on the combination of factors that promise viable and durable arrangements. The agenda – the problem-solving task, resource adequacy, and alignment by key actors (the governing coalition) in the absence of a command system (hence a need for a scheme of cooperation) – these are the elements that need to be brought to strength and aligned for governing arrangements to be viable and stable. The analysis determines the strength level and predicts stability and effectiveness accordingly. In this way governing arrangements can be arrayed along a spectrum, from strong and durable to weak and unstable, with most places in fact likely to fall somewhere in the inter-mediate area because bringing all four elements to full and complementary strength is difficult.

On the surface, this kind of analysis is value-free. A strong and stable regime might serve ends that are reprehensible (note, as an example, the racially exclusionary regime in Dearborn, Michigan, under Mayor Orville Hubbard). However, regime analysis can be used to show how a given set of arrange-ments falls short in scope of representation. . . . Thus, while the internal dynamics of the model pre-dict outcomes, the model itself does not pass a normative judgment (even though regime analysis itself grows out of a normative concern about the consequences of socio-economic inequality). At the same time, the model is also in line with the norm-ative ideal that those people who are most directly affected by a policy should be included in the pro-cess of making policy. The regime model predicts that, if they are not included, the policy problem addressed will not be effectively tackled and the policy effort may falter. From current research, as we examine various efforts to combat poverty and achieve social inclusion in both Europe and the U.S., we see a pattern of modest efforts, limited engagement, and persisting problems. Resources are typically spare and efforts to enlist the poor frequently amount to little more than a triage operation. The urban-regime model predicts that such efforts will not occupy a priority position, will not be scaled up, and can be stabilized, if at all, only as a small and marginal activity commensurate with the limited resources it commands.

V. CONCLUSION

Urban regime analysis does not offer itself as a comprehensive theory and thus does not claim the scope of theories such as rational choice and Marxism. Its aim is more at an intermediate level where attention can be directed toward effective forms of problem-solving. Urban regime analysis does, however, rest on assumptions both about the grounds of micro-behavior and about the nature of society writ large. And the specific view that human beings are more than interest-driven creatures provides a foundation for the important place that purpose occupies in urban regime analysis.

Structures enter urban regime analysis in a framework of structure and agency. The research posture taken is that of seeking to understand how human agents in a given local context (or range of local contexts) see their situation and choose to act on it. Structures lie behind such activities as fram-ing agendas, building coalitions, devising schemes of cooperation, and making use of and sometimes reshaping inter-organizational and interpersonal networks, but the focus is on human agents in action. Even if global capitalism is the overall setting, human agents devise responses and these responses take into consideration factors much more proximate than the international economy.

The parentage of urban regime analysis lies in political economy and the transparent reality that government actions inevitably intertwine with the workings of the economy. In the case of local economic development, much attention is given to matching governmental and business efforts, and public–private partnerships often give formal recognition to complementarity. Yet we need to look beyond this obvious instance. It is no happenstance that civil society and social capital have become major topics of interest. Complementarity extends across the full scope of public policy – efforts in such matters as education, public safety, and disease prevention depend on how government actions mesh with private behavior. In some cases meshing involves little overt coordination.

We should bear in mind that to invoke the word "governing" or "governance" is not to suggest that the process is always effective. Often it is not, or is only partly effective. The heuristic core of urban regime analysis illuminates what has to be harmonized to provide stable and effective arrangements. An urban-regime perspective thus highlights why many social programs fail or never rise above the level of triage operations. To be effective such programs have to enlist active engagement from the target population. However, a long history of neglect and frustration stands as a barrier to effective involvement. This history can be countered only if substantial resources (including skills and efforts) are committed to the task by governmental actors and others outside the target communities. In a society with substantial social and economic inequalities, this is the severe test for social purpose as leverage to advance democratic inclusion. In such a society is it possible to frame an appropriate agenda and build the needed coalition, complete with the cooperation and mobilization of resources, through which social inclusion can be effectively pursued in a sustained manner?

Small efforts cannot change the socio-political context enough to overcome deeply rooted alienation. From past experience we can see that substantial resources invested in community development and community organization are a needed first step toward political inclusion. Contrary to various pluralist assumptions, neither protests nor voting mobilizations are enough to achieve a place within arrangements for governing. Pressure on a governing arrangement is insufficient. A group needs to be in a position to contribute actively to a shared aim of problem solving. If localities are to achieve large-scale successes in urban education, workforce development, countering youth violence, and promoting neighborhood revitalization, then they need to be able to blend the efforts and contributions of lower socio-economic status populations

with those of established institutions. Bland talk about partnership or parent involvement will fall far short unless accompanied by a sophisticated form of enablement.

Political inclusion is thus at least a two-step process. Taking a cue from early work on the "two faces of power," urban regime analysis lays out a systematic argument that governing, as opposed to ad hoc decisions or concessions here and there, rests on a level of politics in which substantial resources, complex capacities to plan and execute, and skills in building cooperation and devising forms of coordination are far beyond the ordinary citizen. Thus urban regime analysis suggests that institutional repair, community development and community organizing, and reshaping civil society are among the steps needed before we can characterize local politics as open and penetrable. That these are necessary steps does not make them sufficient, but this is as far along the road of reforming society as urban regime analysis can take us on its own.

REFERENCES FROM THE READING

Bachrach, P. and M.S. Baratz (1962) "Two Faces of Power." *American Political Science Review* 56, December: 947–52.

Dahl, R. (1961) *Who Governs?* New Haven: Yale University Press.

Jones, B.D. (2001) *The Architecture of Choice.* Chicago: University of Chicago Press.

Kingdon, J.W. (1995) *Agenda, Alternatives, and Public Policies* (2nd edn). New York: HarperCollins.

Moore, M.H. (1988) "What Sort of Ideas Become Public Ideas?" In R.B. Reich (ed.) *The Power of Public Ideas.* New York: Ballinger.

Stoker, G. (1995) "Regime Theory and Urban Politics." In D. Judge, G. Stoker and H. Wolman (eds) *Theories of Urban Politics.* London: Sage.

Stone, C.N. (1980) "Systemic Power in Community Decision Making." *American Political Science Review* 74, December: 978–90.

—— (1982) "Social Stratification, Nondecisionmaking, and the Study of Community Power." *American Politics Quarterly* 10, 3: 275–302.

—— (1989) *Regime Politics.* Lawrence: University Press of Kansas.

Truman, D. (1951) *The Governmental Process: Political Interests and Public Opinion.* New York: Knopf.

"In Search of the Growth Coalition: American Urban Theories and the Redevelopment of Berlin"

from *Urban Affairs Review* (1996)

Elizabeth Strom

Editors' Introduction

American political scientists have sought to use the local as a laboratory for understanding how the larger political system operates. In some cases they confine their conclusions to the cities they study, or to cities existing within the peculiar constraints of the American system. In others, they present models developed within the U.S. context as having more universal application, leading the reader to wonder whether they are really applicable to other national contexts. For example, Paul Peterson's inferences from public choice theory in *City Limits* would seem to be applicable to all cities within a capitalist economy. The structuralist analyses reviewed by John Mollenkopf or Martin Shefter's "four imperatives" (Part 4) would also seem to speak to contradictions endemic to all local governments within capitalist democracies. So perhaps theories developed by American political scientists to explain how power works in cities can be exported globally, like Hollywood films or Nike running shoes.

Or can they?

In the past decade, urban political scientists have become far more engaged in comparative work. Strom's reading primarily asks whether the theoretical models used by U.S. scholars can provide explanations in the study of cities outside of the U.S. Her conclusion is that they may be helpful, but, like those Hollywood movies, they require some translation. Yes, wielders of public authority must strike complex accommodations with possessors of investment capital in all capitalist democracies; the aspirations of citizens, expressed through electoral participation, are always tempered by economic constraints. Governance does indeed depend on a variety of actors bringing together public and private resources to meet agreed-upon goals.

When both private and public sectors in different countries have distinctive forms of organization, growing out of different histories and traditions, however, they are unlikely to interact in the same way. Both decision-making processes and policy outcomes are likely to vary. Explanatory models based on one system, then, will be less effective in illuminating another.

After the fall of the Berlin Wall, Berlin experienced a sort of "gold rush," as developers and corporations sought to establish themselves in parts of the city that had for decades been inhospitable borderlands and were now, suddenly, centrally located in a city about to become the national capital. Seeing this flurry of real estate activity, Strom went to Berlin sure that she would find this city following the pattern familiar to her from studies of U.S. cities: elected officials eager to fill city coffers and create jobs would happily turn desirable property over to developers. She found, however, that even as building cranes popped up across the central city landscape, public officials, citizens groups, architects and planners, and even investors themselves were reading from a very different script than one would have found in a typical U.S. city. Some of these differences were institutional, with public officials and the "expertocracy" (architects, planners, historians) being given a far greater role in shaping even private development. Some were more cultural, as she found different attitudes about the appropriate roles of public and private actors prevailing across sectors.

There is a rich and growing debate about the export of American urban theories, and in particular the regime approach. Articles such as Gerry Stoker and Karen Mossberger's "Urban Regime Theory in Comparative Perspective" (*Government and Policy* 12 (1995): 195–212); Alan Harding's "Urban Regimes and Growth Machines: Toward a Cross-national Research Agenda (*Urban Affairs Quarterly* 29 (1994)); and Scott Gissendanner's "Mayors, Governance Coalitions, and Strategic Capacity: Drawing Lessons from Germany for Theories of Urban Governance" (*Urban Affairs Review* (2004) 40: 44–77) all explore this topic. Books employing this approach include Alan DiGaetano and John Klemanski, *Power and City Governance: Comparative Perspectives on Urban Development* (Minneapolis: University of Minnesota Press, 1999).

Those interested in the case of Berlin will find many books in English about this city's transformation since reunification, including Brian Ladd's *The Ghosts of Berlin: Confronting German History in the Urban Landscape* (Chicago: University of Chicago Press 1998).

In this article, I explore the redevelopment of Berlin since the fall of the Berlin Wall in November 1989. My primary goal is to determine whether American approaches to urban political economy, which focus on the formal and informal alliances between political and economic elites as determinants of local development policies, help illuminate the sudden and complex transformations occurring in Berlin's central business and government district as the city rejoins the system of capitalist cities. In discussing the Berlin case, I also hope to contribute to the debate over whether American-based analyses of urban regimes and governing coalitions are helpful tools in analyzing urban politics outside of the United States. Berlin is, of course, hardly a typical city, but it is one in which current development conflicts offer an excellent laboratory in which to study the interplay of the public and the private in the German context. My conclusion is that American paradigms, such as the urban-regime perspective, can illuminate aspects of German urban politics that are sometimes overlooked in the German literature. The kinds of governing coalitions that emerge, however, are shaped by a different set of political and cultural dynamics than those common to American cities. Analysis of Berlin thus requires an appreciation of the vertical and horizontal ties within the state

and of the more deeply embedded and powerfully defended sense of place than is normally found in cities in American case studies.

[. . .]

The American literature rests on certain assumptions about the nature of the local political economy that constrain the formation and activities of urban governing regimes. First, the national state is characterized as weak, because it is nearly absent as an agent of fiscal equalization and because its intervention into the market is limited and largely indirect. Second, within the local political realm, capitalists have privileged access to important resources (money, access to media, organizational skills), making them the most attractive partner for any lasting coalition. Finally, implied in most American case studies is the near hegemonic acceptance of the role of capital in shaping the local policy agenda and determining the central-city built environment, perhaps a natural corollary to the political strength of capitalist interests. . . .

EXPORTING AMERICAN MODELS

These assumptions lose their power when one seeks to understand urban development outside the United States. To be sure, other liberal democracies are also defined by the division of labor between state and market; therefore, government at all levels involves public–private bargaining. Moreover, at least in the case of the similarly urbanized and developed nations of Western Europe, the same economic restructuring pressures are at work, leading to comparable urban economic conditions and, in some cases, evidence of converging policy responses. Nonetheless, the study of European cities is not simply a matter of applying American theories to new settings. . . . Although many researchers have concluded that the tools of regime theory are appropriate for cross-national analysis, it is clear that the task of the analyst requires some adjustment.

European states tend to have stronger central governments in several senses. They have a greater capacity and willingness to intervene in the market, and they play a more significant role in setting policy guidelines even for such traditionally municipal functions as land-use planning and regional development. This leads to more uniform local policies and less intercity competition; it also makes the relationships within the state, both vertical and horizontal, more salient. Fiscal equalization schemes, in which central government uses its greater revenue-raising power to redistribute resources from wealthier to poorer regions, further enhance the role of the central authority within the intergovernmental system. These centrally directed measures leave local units less dependent on the local market, albeit at the expense of political autonomy. Indeed, Gurr and King argued that local governments experience an inevitable trade-off between dependence on national government institutions and dependence on the market. A city's degree of dependence on (or, to put it more optimistically, autonomy from) the central government on one hand and the market on the other comes to shape local politics, including the way that local governing regimes are constructed. The terms of this trade-off vary across nation-states and across time, conditioned by a number of political, social, and economic variables.

Moreover, parties and labor unions are sometimes politically decisive in European cities. . . . Most significantly, the strength of anti-capitalist parties makes it impossible for the city as *growth machine* ideal to emerge as hegemonic.

[. . .]

THE GERMAN STATE AND POSSIBILITIES FOR BUILDING LOCAL COALITIONS

[. . .]

The state

Even in a federal system such as Germany, the central state can play a far greater role in all aspects of local governance than it can in the United States. As in North America, policy implementation is carried out largely by state and local officials, and the federal government can meet its policy goals only through a system of negotiation with lower-level officials. The German federal system, however, differs from the American in several respects that are especially significant for urban politics. Most notably, tax collection and revenue-sharing procedures are far more centralized, leaving local governmental units less autonomous in their tax policies and less dependent on their town revenue-raising capacities. Within the federal system and within each territorial state, an elaborate system of revenue sharing requires wealthier units to share their revenues with poorer ones, preventing the tax-generating inequalities between regions from producing too great a disparity in service provision. Local governments almost universally control land-use matters, but national planning regulations, fiscal equalization formulae, and regional development policies limit both the autonomy of local officials and their dependence on local economic elites; the key arena for coalition building thus becomes central–local, rather than public–private. Furthermore, the German state is more actively interventionist, with a long tradition of public ownership, regulation, and support of industry and financial markets.

Moreover, the German state has been characterized as comparatively bound by rules and inaccessible, thus reducing the opportunities for coalition building that bridge the gap between state and society. . . . Specialized career bureaucrats are charged with applying detailed public laws, thus reducing the flexibility that American local officials have used to open informal channels of cooperation with business elites. In the area of urban development, a policy domain in which local autonomy is comparatively high, the 400-page federal law addresses nearly every aspect of planning, building, and property disposition. Those working within the bureaucracies see their jobs to be the implementation of this law and, in conversations, refer to its specifications as explanation for government action. This does not mean that ties between those within the state apparatus and outsiders are not possible. There are numerous formal, corporatist decision-making bodies that operate in every policy domain, especially in those concerning the regulation of the economy. Under certain circumstances, those within the government have found it advantageous to promote ties to new, unofficial interest groups; thus some of the social movements of the 1970s and 1980s, including those in the area of urban renewal, have been able to influence government policy. The dynamics of these alliances are conditioned, however, by the aforementioned characteristics of the state apparatus.

Ideology and interests

In Germany, as in other European countries, the hegemony of value free development faces greater competition from several sources. It might be expected that the Social Democratic tradition would challenge the prerogatives of capitalist accumulation, but German Christian Democracy, representing the country's center–right population, is also suspicious of free-market liberalism. It was only in the mid-nineteenth century that a real property market emerged in Germany; capitalist land exploitation thus represents a relatively recent phenomenon in the 500- to 1,000-year histories of many German cities. The city areas that were built during the brief period in which speculative development was scarcely checked are viewed with great ambivalence, and socialist administrations during the Weimar Republic encouraged the creation of not-for-profit housing companies to rid urban development of its profit-making associations. Thus alternatives to capitalist hegemony have found institutional support.

This is not to suggest that German culture is at heart anticapitalist. Nowhere is private property more unassailable than in Germany, where ownership rights are protected in the Constitution, upheld by the courts, and championed by the liberal and center–right political parties. Rather, German tradi-

tions regarding the respective private and public roles in shaping the built environment are contradictory. . . . Property is sacrosanct, but according to Article 14 of the German Constitution, "Property imposes duties. Its use should also serve the public weal," a public weal interpreted and upheld by the state. The notion of the civic arena as the buffer between citizens and the powers that would exploit them, be they absolutist rulers or capitalist developers, retains its resonance. Clearly, it is more difficult to base a local governing regime on public support of profit-seeking ventures when the drive for profit is not universally accepted as a legitimate determinant of the built environment.

Berlin as a case study

Berlin is hardly a typical Germany city. Divided by a heavily armed border for nearly forty-five years, the social, economic, and physical characteristics of the city were shaped by geographic isolation, military presence, and the competition between two hostile systems, both of which used the city as an ideological showcase. West Berlin's borders were secured by the thousands of allied soldiers stationed within the city limits, and the city's economic viability was maintained through a plethora of federal subsidies. . . .

The subsidies had the beneficial effect of preventing the economic strangulation of the city. . . . More subtle effects of the extensive subsidy system, however, continue to shape aspects of the city's social and economic fabric. In comparison with other large cities, Berlin has retained an unusually high percentage of industrial jobs, kept in the city by federal subsidies, and has never developed a strong employment base in private business services. Both the city's public and private sectors have been criticized as suffering for a "subsidy mentality," a vaguely defined illness in which private managers fail to innovate and public managers fail to scrutinize. . . .

Finally, Berlin (and formerly West Berlin) is a city-state and, therefore, differs from typical cities administratively as well. A city-state is, functionally, a city but is treated as one of the nation's sixteen states. . . . It, therefore, has powers, resources, and responsibilities that other cities must share with state governments.

[. . .]

THE PUBLIC–PRIVATE PARTNERSHIP IN WEST BERLIN

Local capital, although not absent from the political scene, was not a significant force [in the postwar period]. Berlin's nineteenth-century dominance in manufacturing and finance declined quickly after the war, as industry sought less troubled, more central locations. Into this vacuum stepped the state. The local state in Berlin has, as the channel for Marshall Plan and later federal aid, acted as the catalyst of most aspects of postwar reconstruction. It was full or partial owner of the major banks, and its publicly owned housing companies built and managed much of the city's housing stock. The national government's Berlin subsidies were so all-encompassing and so generous that there was little room left for a local economic development policy.... The city's economic experts did, in the 1980s, look to other kinds of development initiatives to improve the city's image and to try to improve the economy's structural base toward more innovative, higher value-added production, but the major political battles over Berlin's development continued to be played out not locally but in [the national capital].

In some cities, real estate itself has emerged as a major industry, and real estate developers have come to wield important influence in local politics. But this sector was weak in Berlin. In the east, there was no private commercial market; all building was done at the behest of the state or one of its organs. The West Berlin economy had lost nearly all prewar corporate headquarters and, therefore, never developed the service sector that represents much of the demand for commercial space in other cities. There was of course a real estate "scene" in West Berlin, but it was insular and risk averse. Developers tended to focus on the heavily subsidized housing market. Rents were subsidized down to $4.00 per square meter per month even when actual costs were $13.00 or more; in such a climate, anyone with the patience and connections to deal with the public bureaucracy was sure to be able to make a living. Even commercial development in Berlin lacked the competitive rigor of more typical market conditions. Berlin's credit agency was generous in providing loan guarantees for commercial projects; with this safety net, banks were also less cost conscious. Several costly failures shed light on the sometimes scandalous ease with which public moneys were allocated, but did not alter the nature of local development.

Key to growth politics in American cities has been not merely the fact of private interest but the ability of the private sector to organize itself effectively. In cities like Atlanta, Pittsburgh, or even in other German cities such as Hamburg, the private sector has organized into peak organizations through which corporate leaders have channeled their economic clout into political action.... Lacking the clout of corporate headquarters or a strong financial services sector, the Berlin business community was not a compelling or, given the resources of the public sector, a necessary governing partner. There was a growth coalition active in the rebuilding of Berlin in the postwar period, but it was one linking various arms of the state. Private interests played a minor and subordinate role....

AFTER THE WALL: POLITICS AND PLANNING IN UNITED BERLIN

The fall of the Berlin Wall and the reunification of Germany with Berlin as its seat of government have made Berlin once more a center of economic and political power. These changes have been rapidly registered in the built landscape, in which the central-city space not claimed by the federal government has been quickly grabbed by an array of German and international companies and real estate developers for whom West Berlin had previously held no interest. The phasing out of the various subsidies that had buffered the city from market forces began shortly after reunification. As fiscal pressures forced city officials to cut programs and more aggressively seek revenue and as investors introduced plans to transform Berlin into the current image of a world city, some observers began to wonder whether the city's politics would be shaped into an urban regime resembling the American growth coalition.

The signs of such a development seemed to be in place. First, the all-Berlin elections held in December 1990 resulted in a government much more sympathetic to growth concerns than the previous one. The conservative Christian Democratic Union had become the largest party in the local parliament; they invited the SPD to join them as junior partners in a Grand Coalition (the Coalition). Replacing the left-leaning government of Social Democrats and the Alternative List (Berlin's Greens), the Coalition represented a turning away from the experimentation that had marked the previous period.

The Coalition's early policy decisions indicated their awareness of the city's new competitive position, as is suggested by their approaches to two issues that dominated planning and development debates throughout 1990 and 1991: the campaign to attract the summer Olympics to Berlin in 2000 and the plans to create a new business center at Potsdamer Platz. These two cases are presented as exemplary not because they were routine but because they brought together exactly the kind of key, citywide political and economic interests that, according to much of the American literature, would come to shape a growth-oriented regime. The Potsdamer Platz development was sponsored by several multinational corporations; the Olympic campaign was universally endorsed by various business groups, and the chamber of commerce took it on as its major theme throughout 1992 and 1993. The opposition to both these initiatives also included the most significant alternative groups, including the Alternative List and other environmental movement representatives. . . . As a brief discussion of these two cases will reveal, a growth-oriented coalition [did not] gain dominance over development policy in the city.

Olympia 2000 and Potsdamer Platz

Even before taking office, the new government had made the campaign to become the host of the summer Olympic games in 2000 its priority. The notion of competing for the Olympics had actually first emerged in 1987, when Berlin was still two cities, as a joint project of East and West Berlin; it lay dormant until the leaders of the Coalition co-opted it as a means of mobilizing public and private support for the city's redevelopment. Berlin's leaders saw the games as a way to meet multiple goals. In Berlin's case, the desire to project a new image to an international audience was particularly strong; the last Berlin Olympia (1936) had been a showcase for Nazism. The Olympics promised to provide the foundation for a new public–private partnership:

public investments would be required to improve transportation between proposed sites, but much of the actual development, it was hoped, would be carried out by private investors. Corporate sponsorships, which had footed much of the bill for the Los Angeles Olympics, were seen as an indispensable part of the campaign, and a newly created agency to promote the Olympic candidacy was headed by an IBM executive, a sign that the business community would be integral to this effort. In addition, political leaders predicted the rallying of popular support behind the allure of Olympia, which would help bring the East and West together behind a development agenda.

Berlin was not awarded the Olympic games. That they failed to convince the International Olympic Committee does not alone tell much about the city's internal politics, but an examination of the debate set off in the city by this campaign suggests some ways in which the foundations of a pro-growth coalition in Berlin remain shaky. The quasi-public local Olympic agency was under heavy criticism from all sides before it even got into gear; its private sector manager was replaced amid a barrage of criticism. Berlin businesses dutifully displayed their "Berlin 2000" posters, and the chamber of commerce held Olympia-connected events, but the local press – including the more conservative mainstream newspapers – eagerly reported evidence of financial mismanagement and poor planning.

Perhaps the biggest disaster was the lack of support for the effort among the population. The Berlin Left, mobilized by an Anti-Olympia Committee supported by the Alternative List and its East Berlin allies, took to the streets to oppose the Olympia campaign; [their] protest . . . was carried out with far more imagination and enthusiasm than the pro-Olympic campaign. Even among the apolitical, the prospect of hosting the Olympics was not greeted with enthusiasm. The local Olympic agency, continually polling citizens for signs of increasing support, could find no more than 53 percent who viewed the Olympics positively, even when polling questions stressed the financial and civic benefits enjoyed by other Olympic cities. Other, more objective surveys found majorities of up to two-thirds opposed to the games. . . . Ultimately, Berlin failed to capture the Olympics or to forge a new growth alliance. Pre-unification West Berlin political relations prevailed: Business leaders were largely passive, the police treated political opponents as state enemies, and Berlin's leaders found themselves pleading with their party allies in Bonn for more money.

Another indication that a new kind of politics might be underway was the sale of large, central sites around the vacant Potsdamer Platz to three multinational firms. In 1989, before the breaching of the Berlin Wall, Daimler Benz Chairman Edzard Reuter (son of Ernst Reuter, SPD mayor of Berlin in the 1940s and 1950s) approached Berlin officials about the possibility of locating a new Daimler service subsidiary in the city. Several sites were discussed, and the western half of Potsdamer Platz was agreed upon. Discussions were under way and preliminary agreements had been made when unrest in East Germany made reunification a possibility and metamorphosed Potsdamer Platz from an undesirable border area into the urban core. In spring 1990, Daimler purchased the land from the city for approximately $90.00 per square foot, planning to build the headquarters for its new subsidiary, plus speculative commercial and retail space. A year later, the adjacent site was sold to Sony, which intended to build its European headquarters and additional rental offices.

The announcement of the deal in spring 1990 by the SPD mayor and leading senators nearly ruptured the governing coalition. Technically, it is the finance senator who negotiates land sales, but the failure of the SPD to consult the senator for urban development, representing the Alternative List, in what was a significant planning issue was clearly a breach in the expected collegiality. In addition, the Alternative List objected to the size of the planned development and to the low offering price. The coalition was saved only when the SPD agreed to use a planning device – the architectural competition – that has been the major tool of German development planning for a century. In such competitions, plans for an area or for a specific building are chosen by a jury, usually appointed by public officials, consisting of professional architects and representatives of relevant city agencies. Architects are invited by the government to submit proposals based on the city's articulated goals; it is the job of the jury to choose a winner whose plans will ultimately be realized. Operating within a protocol spelled out in public law and monitored by the Architects' Chamber, city governments very often use competitions to choose plans for significant areas or designs for public buildings. Private investors may also sponsor competitions for their projects as a way to gain public confidence in and support for their plans.

[. . .]

Governing Berlin: a new growth regime?

[Is unified Berlin headed by a growth-oriented, public-private partnership?] If the Olympic effort failed to institutionalize a new working relationship between public- and private-sector leaders, the sale of centrally located sites to multinational corporations does seem more indicative of a progrowth philosophy. In [the case of Potsdamer Platz], public land was viewed as a commodity to be granted to business interests in the hope that their profit-seeking activities would create jobs, tax revenue, and prestige for the city. But even here, the coalition supporting the Potsdamer Platz development appears unsure and tenuous, not the foundation for a new governing regime. Different arms of the state have been in conflict over the future of the central city, and investors have been forced to engage in extensive negotiations with state officials before building plans could be finalized. There is scant evidence that important private investors, at Potsdamer Platz or at other key development sites in East Berlin, are effectively organized. They concentrate on issues specific to their projects rather than a city- or areawide agenda, and they press their demands separately, at whichever point in the government they happen to have the greatest leverage.

Debate about proposed development has remained heated within the government and among political parties and the general public. Opponents have questioned whether the city's center can legitimately be sold to private interests; they have insisted on the right of the state, rather than private owners, to determine design guidelines; they have protested on environmental grounds. In short, capitalist development logic does not dominate. Moreover, opponents who raise fundamental objections to proposed developments are unlikely to be drawn into the kinds of compromises found in some more progressive American cities . . . in which erstwhile growth opponents won a share of the expected benefits of development in the form of linkage payments.

Features of the German State and Coalition Building in Berlin

The boundary conditions in which these deals have been made suggest the salience of forces not usually explored in the American literature. Turning again to the features of the German state relevant to urban political economy, one can see how these characteristics have shaped urban development politics in Berlin. The importance of the strong state is even greater in Berlin than in other German cities. As mentioned previously, the state (local and national) has had to act in place of private capital in many aspects of economic development. National subsidies overshadow local interventions. In other words, if American urban regimes grow out of the many incentives for co-operation between the public sector and selected business interests, it is clear that these incentives are far reduced when the national government or publicly controlled financial institutions are the crucial negotiating partners.

The formalized *Rechtsstaat* and the specialized bureaucracy shape the formation of policy coalitions. National building law provides clear legal guidelines for local planning officials and potential investors that cannot be negotiated away. For instance, the creation of a legally binding construction plan (*Bebauungsplan*) requires planning officials to solicit input from affected interests (as defined by the local government) and to make the plans available for public review and comment for at least a month. They must read, consider, and respond to every comment they receive and be able to prove that they have done so to an administrative court established to hear complaints filed by those who challenge the city's process. This cumbersome planning procedure (which contrasts markedly with the stealthy process through which public land is sold) shapes the politics of planning. Although it could – and sometimes does – lead to lengthy adversarial procedures, more commonly it gives all parties the incentive to negotiate. Investors consult with state officials; city planners consult with those interests who are most likely to take full advantage of their legal veto power. This is very similar to the kinds of negotiations that underpin governing regimes in American cities, but several factors make the nature of the negotiating process and the typical outcomes slightly different.

First, for reasons suggested earlier, the state remains far more central to the process. In New York, for instance, community opponents have at times negotiated directly with developers, with city officials playing only a facilitatory role (although, of course, demanding credit for successful outcomes). State actors in Germany would be unlikely to abdicate their central position, nor would competing interests be likely to seek a mutual accommodation independently. As a rule, investors and their representatives would not be expected to have a natural advantage in gaining the support of bureaucrats with whom they share neither training nor work ethic. . . . Although investors certainly have their important allies and, as the case of Potsdamer Platz shows, they are quite often at the highest levels of the party and bureaucratic hierarchies, most likely other kinds of interests have advantages in their routine dealings with midlevel state officials.

Second, and most striking to the American observer, is the prominence of interests and ideologies that form a cultural barrier to progrowth governing ideologies. Whereas in literature from other European countries authors have noted the importance of socialist parties as counterweights to capitalist development policies, in Berlin the postwar SPD has never played this role. In the 1980s, the Alternative List emerged as a voice for an environmentally based opposition. To be sure, members have mobilized their parliamentary resources quite effectively and have been able to use the legally mandated planning process to raise questions about planned development, thus extending their influence beyond their modest membership. But the most significant cultural and organizational factors are found outside the party system.

In Germany, there is a strong consciousness of city building as a public enterprise that does not exist in the United States, where urbanization and capitalist land development have always gone hand in hand. The state is seen as the legitimate arbiter of the built environment; investors, it is believed, should be considered just one of many interests. Architects and planning professionals, both those within the state apparatus and those outside who are asked to participate in decisions on behalf of the state, play a key role in shaping the discussion of central-city redevelopment. Whereas

Americans tend to view property investors as the heroes of urban development and identify buildings with those who initiated and profited from their construction (e.g., Trump Tower, Rockefeller Center), in Berlin the architects are the celebrities. Foreign investors have caught on quickly, choosing their architects from the pool of local stars in hopes of winning public credibility and quicker building approvals.

Indeed, architects and their values have shaped most of Berlin's planning debates. Conflicts have been interpreted – and resolved – as disputes between conflicting architectural ideals: high-rise versus low-rise, busy street grid versus boulevards and plazas. Public officials have indeed sold land for intensive office development, but the issuing of building permits is made contingent on the developer's willingness to comply with their vision of the traditional European city. Thus have most central-city plans favored bulky, low-rise buildings that mix at least a minimal housing component with commercial uses. This is not a new development in Berlin, where urban renewal, the major social conflict of the 1970s and 1980s, was also perceived as a debate between two schools of urban planning. But given the pace of change today and the uncertainty over where the city is heading, the architectural perspective on urban problems becomes more salient. Architects, of course, control neither political nor economic capital, but their views have such symbolic resonance – the Architects' Chamber is, after all, the official guardian of the built environment – that their imprimatur becomes necessary for public authorities and investors engaging in significant building projects. The architectural community thus bestows legitimacy upon compromises reached between the rights of property and the state-defined public interest.

The influence of architects on actual planning policy is not merely an ideal phenomenon. It is institutionalized through the opinion-shaping power of professional associations and realized through the many state-mandated planning processes. Planning consultants often work under contract to the city: They, in fact, do most of the actual planning, from testing alternative development proposals to carrying out the nuts and bolts of the process for a major area like Potsdamer Platz (including the drafting of the responses to citizens who have raised complaints during their official participation

period). Blue-ribbon panels, advisory commissions, and the like are common features of the political landscape. Upon taking office, the current urban development senator convened a city forum that included the key notables of the architecture and planning world; their recommendations were to help him shape new planning policies, although some now complain that it is more a discussion group than an advisory panel.

Most significant remains the use of planning and architectural competitions to design major development sites, as discussed earlier. The architectural competition as the primary mode of city planning has multiple attractions to local policy makers. The competition approach places architects – as applicants and as judges – at the center of the planning process, where they are believed to belong. The notion of urban development policy making as the search for the technically perfect solution as determined by experts is consistent with a legalistic, bureaucratic tradition of governance. Finally, the competition format brings the legitimacy and presumed public-regardingness of the architectural profession to planning decisions. The formal planning process appears to guarantee that the public interest, represented by the state and supported by the publicly obligated Architects' Chamber, is the guiding light determining the city's built environment, even though property is, in fact, privately owned.

BERLIN IN COMPARATIVE PERSPECTIVE

The regime approach, with its emphasis on the construction and institutionalization of cross-sector governing coalitions, offers a helpful perspective for analyzing the political economy of a city in the midst of transformation. By placing the relationship between wielders of political power and wielders of economic power at the center of an analysis of the city, regime models illuminate aspects of urban development politics in Berlin that have not yet been fully explored. But the relative strength of the coalition partners, the rules under which they operate, and the incentives to which they respond can be – and in the case of Berlin indeed are – markedly different. In Berlin, even in the face of new development pressures, the state remains the political center of gravity, and the most important ties are still the vertical and horizontal links within the public sector. Coalition-building possibilities with private interests are further constrained by values among elites of all political stripes and among members of the general population, who question the legitimacy of private development imperatives as the determinant of city building. This does not necessarily make planning in Berlin more democratic, nor does it guarantee a more attractive city. It simply suggests that the political dynamic driving urban development is fueled by incentives not routinely found in American cities.

Some of these observations should apply to other German cities, and perhaps other European cities as well; others grow out of circumstances particular to Berlin but may still have comparative utility. For example, organized business interests do not play the leading role that they do in American cities anywhere in Germany, where the social welfare, community improvement, and cultural functions sometimes undertaken by legitimacy-seeking businesses in the United States are funded by the state. The extreme weakness of the private sector

in Berlin grows out of specific historical factors and may be less relevant to other Western European cases. Berlin's case may, however, suggest some comparisons with the new capitals of Eastern Europe, in which capitalist property markets and development pressures are emerging.

[. . .]

The cross-national diversity of values regarding by whom and for whom the city should be built needs to be more rigorously explored. The salience of architectural debates in Berlin, a city that has long been held as a political symbol, may be unusually high. That different planning cultures mark different cities, and particularly different nation-states, however, is clear. Such differences can only be appreciated with attention to specific historical factors, such as the timing and circumstances of the emergence of capitalist property relations, and the respective roles of public and private actors in defining the city's landscape in formative years. Comparative studies of historical urban development . . . can shed light on the understanding of these issues today.

PART FOUR

The political economy of cities and communities

Plate 4 Commerce on the Grand Canal, Venice.

INTRODUCTION TO PART FOUR

In 1950, the ten biggest central cities in the U.S. housed one out of every six people in the U.S. – 21.7 out of 151 million people.[1] These urban dwellers were overwhelmingly white (84.3 percent), although blacks had become a significant minority (13.3 percent) and Hispanics were beginning to be present (2 percent). Fifty years later, the population of these ten cities had declined by more than 10 percent, to 19.1 million (their larger metro areas were more successful at retaining their population). While losing two million residents might not seem catastrophic, it caused a precipitous decline in their share of the national population. By 2000, these ten cities held only 6.8 percent of the nation's 281 million residents. Moreover, whites made up only 42 percent of this population, while blacks had grown to 24.7 percent, Hispanics to 22.6 percent, and Asian to 7.4 percent. In other words, the big cities of 1950 not only shrank rapidly as a share of the national population, but their racial and ethnic make-up changed dramatically.

Where did the growth of the national population take place if not in the old central cities? Certainly, the population of the suburbs surrounding the old central has grown enormously over the past five decades, beginning with World War II veterans returning to use GI loans to buy their first homes, the continuation of the exodus that large industrial employers had set in motion during the war, and massive post-war investments in metropolitan freeways that opened up new land to tract housing development. In 1950, 57 percent of the population of the metropolitan areas around the ten largest cities had lived in the central cities, including the overwhelming share of the blacks, Hispanics, and Asians. By 2000, their metropolitan areas had grown to 38.8 million, but the central city share had fallen to only 37.5 percent. (While whites were still more likely than other groups to live in the suburbs, about half of the other groups also lived outside of these large old central cities.)

Much more growth, however, took place in the metropolitan areas outside of the Northeast and Midwest. In 1950, Phoenix was only the ninety-ninth largest city, lagging behind such places as New Bedford and Fall River in Massachusetts. By 2000, it was the sixth largest city, just behind Philadelphia. Many other cities outside the old, industrial Northeast and Midwest regions had also grown rapidly. (Indeed, ten cities rose at least ten places in the rankings between 1950 and 2000 to enter the top twenty cities.[2]) Among the cities rising into the top ranks over the past half century are Houston (rising to number 4 from number 14), San Diego (rising to 7 from 31), Dallas (rising to 8 from 22), and San José (rising to 11 from below the top 100). Called "elastic cities" by former Phoenix Mayor David Rusk in *Inside Game/Outside Game* (Washington: Brookings Institution, 1999), their central cities contain more of their metro area populations than the older cities. Compared to the ten biggest cities, the ten "new cities" had only

---

[1] In order: New York, Chicago, Philadelphia, Los Angeles, Detroit, Baltimore, Cleveland, St. Louis, Washington DC, and Boston. All calculations by authors from the 1950 and 2000 Integrated Public Use Microdata Series: Version 3.0, Minnesota Population Center.

[2] In order of their size in 2000, these "new cities" are Houston, Phoenix, San Diego, Dallas, San Antonio, San José, Indianapolis, Jacksonville, Florida, Columbus, Ohio, and Austin.

3.1 million people in 1950, one-seventh of the biggest cities, and their metro areas 4.6 million. By 2000, their metropolitan populations had surged more than fivefold to 20.7 million. These cities, too, had become increasingly diverse, with whites making up only about half of the metropolitan population.

These massive demographic transformations were accompanied by an equally profound restructuring of urban economies. Blue-collar activities dominated the employment base of the ten largest cities in 1950. Manufacturing accounted for more than one-third of the jobs, transport, communications, and utilities another tenth, and wholesale trade another 5 percent, for a total of half the employment. (More than half of all workers had crafts, operative, or service worker occupations.) Some 58 percent of all the jobs in the metro areas were located in the central cities. By 2000, blue-collar activities employed only one in five, with manufacturing having declined to a mere 8.5 percent of the labor force. Only a third of the workers reported a blue-collar occupation. Moreover, jobs were even more suburbanized than the population in these large, old metropolitan areas. (More than three-quarters of the durable manufacturing jobs were located outside the central cities.) Instead, the mainstay economic sectors of the large, old cities were retail, financial services, professional services, health services, and educational and social services. Almost half the workers living in these cities were professionals, managers, or administrative support workers. The dramatic growth of the new metropolitan areas was driven not only by the growth of service activities, but also by post-war high technology sectors such as computer equipment and software, telecommunications, and aircraft production.

Such massive transformations in demography, population, and location could well be expected to produce a great deal of political turmoil and change within and across the nation's cities. The following six readings outline the main dimensions of these changes. This section looks at what happened within the large, old, central cities as they stood on the precipice of change. John Mollenkopf originated the term "pro-growth coalition" in his study of how the leaders of large firms centered in downtown business districts came to partner with a new generation of mayors and urban developers to use such tools as urban renewal, public–private development partnerships, and highway construction to re-engineer the spatial layout of their cities' economies and neighborhoods. A key motivating factor was their fear of the "blight" encroaching up on central business districts and important institutions such as universities and medical centers in the form of growing black and Hispanic communities. He shows that public policy played the midwife for the emergence of the post-industrial central city economy.

As the populations and electorates of these cities shifted from being predominantly white, ethnic, and blue collar to being characterized both by white managers and professionals on the one hand and by black and Hispanic service and administrative workers on the other, a series of new tensions and political accommodations were struck. Using the fiscal crisis of New York City as a lens on this process, Martin Shefter examines how the imperatives of public management cross-cut one another, particularly in terms of the tension between attracting new forms of investment (and revenues) and serving the growing working-class minority population on the other. Savitch, Kantor, and Vicari provide a comparative perspective on the imperative to promote investment, noting the importance of fundamental institutional arrangements and the ways in which those arrangements strongly promote inter-city competition in the United States. The extremes to which this can be taken in the American setting are parodied in a short selection from *The Onion*.

The drive to transform old central cities by tearing down structures, land-use patterns – and neighborhoods – deemed to be outmoded provoked a large-scale counter-mobilization from community organizations. Peter Dreier, who sees this dynamic not only as a social scientist but as a former housing and development policy advisor to the mayor of Boston, traces out the ways in which grass-roots activism has influenced urban politics in Europe as well as the U.S. The section concludes with Amy Lind's appreciation of how women have arisen as important leaders in such movements in North and South America.

"The Postwar Politics of Urban Development"

from *Politics and Society* (1975)

John Mollenkopf

Editors' Introduction

In "The Postwar Politics of Urban Development," John Mollenkopf brings together a number of analytical threads to shed light on the dramatic rise in grass-roots activism found in central cities in the late 1960s. Mollenkopf's analysis covers the entire U.S. experience, but focuses in particular on four cities in two metropolitan areas: Boston and Cambridge in Massachusetts, and San Francisco and Berkeley in California. All these cities, like so many others, suffered the decline of manufacturing and residential exodus to the suburbs, but, in American cities, these also proved to be among the more resilient. The strength of their university sectors, their surprisingly robust downtown commercial real estate markets, and their regional importance as financial centers prevented them from declining too dramatically. Nevertheless, these communities experienced much of the fiscal stress, economic decline, and political tension found elsewhere.

At the outset, Mollenkopf notes one of the sources of tension found in urban politics. Cities are, first and foremost, economic entities, sites of production and exchange, whose built environments reflect the requirements of economic efficiency for the eras in which they were built. They are valued for their usefulness in carrying out such exchange, and land within the city has real economic value based on the desire of economic actors to conduct business in that particular space. However, despite these overarching economic imperatives, cities aren't just profit-making engines: they are also places where people live, engage in cultural and social lives, and develop communal feelings and expectations that might be unrelated to the city's market functions. Some of the political strife found in cities, both in the 1960s and today, grows from the tensions between the different functions of cities within any society. The significance of these tensions has been further explored in John Logan and Harvey Molotch's 1987 book *Urban Fortunes* (Berkeley: University of California Press).

This reading also demonstrates the importance of the federal urban renewal program on several levels. First, urban renewal represented both a political and a policy response to the decentralization trends that, even by the 1940s, seemed about to render some city centers obsolete. Growth-oriented mayors and their politically astute, aggressive new redevelopment directors teamed with those business interests least likely (or able) to flee the city. Together, they lobbied Congress to create a federal program that would facilitate their making changes to the built environment that, they hoped, would make their cities more attractive sites for post-industrial investment. In many cities, projects carried out with urban renewal funds served to dramatically reshape the urban environment – that was certainly the case in Boston, San Francisco and Cambridge, among those studied in this selection, as well as in other older cities such as New York and Philadelphia.

But the significance of urban renewal springs not merely from its policy "successes," but perhaps even more from the side-effects of these "successes." In city after city, urban renewal was used to clear poorer neighborhoods, especially poor neighborhoods located near potentially valuable central city land. Mollkenkopf demonstrates how few resources went toward relocating those in the path of urban renewal, and how many urban residents were forced out to make way for the corporate centers, university campuses and upper-income housing preferred by those running urban renewal programs. The displacement that accompanied urban renewal was to have a profound effect upon cities. The poor became poorer (and more racially segregated), as the displaced crammed into adjacent neighborhoods, but they also began to mount a more determined resistance to top-down urban renewal policies. Mollenkopf notes how often grass-roots mobilization grew out of community efforts to resist large-scale clearance policies. Those directly and negatively affected by urban renewal were joined in their resistance by others (middle-class preservationists and liberals) who, together, challenged the public–private alliances that had pushed the renewal agenda. By the late 1960s, the pro-growth coalition had essentially broken apart in most cities, urban renewal was in retreat, and subsequent urban policies would rarely depend on large-scale clearance. In short, urban renewal's opponents had permanently altered local politics and federal urban policy approaches.

This article presages Mollenkopf's fuller treatment of the connections between federal policy, local urban renewal programs, and community responses in *The Contested City* (Princeton University Press, 1983). There are a number of good studies of urban renewal and its aftermath. See Raymond Wolfinger, *The Politics of Progress* (Englewood Cliffs: Prentice Hall, 1974); Martin Anderson, *The Federal Bulldozer* (Cambridge, MA: MIT Press, 1964), and James Q. Wilson, *Urban Renewal: The Record and the Controversy* (Cambridge, MA: MIT Press, 1966). There are also countless studies of the impact of urban renewal on specific cities. Among these are: Harold Kaplan, *Urban Renewal Politics: Slum Clearance in Newark* (New York: Columbia University Press, 1963); Thomas O'Connor, *Building a New Boston: Politics and Urban Renewal 1950–1970* (Boston: Northeastern University Press, 1993). Clarence Stone's *Regime Politics* (Lawrence: University Press of Kansas) offers a case study of urban renewal as a key project of the city's regime in Atlanta, and Robert Caro's *The Power Broker* (New York: Vintage, 1974) presents a highly detailed, journalistic account of Robert Moses's leadership in New York City's urban renewal program.

During the 1960s, a dual political crisis unfolded in America's big cities. Across the country, neighborhood groups mounted campaigns to halt urban renewal, provide decent housing for themselves, and reclaim the public institutions. These groups squatted in vacant housing, stormed public hearings, mobilized thousands for marches and mass meetings, sat in front of bulldozers, and generally made life miserable for public officials. Starting in 1964, ghetto rebellions also swept from Cleveland, Watts, Newark, Detroit and Washington to dozens of other cities, generalizing this community-based political crisis. To make matters worse, the national government found itself increasingly unable to sustain support for imperialism abroad and liberal myths at home. The revolt of the neighborhoods thus became part of a widespread, though ill-defined culture of insurgency.

While these waves were beating against City Hall, pressures began to undermine the mayoral coalitions that had structured urban politics since the mid-1950s. Increasingly militant civil servants pressed for higher pay and better working conditions. Neighborhood property owners insisted on improved services at lower tax cost, igniting a "taxpayer's revolt." Reformist critics took well-aimed potshots at the customary ways of doing city business. Mayors found their habitual solutions to such conflicts, namely increased federal funding, increased taxes, reduced services, and stimulated economic growth, to no avail.

Big city mayorships were never as secure as other electoral offices, but in 1969 alone such prominent mayors as Lee (New Haven), Cavanagh (Detroit), Naftalin (Minneapolis), and Allen (Atlanta) left office. . . . The prominence of those deposed suggests that the late 1960s were a time of growing instability in city politics.

Why did these two developments take place? What is their significance? Urban violence and protest generated much "instant research" but few convincing explanations. Some blamed these results on the civil rights movement and [the federal Office of Economic Opportunity's] "stirring up trouble," while others focused on "alienation," "frustrated expectations," "pervasive racism," and similar attitudinal arguments. Some attributed the destabilization of city political coalitions to white ethnic reaction to black neighborhoods' transgressions of tacit boundary lines.

Such views are both conceptually incomplete and empirically inadequate. Why did people develop these ideas and attitudes? What specific mechanisms propelled them to act in the mid-1960s, rather than ten years earlier or later? Was this activity, as Bell and Held argue, merely an over-heating of the democratic cauldron? To answer these questions, a larger causal context must be established. The dual political breakdown of the 1960s was part of a larger dialectic between local government's attempt to solve a central city land value and revenue crisis and its striving to manage the unanticipated, but sharp, political consequences of the "solution" chosen. In short, the dual political crisis grew out of a struggle over the nature of urban development.

A FRAMEWORK FOR UNDERSTANDING URBAN POLITICS

First and foremost, the city must be analyzed as the main location in which production, distribution, and the accumulation of wealth take place. This strongly colors urban institutions which, in turn, exert their particularizing influence on urban economic activity. The physical layout of cities reflects, albeit in a complex and incomplete manner, the requirements of efficient production and exchange. Gary, Indiana, and Pullman, Illinois, were laid out with such ideas specifically in mind. Although the city planning profession never lived up to the dreams of Le Corbusier, it sought to reduce economically "irrational" elements of older, more haphazardly-developed cities.

But cities are more than networks of streets, sidewalks, office buildings, and other public and private investments designed to produce maximum returns. The people who live in them inevitably express their human needs and pleasures in cultural, social, and geographic ways. By turning their environments to human as well as economic ends, they develop a sense of communal enjoyment.

Such communal feelings can collide with the demands of production. The neighborhood saloon, with its time-honored rowdiness, clashed with the orderly, sober work habits desired in the factory. In the Midwest, an entire corporate-based urban reform movement was launched to "Americanize" immigrant workers. The battle to control Pullman, Illinois led to a nationwide rail strike in 1894. As circumstances dictated that cities be refashioned after World War II, such struggles became increasingly prominent.

Between 1945 and the 1970s, cities experienced a transition from an industrial to a post-industrial economy, converting the central city from a location of industrial work to one dominated by office-based command and control activities. This has been reinforced by the emergence of world cities. It seems clear that the modern corporate-based central city requires more city services (i.e., transportation, traffic management, utilities, economic development and manpower training among others) than its nineteenth century counterpart.

The location of residences and economic activities has also changed substantially since 1945. Suburbanization has reshaped the urban political arena, stimulated new political conflicts, and posed unexpected difficulties for metropolitan economies. Migration of rural, poor minorities to the central city, the dispersion of manufacturing, warehousing, and sales activities to the metropolitan periphery, and the relatively declining central city tax base all played a part in the 1960s community mobilization and political instability.

Local and national government actively promoted these contradictory changes. By setting up the Federal National Mortgage Association, the Home Loan Bank Board, and the FHA and VA loan programs, and subsidizing home ownership through the tax laws, as well as spending on urban renewal, urban freeways, and mass transit, the federal government financed these changes and provided individuals with a strong incentive to undertake them. One result, however, was to jeopardize central city property values and to precipitate a chronic central city fiscal crisis, leaving the cities and the federal government to reinforce downtown development and bridge the revenue gap while paying to maintain central city social peace.

As a result, the government of metropolitan areas became increasingly fragmented and segmented.

The nineteenth century reform movement took many elements of local government activity out of the public arena by establishing non-partisan, business like city charters, appointed commission, and the like. Independent authorities (e.g., redevelopment authorities, port authorities, and transit authorities) proliferated in the period after 1930. With the end of annexation, suburban governments also proliferated.

The interaction of these four tendencies shaped the decline of political stability in the big cities. National changes in the spatial structure of economic activities and federal policies aimed at converting the city from production to consumption precipitated a crisis in central city real estate shortly after the second world war. One aspect of this crisis was "the cancerous effect of the slums." This set the stage for a new breed of mayors, bureaucrats, large corporations, central business district real estate and merchant interests, and the construction trades to forge pro-growth coalitions. These coalitions pushed nationally for a strong urban renewal and highway program, and locally for downtown redevelopment. Innovative mayors like Richard Daley, Richard Lee, Keven White, and Joseph Alioto, joined with strong, not to say dictatorial, renewal administrators like Robert Moses, Edward Logue, and M. Justin Herman to change the skylines of every major city.

Unfortunately, their successes in reinforcing the command and control functions of the central city imposed tremendous costs on central city residents. They displaced stable communities, exacerbated racial tensions, imposed heavy taxes on those least able to pay, and increased commuting time, congestion, and pollution. We can trace the dual urban political crisis of the 1960s directly to the consequences of pro-growth policies.

The remainder of this essay uses the case studies of Boston, Cambridge, San Francisco, and Berkeley to analyze the post-II World War politics of urban development. These metropolitan areas are centers of finance, high technology, high-level service activities, and corporate headquarters. The satellite cities of Cambridge and Berkeley house major centers of education, research, and development, but, like the central cities, they retained a legacy of factory work and ethnic diversity in the years following World War II.

THE CENTRAL CITY FROM 1940–1956: CRISIS AND RESPONSE

In 1950, "urban crisis" referred to the central cities' loss of population to the suburbs, their growing minority populations, their expanding black slums, and the threats to their property values. By 1970, the term had expanded to cover social problems like racism, poverty, crime, and poor housing, as well as political chaos and growing doubts about whether cities could overcome "negative externalities" like traffic congestion, pollution, and real estate disinvestment. It was, in short, a political crisis as well as a fiscal crisis.

Central city property values peaked about 1930, only to be knocked down during the Depression. Mayors and central city bond-holders who had narrowly averted default during the Depression and downtown business interests had ample reason to worry about the post-war future of the central city. During the 1940s and 1950s, central cities experienced a growing gap between their expenditures and their revenues. . . . To finance rising expenditures for existing services and compensate for a declining tax base, Boston boosted assessments on commercial and industrial property to 100 percent of true value and raised the tax rate to $86 per $1,000, leading the business community to complain that a further increase "beyond this level will spell disaster."

The tax base in all four case study cities peaked around 1930 and did not recover until much later. While they made small gains during the 1950s, assessed values did not turn sharply upwards until the late 1960s. Immediately after the war, rising costs for city services and the need for capital investment butted up against limited tax bases. As a result, all four had to raise tax rates sharply.

At the time, analysts saw this fiscal and property value crisis as a result of metropolitan dispersion and the growth of poor black and other minority neighborhoods that required expenditures rather than contributed revenues and "threatened" neighboring property values. The appropriate policy seemed clear: eliminate "blighting" slums, stimulate investment in the central business district, and provide the transportation infrastructure necessary to keep the CBD "viable." And the only source of this policy, it seemed equally clear, was the federal government.

In his testimony before the House Banking and Currency Committee on the 1949 Act, Housing and Home Finance Agency Director Raymond Foley stated that urban renewal should be enacted because:

> The mayors and other city officials daily face the problem of heavy municipal expenditures for essential municipal services in slum areas which far exceed the taxes derived from those areas. As new building is forced to the periphery of cities and the tax base in cities and the tax base in the central city areas decreases, they face the problem of constantly increasing municipal outlays for capital improvements and additions required to serve the newly developed areas. They do not have access to the financial resources required to absorb the full costs of the necessary write down in anything approaching the volume that is required for effective slum clearance operation.

Urban renewal was to throw a wall around the growing social problem of the minority urban poor to preserve and enhance central city land values. It was no wonder that the program was frequently executed with racist overtones.

In September, 1954, President Eisenhower appointed a five person Advisory Committee on a National Highway Program.... Their report, "A Ten-Year National Highway Program," led to the passage of the $56 billion Interstate and National Defense Highway Act in 1956. Though this program was largely administered through State Highway Departments, planning for urban freeway construction was a highly political affair. Not only were congressmen and local political officials influential in helping to determine route and interchange locations, but so were big city chambers of commerce. Together with auto-oil-highway construction-trucking interests in the national economy, they formed a formidable alliance for the construction of urban freeways.

All four case study cities embraced these two programs, but it was not clear that they were taking doses of the right medicine. These programs might be effective for real estate owners, but it is highly doubtful that they could resolve problems like crime, delinquency, or the demand for social services.

THE DEEPER ROOTS OF THE POST-WAR URBAN CRISIS

[...]

Boston, San Francisco, and their satellites Cambridge and Berkeley, are long-developed, older central cities. Though the latter contain large student and professional populations, they also contained industrial areas and significant black populations. Cities of this type were hit hardest by the national shift from production to services, and all four lost jobs in construction, manufacturing, transportation, and sales during the 1950s.

Boston and San Francisco are major capital markets and house numerous corporations, banks, and insurance companies. Cambridge and Berkeley contain large educational, technology, and research and development establishments. These sectors made possible strong employment gains in clerical work, government, and professional occupations, as well as finance and real estate, especially after the 1960s. But these trends were not accidental. They resulted from the plans of the pro-growth coalitions.

[Raymond Vernon and other regional economists] hold that technological and transportation forces dictated suburban development. Similarly, urban historians traced suburbanization to streetcar line extensions, speculative home building, and the decline of annexations in the late nineteenth century.

These forces are undoubtedly real, but suburbanization was also driven by firms' search for advantage over their workforces and their milieux. This characterized metropolitan dispersion since the mid-nineteenth century, but it seems to have accelerated in the period after World War II. While the evidence is fragmentary, it seems that relocation production and housing to the suburbs provided a way to absorb the productive capacity developed during the war and prevent a return to the Depression. The war years had also led to serious labor problems, overcrowded housing, pent-up housing demand, and new racial antagonisms. The location of large manufacturing establishments in working class neighborhoods beset by all manner of problems, spurred corporations to set up new small plants in safer suburban locations.

The role of these concerns in the Bay Area was clear in a 1948 debate published by San Francisco's

Commonwealth Club. Chaired by the ex-president of the National Association of Home Builders, this session discussed reasons for urban decentralization. As the vice-president of the American Trust Company stated:

> Labor developments in the last decade may well be the chief contributing factor in speeding regional dispersion of industry, and have an important part in the nationwide tendency toward industrial decentralization. In this period good employee relations have become a number one goal. Labor costs have expanded markedly. Conditions under which employees live, as well as work, vitally influence management–labor relations. Generally, large aggregations of labor in one big [central city] plant are more subject to outside disrupting influences, and have less happy relations with management, than in smaller suburban plants.

The report quotes the California State Reconstruction and Reemployment Commission's 1946 report "New Factories for California Communities:" "Workers could own their own homes and enjoy contentment, leisure, lower living costs, and better health. . . ." The managers of many large and small plants which have located in Santa Clara County testify that their employees are more loyal, more cooperative, and more productive workers than those they have had in the big cities. Another manufacturer said even five hundred employees was too many to have in one place. "We are going to get away from tenements, traffic jams, high taxes, crowded streetcars, and transient labor, with all the economic waste and irritation those things involve. The development of the electronics and aircraft industries on the San Francisco Peninsula reflected these feelings.

Similar patterns developed in metropolitan Boston. Cabot, Cabot and Forbes effectively stimulated development of the electronics industry along Boston's Route 128. The interlocking directorates, financing, and the presence of old Brahmin families in some of these firms suggest that Boston's corporate establishment quite consciously planned this development.

Black migration to the urban north began around 1920, with the reduction of European migration. During the 1940s, 1.6 million blacks migrated city-ward, more than the total of the previous thirty years, in response to wartime labor demands. This inserted blacks into the Northern urban economy, but then deprived them of jobs and housing as veterans returned in the post-war period. In the 1950s, mechanized agriculture forced even more blacks towards the north. Between 1949 and 1952, the demand for unskilled agricultural labor in the Mississippi Delta counties dropped by 72 percent. The 1950s were thus a time when northern urban blacks struggled to establish a sense of community, elaborate communal institutions, and carve out an economic niche. By 1968, blacks were more urbanized than whites and formed an increasingly large component, along with other minority populations, of the central city citizenry.

All four case study cities developed black neighborhoods and they were expanding more by natural increase than migration by the 1960s. By the time that the influence of the pro-growth coalition was most devastatingly felt, these black communities had passed a generation from their conservative Southern roots and had begun to establish a more independent, community-based leadership.

In the period after World War II, all four cities lacked strong political leadership. Local government did not seem likely to live up to the demands being placed on it. Instead, they were dominated by haphazardly corrupt, functionally isolated bureaucracies with no particular goal except muddling through the deepening crisis. This set the stage for a new turn in city politics. National policy had responded to central business district needs with two key tools: urban renewal and freeways. Though central city corporations, real estate and retailing interests were organized and well aware of the need for action, they had not made connections with big city mayors.

THE RISE OF THE PRO-GROWTH COALITIONS, 1955–1970

Big city politicians and businessmen forged new alliances around central city development between 1955 and 1970. Robert Salisbury argued that a "new convergence of power" was developing in city politics, based on "an executive-centered coalition" of businessmen, progressive mayors, "and planning-oriented" technocrats. One common element proved to be strong leadership from renewal executives, who learned how to mass forces in secret and launch lightning attacks on the chosen territory. As Robert Caro's widely-heralded biography shows, Robert Moses invented the basic techniques, the independent authority, secret planning, well-timed deadline manipulation, and good Washington contacts. Edward Logue, first renewal director under Lee in New Haven, then under Collins in Boston, and finally director of New York's Urban Development Corporation, refined these techniques and endowed them with political sophistication. He trained a bevy of administrators who went on to direct renewal in Washington, D.C., Miami, New York, and Cleveland. These leaders developed large new bureaucracies commanding money, technical expertise, and manpower.

As Roger Friedland has shown, nearly every major city developed also developed a corporate-based urban planning body during the 1950s. More often than not, these groups included executives from the city's largest corporations; many raised funds to hire staff members and conduct studies of "proper" urban development patterns.

[...]

A third element in this alliance proved to be a new generation of growth-oriented mayors. Richard Daley was elected mayor of Chicago in 1955. As Mike Royko points out, "the fastest way to show people that something is happening is to build things" and Daley did. New Haven's Mayor Lee was elected in 1953 on a campaign of modernizing the city. He too succeeded a relatively weak mayor, and managed to forge decent working relations with the city's Democratic organization. Lee, a former Yale and Chamber of Commerce public relations officer, subsequently piled up larger election margins upon putting renewal into action.

These mayors have a liberal, technocratic outlook, come from professional careers, and have strong ties to the local business community. Yet they also come from ethnic backgrounds and bank on ethnic appeal. Almost without exception they are Democrats, and if they begin their careers without the support of old-line city bureaucracies and regular party organizations, they have usually been able to build ties with them. They have introduced techniques like program budgeting, program planning, and computerized management systems. They

have skillfully shaken the federal money tree for resources to build parallel bureaucracies, often outside of city civil service, as a political base.

Other elements also joined this coalition: city labor councils and construction trades councils, regular party organizations, realty interests, and good government groups. Their participation, however, was more variable and less decisive. In the late 1950s and early 1960s, a growth platform could be touted as a panacea. Since the negative consequences had not yet fully emerged, attention could be focused on rising tax bases, construction jobs and contracts, new housing, and expanded central city institutions. Local business could see the wide range of benefits. Renewal, therefore, commanded, for a time, a working if not general, consensus.

The four case study cities embraced the pro-growth coalition during the late 1950s. Boston's first pro-growth mayor was John Hynes, who defeated James Michael Curley in 1951 with the aid of the New Boston Committee, a business-based reform group. Faced with a deepening fiscal crisis, bankers forced Hynes to cut city employees by 5 percent when he attempted to secure loans for city operations in 1957. Hynes initiated the now-infamous West End Project, one of the first massive slum clearance projects in the country. It displaced over 2,600 families. He also put together the Prudential Center project, comparing it to rolling away of the stone in front of Jesus' tomb, with support from one of Boston's leading real estate firms, Cardinal Cushing, and Charles Coolidge, partner in Boston's most prestigious law firm.

In 1959, a second important election occurred. One candidate, machine Democrat John E. Powers, threatened, if elected, to declare bankruptcy to get the city out of its fiscal jam. This threw shivers down the spines of the city's Brahmin bond-holders, and a business group, soon dubbed "The Vault," organized to put John Collins into the mayoralty. Originally conceived as receiver should the city default on its obligations, the group raised a substantial amount of money to back Irish Democrat Collins. He won.

Among the Vault's members were Gerald Blakely (prime mover behind Route 128 for Cabot, Cabot and Forbes), Ralph Lowell (retired chairman of the Boston Safety Deposit and Trust Co. and director of numerous Boston corporations), Carl Gilbert (chairman at Gillette and Raytheon director),

Lloyd Brace (chairman of the First National Bank of Boston), and various other bankers, insurance company executives, retailers, and utility executives. With their backing, Collins improved city administration and trimmed 1,200 more employees from the payroll. But most of all, he hired Ed Logue as redevelopment director and initiated the "New Boston" renewal program that ultimately subjected 10 percent of the city's land area to redevelopment.

In Cambridge, the universities and their high-technology industrial allies provided the corporate organization for urban renewal. The city's ethnic politicians pushed M.I.T.'s plans to develop Tech Square and Kendall Square in an effort to "Blitz the slums," as then-Mayor Edward Crane put it. The universities provided technical assistance for renewal applications, set up a non-profit corporation to spur housing development, and undertook their own substantial development programs. The net result was to reshape Cambridge's composition by the end of the 1960s. Most industrial plants, including Riverside Press, Simplex Wire, and Biltrite Rubber, were bought by the universities and developed for new university and high technology research and development uses. Not only were many families and small businesses displaced, but incredible stresses were placed on the working class housing stock.

San Francisco's Mayor Joseph Alioto was a comparatively late arrival, having been first elected in 1967, but had done duty as chairman of the redevelopment agency in the late 1950s. Alioto put together a coalition based on big labor, big real estate, and big corporations, with substantial minority neighborhood support for good measure. This alliance gave strong backing to renewal during the mid-1960s, but it is necessary to go back much further to establish the essentially non-partisan sources of San Francisco's pro-growth coalition.

In 1945, leaders of the region's major corporations founded the Bay Area Council, upon which membership was open only to chief executives, including those of the Bank of America, American Trust Company, Standard Oil of California, Pacific Gas and Electric, etc. The BAC concentrated primarily on two issues: regional transportation, which had been in a shambles during World War II, and industrial location, including urban renewal. It issued a number of important studies,

and developed business consensus most importantly for the development of the Bay Area Rapid Transit District, or BART. BART's impact on the Bay Area is a complex story of government action at corporate behest. BART's $1.6 billion dollar capital investment will influence Bay Area development for decades alone, and strongly reinforces the San Francisco CBD.

Some of its San Francisco members founded a second committee in 1956, the Blythe-Zellerbach Committee, to back urban renewal. It in turn set up a broader group, S.F. Planning and Urban Renewal Association (SPUR), to build support largely among the city's professionals for urban renewal. In the late 1950s B-Z gave money for renewal related studies to the City Planning Department, and in the early 1960s, SPUR was designated the official citizen participation unit for renewal in San Francisco.

As a result of these business initiatives and the resulting alliance they forged with Mayor Alioto and his two predecessors, San Francisco, like Boston and Cambridge, launched into massive urban renewal efforts. During all three administrations, M. Justin Herman, previously a HHFA official responsible for overseeing renewal in the western regions, provided strong leadership for renewal. Indeed, community spokesmen often berated him as a dictator, a charge Herman's usual response did little to disprove. With backing from B-Z, SPUR, major city property owners, and most trade unions, Herman undertook to redevelop large areas adjacent to the central business district, the city's produce market, its Japanese neighborhood, its major black neighborhood, and a variety of other sites. Though community opposition prevented entry into the Mission District, San Francisco's Latino neighborhood, in 1967, these other projects were destined for completion.

Berkeley established another pro-growth orientation. Unlike the other three cities, Berkeley is not a center of corporate activity or banking. Rather, it is an odd combination of middle class suburb, working class black neighborhood, and university "ghetto." Berkeley has the most "professionalized" bureaucracy of all four cities and the most ideologically charged electoral politics. A city-manager city with a strong progressive history, there were no machine politicians in Berkeley to build a corporate–ethnic alliance on behalf of renewal. Instead, Berkeley had for most of the 1960s a self-made millionaire engineer for a mayor. This official, Wallace Johnson, stood slightly to the right of Barry Goldwater and espoused non-political, business like administration. He achieved office during a highly-divisive 1963 campaign concerning open housing (which Berkeley defeated). During the student riots in the late 1960s he advocated "riot wardens" drawn from the city's more responsible classes to keep order.

A vigorous opponent of Berkeley radicalism, Johnson backed private enterprise and renewal. He provided political support for the controversial West Berkeley Industrial Park, which would have displaced a number of black families, and a bayshore shopping center and marina, strongly opposed by environmentalists. More importantly, Johnson spearheaded a campaign to bury BART lines in Berkeley rather than have them run on elevated structures. This campaign resulted in a cut-and-cover operation which displaced substantial amounts of housing; BART itself, of course, proved to be a major spur to Berkeley highrise development. Johnson and Berkeley Real Estate Board allies proved to be among the strongest supporters of BART.

As in Cambridge, the university acted as a stimulant to development. Under Clark Kerr, University enrollments rose rapidly after World War II. With them came university construction, pressure on the housing market, and a shift in merchandising to the student market. Like Cambridge, the university promoted high-technology and research and development firms. And as in Cambridge, Lawrence Radiation Laboratory and other university efforts were linked with military-industrial activities. Unlike the other cities, however, no full-blown pro-growth coalition ever developed.

These cities suggest certain basic themes. One is the clear corporate planning initiative, which develops the basic renewal scheme, often with corporate resources, and then sells it to the right bureaucratic and political figures. It may even recruit the executives to operate the renewal program. The second theme is the emergence of growth-oriented mayors who seize the corporate-inspired development plans as a program on which they can create a strong organization. The developers and central business district interests which benefited from urban renewal proved to be the largest campaign contributors to the growth-oriented politicians

in each city. But more was involved than a simple graft relationship – pro-growth mayors were able to build a much broader base for their regimes. Renewal provided manpower, benefits, resources, and latent political support, for any mayor audacious enough to reach for them.

The initiative for the pro-growth coalition thus came from two sides: corporate planning interests on the one, dynamic and aggressive politicians on the other. The extent to which these jelled determined how massively a city moved into renewal. Three of the four case study cities produced a solid coalition. Boston, Cambridge, and San Francisco rank 4th, 7th and 10th nationally, after New Haven, in terms of renewal funds per capita. Boston and San Francisco contain the two largest residential renewal projects in the country. These cities, as we have seen, were governed after the mid-1950s by a strong pro-growth coalition. In Berkeley, where the material conditions for such a coalition were weaker, the thrust for renewal was weaker, but not absent.

URBAN STRUGGLES AGAINST THE CONSEQUENCES OF GROWTH, 1965–1975

The pro-growth coalition engineered a massive allocation of private and social resources. Between 1949 and 1971, the federal government committed over $8.2 billion in direct outlays and more than $22.5 billion in bonded debt. By 1968, private investors had sunk an estimated additional $35.3 billion into 524 renewal projects across the country. In addition, some $70 billion has been expended on interstate highways, a substantial portion of which went to high-cost urban areas. As a result of these investments, more than a quarter of a million families have been displaced each year. They have received only $34.8 million in relocation payments, or less than 1 percent of the direct federal outlay!

Even those who defend the renewal program on other grounds admit that "the result is a regressive income redistribution," according to one analyst, "with lower-income groups who consume at the lower end of the housing stock suffering the most." More than one million people per year were being displaced in the late 1960s. Assuming an equal chance for everyone, and no one displaced twice, this means that in a ten-year period, fully 6.3 percent of the urban population was displaced. Of course chances are not equal, so residents of poor neighborhoods near central business districts probably are three or four times more likely to suffer. Displacement carries out of pocket expenses and leads to higher rents, psychic trauma, sundered friendships, and more crowded low-rent housing. Much of this damage was done in the cities under examination.

The Western Addition A-1 and A-2 projects involve $100 million in public money, and have displaced eight thousand people; the Yerba Buena convention center has also displaced thousands. Boston's South End redevelopment area, rated at $37 million in public money, ranks among the top three residential projects and has also displaced thousands of people. In Berkeley BART's construction has displaced about a thousand low income people. In all of these areas, renewal and highway construction demolished far more low rent housing units than were ever replaced. Equally important, the re-uses by and large served the central business district and dominant government and educational institutions.

In Boston, renewal in the South End was designed to produce "maximum upgrading," to use Edward Logue's words, in a housing stock adjacent to the CBD, a hospital complex, and the newer office developments in the Back Bay. Other large Boston renewal projects cleared land near Massachusetts General Hospital, for a new Government Center office complex, and for the Prudential Life Insurance Company. In San Francisco, renewal made possible the Golden Gateway Center, upgraded housing near City Hall, and a large-scale convention and sports center/office building complex. Renewal in Cambridge cleared white working class neighborhoods to make way for office building complexes sponsored by M.I.T. and Cabot, Cabot and Forbes. In Berkeley, renewal removed low-rent housing in West Berkeley for an industrial park.

This shift of territorial organization triggered a whole chain of disruptive consequences for the central city, most of which fell upon poor neighborhoods or pitted newer minority areas against older ethnic communities. Not only did growth mobilize those directly and adversely affected, but

it stimulated racial antagonisms as the displaced sought housing in other areas.

The neighborhood response

As time passed, the seamy side of urban development started to seep out from under its glossy covering. As a sign on a vacant lot created by renewal in Cambridge said,

No War Declared,
No Storm Had Flared
No Sudden Bomb So Cruel,
Just a Need for Land,
A Greedy Hand
And A Sign
That Said,
"Urban Renewal"

Though Boston's West End residents failed to mobilize against that city's first massive clearance, they banded together in the many neighborhoods to which they were displaced, warning that renewal portended destruction for working class neighborhoods.

Almost every instance of neighborhood mobilization has its roots in struggles over growth. Most of the community turbulence of the 1960s was directed against urban renewal, highway construction, the declining availability of decent, inexpensive housing, expansion of dominant institutions, and city bureaucracies tightly dominated by ethnic groups being displaced in the urban population by minority newcomers.

The four cities demonstrate the importance of struggles over growth. In Boston, community protests arose against urban renewal, construction of the Inner Belt highway, inadequate housing, hospital expansion, and rent control. In Cambridge, the issues were largely the same, coupled with opposition to Harvard's and M.I.T.'s expansion. In San Francisco, protests emerged against urban renewal in the Western Addition and the Mission, against housing quality, in favor of community control over housing development, and against racism in the education and law enforcement bureaucracies, and in various firms' hiring practices. Finally, in Berkeley movements developed against renewal, displacement from BART, community control over development of areas around BART, police brutality, and political control of the city as a whole (with students and blacks pitted against suburbanites).

The same factors contributed to the wave of riots across urban America between 1964 and 1968. Virtually all of the riot areas were sites of major renewal efforts; quite frequently struggles over renewal lurked behind the riots, as in the case of Newark's planned new medical school. On a-city-by-city basis, there was a strong correlation between city expenditures for renewal and the incidence of riots between 1964–1968. Where riotous violence occurred in the four cities, it was in neighborhoods strongly affected by urban renewal.

These movements had an alternative conception of how urban development might occur, at least within their neighborhoods. Although this conception never completely jelled, it tended to have the same elements. Opposition to market allocation of private housing, opposition to planning for businesses rather than people, desire for community control of major government services, calls for rent control and government-subsidized housing, and experimentation with local self-development were repeated in all four cities. In battling development, new neighborhood-oriented institutions like tenant unions, advocacy planning bodies, tenant self-management corporations, and social service advocacy organizations grew up. Though many failed or were drawn into the mechanics of traditional city bureaucracies, many retained an important role in the neighborhood and modestly yet obviously pointed towards how a whole society might be organized along alternative lines.

THE BREAKDOWN OF THE PRO-GROWTH COALITIONS

The second part of the mid-1960s dual political crisis, namely the internal breakdown of the pro-growth coalition's network of alliances, has three basic parts. First, neighborhood mobilization slowed down government decision-making and threatened electoral consequences. In response, mayors, realizing they needed a more sophisticated approach, began to allow groups a mostly symbolic part in the policy process, and to set up parallel but for the most part powerless bureaucracies like Model Cities. This introduced discordant elements into city hall.

Second, constituencies which previously supported growth began to swing into opposition. Fiscally conservative elected officials urged that commuters be taxed more fully for the costs of growth, while urban upper middle class professionals opposed the transformation of "their" cities and mixed neighborhoods. The climate of public opinion, once unified, became divided. Middle class blacks in Boston, Harvard liberals in Cambridge, and ecology-oriented activist professionals in Berkeley and San Francisco all mounted anti-growth campaigns.

Finally, business and labor both withdrew support for the pro-growth coalition because both disfavor, and to some extent fear, the politicization of their activities. Neither wished to accept the public control, environmental impact reports, neighborhood vetoes, and affirmative action programs which were emerging from the battles against growth.

From 1969 onward, these forces clearly sapped local pro-growth electoral coalitions. In general, they have survived only where growth's negative consequences were relatively slight (perhaps because development was more sophisticated politically), or where pro-growth mayors have demonstrated exceedingly clever political skills. To mention some famous cases, "cop" mayors took over in Minneapolis, St. Paul, and Philadelphia, and law-and-order candidates came close to winning in Newark (Anthony Imperiale), Boston (Louise Day Hicks), and New York (Mario Procaccino). Blacks managed to win in Gary, Cleveland, Newark, Detroit, Los Angeles, and Atlanta. Shrewd pro-growth mayors, particularly Alioto in San Francisco, White in Boston, Uhlman in Seattle, and Landieu in New Orleans, managed to reconcile opposing sides and avert losses to blacks, middle class anti-growth exponents, or law-and-order advocates, but they often did so by the skin of their teeth.

White and Alioto both employed Model Cities, OEO, job programs, and similar federal patronage resources to sustain minority support while acting tough on disorder and boosting growth. They shrewdly allowed some "community control." In the last analysis, however, their staying power remains in doubt. In Cambridge and Berkeley, an alliance of poor neighborhoods, radical students, and middle class people upset by the depredations of growth, united to elect "radical" mayors and city councilors.

What are the consequences of this declining potency of the pro-growth coalition? Aside from the fact that city politics has been opened up to all manner of electoral ventures, the main consequence has been to throw up barriers against growth. In Berkeley, this influence has accomplished this quite concretely, as it were, by building barriers across many of the city's streets.

In response to neighborhood protest in the mid-1960s, federal policy and local practice with respect to renewal changed substantially. Relocation rights were strengthened, neighborhood advisory committees required, the amount of subsidized housing increased, affirmative action hiring propounded, and many renewal projects delayed, restructured, or killed altogether. New zoning and other impediments to development were thrown up in various neighborhoods in the cities under consideration. To summarize the matter, quite real and serious obstacles were developed to thwart some of the worst consequences of growth. While many of them were designed simply to "buy off" neighborhood opposition and allow the realization of corporate ends, others have proven to be substantial impediments.

[. . .]

CONTINUING TENDENCIES

The pro-growth coalition took on its most public role in cities where state action was most necessary to promote large-scale urban changes. These cities are primarily large, old, and highly ethnic and blue collar, with large but worn capital investments which needed to be replaced. The newer, fast-growing cities of the South and Southwest, by contrast, had no outmoded built environment that needed to be dismantled. In Phoenix, Houston, Los Angeles, and San Jose, growth interests needed only an efficient, business-minded city government, and not even a particularly powerful one. Alliances among major corporations, banks, and developers proved to be all that were needed, and they could operate largely outside public scrutiny. In these cities, growth-oriented mayors survived much more easily than their Northeastern colleagues.

In the cities which experienced the biggest reaction to pro-growth policies, a variety of outcomes appeared. Some mayors managed to paper over the divisions between growth proponents, neighborhoods suffering from growth, and unruly public workers. In other cases, conservative, white ethnic, small property-owning candidates won office on the basis of a revolt against rising taxes, neighborhood disruption, and growing black influence. These "law and order" and "anti busing" candidates made an important but probably not lasting imprint on urban politics.

A third tendency includes black mayors who, having foregone support from (white) regular organizations, must find it in the same platform as their pro-growth predecessors no matter how much black neighborhoods might oppose renewal. When all is said and done, renewal can be used as an important source of patronage for black elites. Black mayors in Newark, Detroit, Cleveland, Atlanta, Gary, and Los Angeles found it necessary to elicit support from their business communities, if only to keep firms from leaving their cities. Thus though black mayors have opposed the depredations of renewal and white ethnic-dominated bureaucracies, their abilities have been strictly circumscribed.

The fourth tendency rejects both growth and reliance on orthodox local political organization based on public employees: so-called "radical" candidates. Aside from the university-dominated cities, environmentalists have sometimes joined with threatened neighborhoods to overcome both business and bureaucracy. Because they were not rooted in the patronage structure of the cities' bureaucracies, however, these candidates also found themselves unable to substantially change the course of local government. At most they could veto widely opposed projects; positive achievements were another matter altogether.

[. . .]

Social scientists have attempted to explain metropolitan decentralization with market models. For them, individuals simply made a choice about lifestyles given a range of means and opportunities. This perspective has difficulty evaluating the content of urban policy and tends to think markets would do a better job (alas, if they could only be introduced!).

This chapter rejects such views. Market explanations remain essentially circular unless we specify the context in which the market operates. In this case, markets function in the midst of a basic conflict between using urban form for human purposes and using urban form for efficient production. New transportation technologies or plant locations are introduced not merely to reduce marginal costs, but to increase control over the environment of work. The up-grading of inner city neighborhoods occurs not simply because young lawyers and business-people find them quaint, but because city economic systems are predicated on concentrating command functions and making them accessible to a relevant workforce.

Of all the contextual factors which structure markets, the state is the most important. Land use patterns are inherently public matters. They cannot be set up without roads and sewers, police and fire protection, and regulations about how owners use their property. As a result, land use questions inevitably tend to become political. The politics of land use is perhaps the central theme of U.S. urban political history.

In order to promote land values and city revenues, business, government, and to a lesser extent organized labor forged a pro-growth alliance. This alliance achieved a class-based transformation of land uses, which then triggered severe tensions within and outside the growth alliance. . . .

That situation provid[ed] an opening for a new, more progressive coalition within city politics. . . . On what grounds could such a new alliance be built? The missing link has been a political movement explicitly based on putting new spending priorities and land use patterns into place rather than merely redistributing patronage. An alliance between public service producers and consumers would encourage each to be directly responsive to the other rather than insulated by brokerage-oriented politics. It would seek to change the framework of decision-making and the values implicit within it rather than reducing expenditures here and increasing them there. Whether such an alliance can emerge remains to be seen, but when all is said and done, this remains the task of city dwellers who would like to live in a truly humane city.

[. . .]

SELECT BIBLIOGRAPHY

Bell, Daniel and Virginia Held. 1969. "The Community Revolution." *Public Interest* 16 (Summer), 142–177.

Friedland, Roger. 1982. *Power and Crisis in the City.* (London: Macmillan).

Salisbury, Robert. 1964. "Urban Politics: the New Convergence of Power." *Journal of Politics* 26 (November), 775–797.

Vernon, Raymond. 1967. "The Changing Economic Function of the Central City." In James Q. Wilson, ed. *Urban Renewal* (Cambridge: M.I.T. Press).

FOUR

"Political Economy of Fiscal Crisis"

from *Political Crisis/Fiscal Crisis: The Collapse and Revival of New York City* (1985)

Martin Shefter

Editors' Introduction

"Ford to City: Drop Dead" read a famous October 29, 1975 newspaper headline. The newspaper was the New York *Daily News*, and the headline was a crude summary of President Gerald Ford's response to New York City's request for aid in the face of impending bankruptcy. New York is hardly the only major U.S. city in modern times to find its expenses outstripping its revenues: Philadelphia faced a major budget crunch in 1991 (see Bissinger's reading in Part 6), and cities like Pittsburgh, Buffalo and Detroit routinely confront rising expenses and lagging revenues. Even European cities, although they are far less dependent on their own revenue-raising capacity, face pressures to remain fiscally solvent, experimenting with privatization schemes and cuts in traditional municipal services to reduce costs (see "Municipal Mayhem," *The Economist*, August 16, 2003, p. 31 for some German examples).

Fiscal pressures, and in particular conflicts between the state's need to remain economically healthy while also providing increasingly generous social benefits, are not unique to cities: there is a rich literature on the fiscal crisis of the state that looks at the structural underpinnings of these endemic tensions (see Mollenkopf's reading in Part 3 and Claus Offe, *The Contradictions of the Welfare State* (Cambridge, MA: MIT Press, 1984); James O'Connor, *The Fiscal Crisis of the State* (New York: St. Martin's Press, 1973)). But in this literature, "fiscal crisis" remains a fairly abstract concept – advanced capitalist nation states experience pressures that lead to political and social conflict, but they are not actually at risk of going broke. For cities, the day of fiscal reckoning is very real, and often very close. After all, as Paul Peterson's reading in Part 3 makes clear, national governments can better control the levers of economic policy (money supply, tax rates, immigration flows), and ultimately can run deficits if necessary.

Shefter's work picks up on Peterson's analysis of the fiscal and economic factors limiting city behavior, but it adds a rich political analysis to help us understand urban fiscal woes as a complex and perhaps inevitable outcome of capitalist democracies, at least in its American variation. He outlines four "imperatives" that city officials must heed if they want to retain power. Two are economic in nature: they must maintain the health of the local economy, and preserve the city's credit. Two are political: they must generate votes and contain conflicts. Shefter compares the effort to heed these four imperatives to a difficult juggling act, in which any shift in the environment – layoffs from a large local employer; an influx of new immigrants, a national recession – can destroy the equilibrium that keeps those four balls in the air. What happens to leaders who fail to observe one or more of these imperatives? At the very least they are voted out of office. In the worst cases, their cities suffer the consequences of job loss, service cuts and/or social disorder – and *then* they are voted out of office. Heeding these imperatives, however, is not simple. Cities may need to cut services or raise taxes to balance their budget, but the mayor who adopts this plan is likely to encounter political resistance. Election-seeking officials may be tempted to make costly promises to win votes, but fulfilling these promises can lead to budget shortfalls.

Fiscal Crisis/Political Crisis takes us through New York City's shifts between machine governance and political reform, and shows how this political dynamic helps explain the city's frequent battles with fiscal stress. Certainly, as Shefter's history shows, fiscal crises do have structural origins but the city's swings between surplus and deficit have a clear political dynamic. Although Shefter describes himself in one of the book's footnotes (not reprinted in this excerpt) as more a structuralist than a pluralist, his analysis leaves far more space for political contingency than we find in the work of Paul Peterson or Claus Offe. City leaders can find many ways to appeal to voters, and they can adopt different strategies to piece together a winning electoral coalition. Once they are in office, they make many decisions about how to keep those juggling balls in the air – and their choices will undoubtedly reflect the interests of their electoral base, and their calculation of which groups they can afford to alienate.

Martin Shefter is a Professor in the Department of Government at Cornell University. His research addresses the development of American political institutions at the local, state, and national level. Most recently, Shefter has become interested in the influence of international forces in shaping U.S. political institutions, and he has co-edited (with Ira Katznelson) *Shaped by War and Trade: International Influences on American Political Development* (Princeton University Press, 2002).

Fiscal crises are distinctively urban phenomena. Although the federal government's expenditures have exceeded its tax revenues much more often than not since the 1930s, Washington has never faced the problem of being unable to finance its deficit. That American cities have confronted this problem many times raises the question of why there is such a difference in the nature of the fiscal problems faced by Washington and those faced by urban governments. Moreover, because municipal fiscal crises erupt periodically, a pair of puzzling questions are raised. If cities periodically face severe fiscal problems, why aren't these confronted on an ongoing basis? And, if cities are able to maintain a balance between their expenditures and revenues during the decades between fiscal crises, why does this equilibrium periodically break down?

The answer to these questions lies in recognizing that city officials are subject to a variety of imperatives. The necessity of winning votes and maintaining civil order can lead city officials to increase public expenditures at a rapid pace. However under normal circumstances the necessity of preserving the municipal government's credit and promoting the health of their economy compels them to restrain the growth rate of locally financed municipal expenditures.

Because cities have considerably less autonomy than does the national government, economic and political developments occurring beyond their borders can easily upset the balance between municipal expenditures and revenues, as can changes in the structure of local political coalitions and organizations. Since municipal governments, unlike the national government, cannot print money, a city can only finance its deficits as long as investors are prepared to purchase its securities. If investors fear that a municipal government lacks the economic ability or political capacity to redeem its securities, they will not lend it more money – precipitating the sort of fiscal crisis New York City has confronted a half-dozen times in its history.

Fiscal crises characteristically discredit the city's top elected officials and lead to their defeat. The shock of defeat often convinces local politicians that they must acquiesce to changes in the political practices and public policies that enabled their opponents to triumph. The political alliances local politicians engineer, and the fiscal policies they pursue to regain and retain control over the city government, can then endure until further changes in the national or international economic and political systems again tempt them to increase municipal spending more rapidly than municipal revenues, which will spark yet another fiscal crisis.

THE IMPERATIVES OF URBAN POLITICS

If they are to gain and retain power, city officials must heed a variety of imperatives. Urban politicians have compelling incentives to pursue policies that will (a) generate votes, (b) maintain the health of the local economy, (c) preserve the city's credit, and (d) regulate and contain conflicts among the city's residents. These are not imperatives in the sense that public officials cannot but heed them, but the penalty for failing to do so can be severe. The imperative of vote generation is the most obvious of these – elected officials must win more votes than their opponents if they are to gain control of the municipal government and hold on to the perquisites of power for themselves and their political allies.

Another imperative for an urban regime is securing sufficient revenues to finance the operation of the municipal government. Local taxes are a major source of municipal revenue, and their proceeds vary with the health of the local economy. If local businessmen find that municipal taxes are too high, or if the public facilities and services the municipal government provides to their firms (or to their customers and employees) are inadequate to make the city an attractive place in which to do business, they may close down or leave town. Other firms, for the same reasons, may not take their place. Consequently, the city government may be deprived of revenues sufficient to finance its current activities, a problem in itself for public officials and a threat to their ability to continue providing their political supporters with the public benefits they expect. Beyond this, employment opportunities, income, and the general well-being of city residents are tied to the vitality of the local economy, and voters are likely to reward elected officials who can claim to have contributed to that vitality. In seeking to serve a variety of interests, then, public officials are driven willy-nilly to pay heed to the interests of the city's most substantial taxpayers and its major employers.

Similar considerations lead mayors to be concerned with whether their city is regarded as creditworthy by the municipal capital market. Cities sell notes and bonds for two crucial purposes. One is to cope with short-term divergences between municipal revenue flows and expenditure obligations. Municipal tax revenues and grants from the state and federal governments generally flow into a city's treasury at widely spaced intervals – in many cases only once a year – whereas every month cities must pay their employees, suppliers, and often their creditors. To meet these obligations, cities sell short-term notes backed by the revenues they are scheduled to collect later. Money is also borrowed to finance the construction of public works (such as bridges, water and sewer systems, and schools) and the purchase of capital equipment (such as police cars, fire engines, and sanitation trucks). Selling long-term bonds for this purpose enables a city to pay for these projects and equipment over the life of the items in question. Investors, of course, will only purchase a city's notes and bonds if they are confident that any money lent will be repaid, which requires public officials to conduct the city's financial affairs in a manner that will instill such confidence. Failure to heed this imperative would deprive the municipal government of the credit needed to pay its monthly bills and acquire facilities that make the city a place in which families can comfortably live and firms can profitably do business.

City officials must also contend with conflicts arising from disagreements concerning the entitlements and obligations of members of the community vis-a-vis their fellow citizens and the municipal government. These conflicts can be fought in the streets – in the form of crime or riots – or in the electoral and policymaking institutions of government. City officials generally seek to control conflicts occurring outside the established institutions of government because they generally regard the protection of life and property as a primary mission of government, and because their failure to do so would alienate many voters and investors and encourage higher levels of government to intervene in the city's affairs. City officials also often have an incentive to reach compromises with their opponents that will limit the intensity of conflicts occurring within the institutions of government. By doing so, officials can reduce the costs and uncertainties inherent in all-out electoral battles and discourage the losers in policy conflicts from calling upon outside authorities to reverse decisions made locally. Finally, because urban regimes characteristically do not incorporate every interest in their city as full partners, their stability depends in part upon their ability to forestall or cope with opposition from these excluded groups.

THE OPEN CITY

Municipal officials can find it difficult to meet these imperatives because cities lack autonomy judicially, economically, and politically. In the juridical realm, cities do not have the legal authority to issue visas or impose quotas on immigration, to enact tariffs or manipulate interest rates, and, of course, to use military force to defend themselves against outsiders who wish to intervene in their affairs. Because they cannot insulate themselves from various national or international developments that have important implications for the interests of city residents, the claims made by city dwellers upon municipal governments in an effort to defend their interests are greatly influenced by such external developments. For example, because municipal governments cannot regulate the flow of people across their borders, events occurring elsewhere in the nation or world may alter the very composition of a city's population. The potato famine in Ireland in the 1840s, the anti-Jewish pogroms in Russia in the late nineteenth century, and the mechanization of agriculture in the American South following World War II, led millions of Irish, Jews, and blacks to move to American cities. This had major consequences for the size and character of the labor force available to employers in large cities, for the competition faced by long-term residents for public jobs and services, and for the problems with which municipal officials had to contend if they were to maintain public order. The efforts of both old-timers and newcomers to protect or advance their interests in the face of these developments shaped the agenda of politics in America's largest cities for well over a century.

Commodities and capital, as well as people, can move easily across the boundaries of cities, and the limited economic autonomy of cities can have important consequences for urban politics. A substantial proportion of the business firms in any large city compete in national or international markets, and the prosperity of even those that do not is dependent upon sales to firms (or to the

employees of firms) in the city's "export" sector. As conditions in national or international markets change, the businessmen operating in these markets are likely to want to alter municipal policies in directions that will enhance their firms' competitiveness. They may seek improvements in the facilities for transporting goods and people to and from work sites, shifts in the character of the services the municipal government provides to business firms and their employees, reductions in municipal tax burdens, and so forth. But even in the absence of any overt pressure, city officials can ignore the effects of municipal policies upon the standing of local firms in wider markets only at the peril of seeing these firms leave town or go out of business, depriving the city's residents of jobs and decimating the municipal government's tax bases.

The political system of a city is no more autonomous than its economy. Participants in local politics often seek power in cities in an effort to increase their influence in national politics. Conversely, groups that have allies at the state or national levels may be able to draw upon the legal authority or material resources of these higher levels of government to enhance their power locally. Finally, the views of various actors in city politics concerning the purposes and the proper domain of government, and their entitlements relative to it, are shaped at least as much by national political currents as by local ones. Cities are components of a broader political regime, and there is a two-way flow of political ideas across their boundaries, just as there is a flow of people, commodities, and capital.

POLITICAL COALITIONS AND REGIME CAPACITY

Politics influences a city's response – or lack of response – to external forces and imperatives in a number of ways. Although city officials may face dire consequences if they fail to heed these imperatives, some officials lack the political vision or the ability to alter established patterns of behavior, and hence suffer those consequences. If the incumbent political leaders in a city fail to satisfy a majority of the electorate, they will be defeated at the polls. If they fail to pursue policies that enable local firms to adjust to market changes or policies that encourage new firms to set up shop in their city, the local economy will decline. If they are unable to preserve public order, higher levels of government may intervene and reduce the authority of the municipal government. And if the municipal government does not remain solvent, it may be placed under some form of receivership.

Political factors influence not only whether municipal officials will successfully respond to the imperatives confronting them, but also the precise manner in which they will respond. There are a variety of ways to match municipal expenditures and revenues, establish a balance between the local public and private sectors, and resolve conflicts among a city's major social groups. Opting for one or another of these alternatives will win officials the approval or the enmity of different interests in the city. The choices public officials make among these alternatives are influenced by their judgment concerning whom they can least-or most-afford to alienate. These decisions are influenced by the leverage that various interests are able to exercise in their dealings with the city government – for example, the number of voters sharing that interest, or the assistance that interest is able to provide to public officials, or the injury an interest is able to inflict upon City Hall by virtue of its control over capital or credit.

There is one qualification that must be made to the last statement, but it is one that, if anything, further emphasizes the important role that political choices play in the fiscal affairs of cities. Although elected officials must win more votes than their opponents if they are to gain and retain power, they have some leeway in determining how to put a winning coalition together. A voter can perceive his or her interests in a variety of ways – as a consumer of this or that municipal service, as someone who is saddled with paying one or another local tax, as a member of an ethnic group or social class, or as a resident of a particular neighborhood. Politicians thus have some freedom in deciding how they will appeal for the support of a given group of voters as well as in determining who to appeal to for political support. They can assemble majority coalitions by seeking to win the support of any one of many possible combinations of voting blocs, while conceding the others to the opposition.

The responses of an urban regime to the imperatives confronting it are a function not only of the strategic choices made by its top leaders, but also of the regime's institutional and organizational capacity to enact and implement decisions made at the top. If a regime is to enact a consistent set of policies it must be able to overcome, regularly and reliably, the dispersion of authority among public officials. To implement these policies, it must be able to influence the behavior of the "street-level bureaucrats" – policemen, teachers, clerks, and others who actually perform the tasks of municipal government – and give them the legal authority and material resources necessary to perform their jobs. Finally, if a regime is to mobilize a reliable electoral majority, it must have the means to communicate with voters and induce its supporters to go to the polls on election day.

It must be emphasized that these preconditions are not always fulfilled. Officials in policymaking positions may regularly fall into deadlock; they may be unable to control the behavior of their nominal subordinates; street-level bureaucrats may be incapable of performing the missions they are assigned; and elected officials may be unable to turn out a reliable majority of voters on election day. In any of these events, the behavior of a municipal government will be influenced by whatever pattern of relationships does emerge among the city's political actors, institutions, and organizations. Such a regime, however, is not likely to navigate easily through the cross-currents that could capsize it.

CITY POLITICS AND URBAN FISCAL CRISES

For all of these reasons, those who govern a city must engage in a delicate juggling act. This act is not impossible to pull off if voters recognize that they reap benefits, such as jobs and lower taxes, from policies designed to bolster the local economy, and if members of the business community recognize that their ability to recruit skilled employees and to attract prosperous customers depends in part upon the municipal government's providing a level of public services that makes the city an attractive place for such people to live. Moreover, in cities that rely upon the real property tax for the bulk of their locally generated revenues and in which home ownership is widespread, voters may be, if anything, more tax-conscious than local businessmen. Local officials are likely to find that by keeping municipal taxes and expenditures at a minimum, they can simultaneously please the city's voters, businessmen, and creditors, and avoid sparking any rancorous political conflicts. But even when there are conflicts among these constituencies, municipal officials may arrange compromises that the major social forces in the city may not regard as ideal but are prepared to live with. Under certain conditions, however, it may be difficult for city officials to keep aloft all the balls they juggle. A possible consequence of this is a municipal fiscal crisis.

Cities face fiscal crises when two conditions prevail. The first is when changes in the economic, demographic or political environment make it difficult for local officials to meet the most immediate of their imperatives without spending more money than they collect in taxes and intergovernmental aid. The second condition is when the city's creditors refuse further financing of the resulting budget deficits.

Cities are prone to run budget deficits when cyclical downturns in the a national economy, structural changes in a city's economy, or the emergence of competing centers of manufacturing or commerce lead to widespread unemployment. Municipal officials may then find it politically advantageous, or necessary for the sake of preserving public order, to provide some form of relief to the unemployed and their families. The very economic conditions that lead to such expenditures, however, may make individual taxpayers and local businessmen loathe to pay higher taxes to finance them.

Unless local officials are able either to get the state and federal governments to finance relief for the unemployed or to make compensating cuts in expenditures for other municipal programs – which is difficult to do even during a depression – the combination of higher expenditures and declining revenues will produce a budget deficit. New York provides a striking example of how difficult it is to cut existing expenditures, even during a severe depression. Although the real income of the city's employees rose as prices fell in the early 1930s, the administration found it politically impossible to cut the nominal wages of municipal employees in order to cover some of the costs of unemployment relief. Indeed, cutting municipal expenditures is so difficult that declining revenues can lead to budget deficits even in the absence of a municipal effort to provide relief to the unemployed.

Another circumstance in which municipal expenditures increase more rapidly than revenues is when local officials, in an effort to mobilize political support (or to forestall opposition) among an ethnic or racial minority whose members previously had received less than their proportionate share of public benefits, increase the flow of benefits to the group in question. Alternatively (or simultaneously), a change of national administration may enable newly powerful political forces to press City Hall to channel additional public expenditures to their local allies. In either of these cases, the political weight of the group receiving benefits may not be great enough to enable local officials to finance these increased expenditures by reducing expenditures that benefit other constituencies or by raising taxes. At this point, City Hall might be tempted to engage in deficit spending.

Under a number of circumstances, then, city officials may want investors in municipal securities to finance some of the costs of meeting the imperatives of vote generation, order maintenance, and protecting the strength of the city's economy. Under what conditions are potential investors likely or unlikely to play this allotted role? The limited judicial autonomy of cities makes this more of an open question for municipal governments than it is for the national government, which has the exclusive authority to create legal tender. The federal government can borrow whatever is needed to finance its deficits, because if necessary it can print money to prevent a default on United States Treasury securities. By contrast, because cities cannot create legal tender, they can default on their loans. If potential creditors doubt that a city will be able to repay the money it borrows, they may refuse to lend the funds the city may require to continue functioning.

What occurs – or fails to occur – at each step in the sequence of events precipitating a municipal fiscal crisis is influenced by the structure of a city's political organizations and institutions, the strength and composition of prevailing political coalitions, and the content of prevailing political ideologies. It is not the case, for example, that all cities will respond to an increase in unemployment by providing relief financed with borrowed funds. Whether a rise in unemployment will generate significant pressure for a relief program depends on the views of the unemployed regarding the government's obligations to persons unable to work and their capacity to disrupt civil order or defeat elected officials if these obligations are not fulfilled to their satisfaction. Whether an urban regime finds it necessary to respond to demands of the unemployed depends on the ability of its top leaders to prevent their associates or subordinates from taking up the cause of the unemployed, the willingness of other political forces to coalesce with those advocating a relief program, the vulnerability of incumbent officials to an electoral challenge by the proponents of the program, and, in extreme cases, the municipal government's ability to quell disruptions by the unemployed.

Whether local officials finance relief by increasing taxes, reducing expenditures on other programs, or borrowing money is determined not only by the legal and financial feasibility of each of these courses of action, but also by the relative political costs they would incur by pursuing each alternative. Finally, whether potential creditors are prepared to finance a city's deficits depends upon their estimation of the likelihood that the municipal government will repay the money it borrows. This judgment inevitably is colored by politics. If investors believe that local officials would find it politically impossible to slash public services and cut the salaries of municipal employees for the sake of repaying them, they will refuse to purchase that city's securities, precipitating a municipal fiscal crisis.

A fiscal crisis usually convinces economic elites that the politicians in power are misgoverning their city and discredits incumbents in the eyes of many voters. This often contributes to the formation and triumph of political coalitions that attack these politicians in the name of reform. In most of the nation's largest cities, however, reform administrations have generally been short-lived, because losing control of City Hall gives urban politicians a strong incentive to come to terms with various supporters of the reform coalition. A typical component of such a "post-reform accommodation" is a renewed commitment by these politicians to balance the municipal budget and maintain the city's credit. Subsequent changes in the wider economic and political systems in which cities are embedded, however, can upset this fiscal equilibrium. If it again becomes difficult for municipal officials to meet the imperatives of vote generation and order for maintenance without accelerating the pace of municipal spending, the stage may be set for another fiscal crisis.

"The Political Economy of Urban Regimes"

from *Urban Affairs Review* (1997)

H.V. Savitch, Paul Kantor, and Selena Vicari

Editors' Introduction

It is not uncommon to pick up the newspaper and read that a company is planning to build a new facility or relocate an old one and is choosing from among several potential locations. You will then learn that the candidate locations, eager for investment and employment, are doing all they can to persuade the company to choose them. State or provincial government might be offering a package of tax incentives. City or county government might make land available at below market costs, or promise to fund worker training programs. For example, when Boeing announced in 2001 that it planned to move its headquarters out of Seattle, contending cities rushed to offer packages that would attract the aerospace giant (the Chicago/Illinois package was valued at $64 million over twenty years). Sometimes the "company" in question is a professional sports franchise: in such cases, cities and states are quick to offer new, publicly funded stadiums complete with infrastructure in exchange for the promise of hosting a major league sports franchise.

Outside of the U.S., interregional competition has different characteristics, but is still intense. European cities and regions, for example, have less flexibility in setting tax rates, or determining labor and environmental standards so they are less able to differentiate themselves along these lines. (European Union rules further limit differentials among EU nations.) But cities still compete: for private investment, for quasi-public bodies such as the European Central Bank, for the expensive privilege of hosting the Olympic Games or being named a "European City of Culture." It is clear that some cities tend to fare better in such competition than others. Investors flock to the most desirable cities, paying high rents to locate in their corporate centers even in the absence of significant subsidies, while less privileged cities offer expensive inducements and still have trouble landing new investment.

In this reading the authors use the concept of the urban regime (see Stone's reading in Part 3), and suggest that cities may be characterized by different sorts of regimes, due not merely to ephemeral coalition politics, but also to deeper institutional patterns that structure relations between the public and private sector. Savitch, Kantor, and Vicari draw upon two theoretical approaches. First, they cite Charles Lindblom (*Politics and Markets* (New York: Basic Books, 1977)) who stressed the division of labor between public and private sectors at all levels of government. Second, they bring in the work of Tedd Gurr and Desmond King (*The State and the City* (University of Chicago Press, 1987)) which notes that local government's relationship to the private sector is conditioned by the type and amount of intergovernmental aid available, so that cities within nations that offer a great deal of intergovernmental support have greater autonomy from market forces. Adding an analysis that includes the importance of an engaged population, and that considers the health of the regional economy, the authors establish the idea that some regimes (notably those with strong economies and clear national support) can be much more directed about their development, and more assertive in their bargaining with outside investors. Cities with weaker economies and less generous national governments are more dependent on private investors and poorly positioned to bargain to their own advantage.

Their case studies cover cities in four different nations; the regime found in each grows out of the particular combination of market position, political efficacy, and intergovernmental position found in each city. The authors show that these differences have real policy implications, with the most empowered regimes (e.g., Paris) able to shape development far more than the least empowered (e.g., Detroit). Savitch and Kantor have elaborated upon this analysis, adding additional cases and more analytical variables, in their award-winning 2002 book, *Cities in the International Marketplace* (Princeton: Princeton University Press).

Paul Kantor is Professor of Political Science at Fordham University, Bronx, New York. He is the author of numerous articles, reviews, and books in the fields of urban politics, public policy, and political economy. His books include *The Dependent City Revisited: The Political Economy of Urban Development and Social Policy* (Boulder, CO: Westview, 1995) and *American Urban Politics: A Reader* (with Dennis R. Judd, New York: Longman 2001). Hank V. Savitch is the Brown and Williamson Distinguished Research Professor at the Department of Urban and Public Affairs at the University of Louisville. He has authored numerous books and articles on the subjects of neighborhood politics, national urban policy, and comparative urban development. His most recent work has focused on regionalism (he is the co-editor, with Ronald Vogel, of *Regional Politics: America in a Post City Age* (Thousand Oaks, CA: Sage, 1996)) and cross-national urban policies, with his 1988 *Post-industrial Cities* (Princeton University Press) and the co-edited (with Robin Hambelton and Murray Stewart) *Globalization and Democracy: International Perspectives* (Basingstoke: Palgrave, 2002). Serena Vicari teaches urban sociology at the Bicocca University of Milan. Her fields of interest are urban development and technology and society. She is the author of *La città contemporanea* (Bologna: Il Mulino, 2004), as well as many chapters and journal articles on urban sociology and politics.

In recent years, regime theorists have added realism to scholars' understanding of urban development because they have shown how local political choices matter. Regime theorists focus on the process of inducing cooperation between public and private sectors through the formation of governing coalitions. Regimes are the city's linchpin for catalyzing this cooperation and realizing tangible goals – or as Stone (1989) described it, bringing different sectors together, across institutional lines, to engage in social production.

Yet regime theorists offer few propositions about how specific political and economic contexts might matter in shaping the regimes themselves. Are some types of regimes viable in certain governmental and economic contexts but not in others? Although regime theorists concede the importance of contextual factors in regime politics, they focus on the dynamics of internal decision making to explain economic development policy, leaving the socioeconomic environment as "a source of problems and challenges to which regimes respond" (Stone 1993, 2).

Unless it is specified how different national and international circumstances shape regimes and their policy biases, regime theory has limited comparative utility because the importance of agency may be exaggerated to the neglect of institutional structures in explaining public policy. Changes in policy are explained as a result of regime activities rather than as a consequence of changes in the economic or political environment that constrain regimes. The result is that it is impossible to say how and when politics (in the sense of decision making by governing coalitions) decisively influences public policy. Ironically, this agency bias actually hinders achieving the very thing that regime theorists set out to do – namely, to demonstrate the scope for political agency in urban politics. Efforts to remedy this gap have begun, but investigators are searching for a theory that can link regimes to a wide variety of cities and circumstances.

We pursue this task in our analysis by addressing the difference between agency (autonomy or leeway) and structure (underlying conditions, institutional constraints). We suggest a theoretical framework that relates specific characteristics of local governing regimes to particular political and economic environments within the Western industrial system of advanced democracies. We conceptualize regimes as bargaining agents within different types of liberal-democratic political economies. The framework hardly accounts for all possible types of regimes, but it does signal circumstances that are supportive of particular kinds of governing coalitions, policy agendas, and strategies for realizing regime objectives.

A BARGAINING PERSPECTIVE: DEMOCRACY, MARKETS, AND INTERGOVERNMENTAL FORCES

In our framework, rather than focusing on the internal dynamics of governing regimes, we view regimes in reference to their bargaining contexts. Regimes are treated as governmental agents that function to bargain out the terms of cooperation between the public and private sectors in a liberal-democratic political economy. This means that government must bargain over the conditions for inducing capital investment from private-sector markets to achieve economic goals that are shared by dominant political interests (Lindblom 1977). At the same time, however, government also bargains for political support within popular control systems in which elections, referenda, and other democratic mechanisms require public approval of programs.

From this perspective, what is crucial to understanding regime politics is the distribution of bargaining advantages between public and private sectors in particular places in three respects. First, local communities vary in respect to the democratic conditions within which regimes bid for support to advance public programs. Second, they also may differ in their ability to induce private investment as a result of variation in their market position or competitive position with other localities in inducing capital investment. Finally, cities vary in respect to their intergovernmental environments, with some providing greater assistance than others in regulating the local capital investment process.

As public- and private-sector players compete to determine their terms of cooperation, they take into account the distribution of bargaining advantages – the rules of the game, as it were – that structure their encounters. Over time, democratic conditions, market conditions, and intergovernmental environ-

ments of cities influence the character of dominant regimes as players try to make use of the bargaining resources and opportunities afforded by their circumstances (or the bargaining game). Ultimately, these factors constrain the very composition of governing coalitions (encouraging who does and who does not participate), the mode of bringing about public–private cooperation, and dominant policy agendas (collective versus selective benefits, residential versus commercial development, downtown versus neighborhood investments).

How democratic conditions shape regimes

Democratic local political conditions affect critical dimensions of regime politics. Well-developed popular control systems should motivate (via political competition) the formation of inclusive governing coalitions that limit the power of any one group, especially business interests. The need to mobilize representative coalitions of players and broad citizen support also motivates officeholders to promote policy agendas that will generate diffuse public benefits rather than only the distribution of selective rewards. From the bargaining perspective, well-developed democratic systems raise the cost of securing public approval for governmental cooperation with the private sector by making public support for deals struck by business and governmental elites more uncertain while making elites more accountable.

Nevertheless, democratic conditions alone do not have a determinative impact on regime politics. Even if democratic pressures motivate officeholders to organize open governing coalitions and ambitious social agendas, these goals may be frustrated by other circumstances, especially lack of resources for getting business to cooperate.

How bargaining in the capital investment process influences regimes: market conditions and intergovernmental systems

The ability of a city to induce capital investment is dependent upon its market position as well as assistance from national, provincial, or regional governments. Market position consists of circumstances or forces beyond local control that make cities more or less appealing to private investors. Cities establish stronger or weaker niches in regional, national, or international capital markets. The strength of that niche is determined by location, transportation, or concentrations of specialty services (high technology, corporate finance, and headquarters).

Local governments also face capital markets with varying degrees of intergovernmental assistance made available through fiscal equalization, planning, or administrative requirements. National governments can influence bargaining by providing direct aid, political access, or planning regulations. The existence or absence of these factors have a profound impact on terms of cooperation with private investors.

Some intergovernmental systems are highly interventionist. They impose restrictions on capital mobility and provide fiscal support or other subventions to local government. French governments use regional councils to equalize development in metropolitan areas and to plan infrastructure. Most European cities are integrated within a national structure of budget support and fiscal equalization. French and Italian cities receive more than 90 percent of their revenues from their national governments and rarely are concerned about bond markets or bankruptcy. In contrast, the very limited degree of intergovernmental fiscal assistance from state and national governments in the United States (averaging less than 40 percent of spending) leaves cities more sensitive to local capital movements.

Expenditures by central and regional authorities make cities less dependent upon private capital and can offset business bargaining advantages. Italy's Mezzogiorno (South) has been aided for decades by huge levels of public infrastructure and employment programs sponsored by the national government. In France, national authorities have imposed strict regulatory controls, such as freezing land prices and restricting alternative development, to force private-sector investment into target communities.

Intergovernmental assistance also influences the distribution of bargaining advantages between local governments and business through center–periphery political networks. These networks are

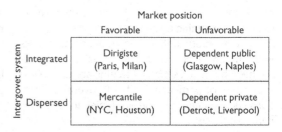

	Market position	
	Favorable	Unfavorable
Integrated	Dirigiste (Paris, Milan)	Dependent public (Glasgow, Naples)
Dispersed	Mercantile (NYC, Houston)	Dependent private (Detroit, Liverpool)

Figure 1 Bargaining context in eight cities.

similar to what Page and Goldsmith (1987) refer to as local access opportunities to national officials and politicians. Such access opportunities include informal networks, multiple electoral mandates, and party organizations that link public elites at local, regional, and national levels.

. . . [T]hese factors [combine to create] four different bargaining environments: dirigiste, mercantile, dependent public, and dependent private. The most advantaged bargaining context is dirigiste. Cities functioning in this context enjoy favorable market positions as well as the support of an integrated intergovernmental environment. The least advantaged bargaining context is dependent private. Cities operating in this context face a weak market position and lack substantial intergovernmental networks. In between are bargaining contexts that provide mixed assets and liabilities and that act upon cities in different ways. Cities in dependent-public contexts have weak economies, but they are able to draw upon substantial resources and access provided by central and/or regional governments. By comparison, mercantile contexts allow cities favorable market positions, but these same localities lack intergovernmental support and compete intensely for capital investment.

Explained somewhat differently, these contexts provide a background that shifts initiative and choice. Dirigiste contexts allow public-sector elites to lead planning and investment and to set priorities. Dependent-private contexts allow cities few choices and instead put initiative with business. Dependent-public contexts keep initiative within public or party organizations, and bargaining is often intragovernmental. Mercantile contexts allow public elites moderate levels of initiative and maximize bargaining on all sides. Thus the dynamics of regime formation are linked to the interplay of democratic conditions, market positions, and

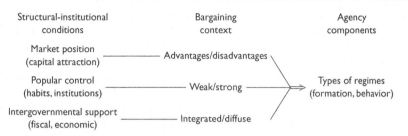

Figure 2 Dynamics of regime formation.

intergovernmental environments, as suggested in Figure 2.

When differences in democratic conditions are considered in reference to the four types of capital bargaining environments, eight different types of contexts emerge that empower regimes that differ in respect to their dominant governing coalitions, their modes of public–private cooperation, and their dominant agendas (see Figure 3). Dirigiste systems yield planner and distributor regimes; dependent-public environments have grantsman and clientelist types; dependent-private bargaining yields radical and vendor cities; and mercantile bargaining leads to commercial and free-enterprise regime politics. These are but ideal types that derive deductively from our bargaining model; undoubtedly, there are other real-world regime types, hybrids as well as some we cannot anticipate because of the limited variables we are considering in our framework.

Although we cannot systematically test our propositions, we are able to illustrate them by drawing upon the experience of five cities in Western Europe (Paris, Glasgow, Milan, Naples, and Liverpool) and three cities in the United States (New York City, Detroit, and Houston) between 1970 and the early 1990s. (2) These cities hardly represent all regime types, but this group of cities does illustrate the range of the critical variables in our framework.

This group includes cities that have well-developed local democratic political systems (Paris, Glasgow, Liverpool, New York City) and cities that are weaker in this respect (Milan, Naples, Houston, and Detroit). (3) The cities also illustrate all four types of bargaining environments because they differ in regard to their market conditions and intergovernmental systems. We will now examine how each of these cities plays in very different bargaining environments that, in turn, influence their regime politics.

DIRIGISTE BARGAINING: PLANNERS AND DISTRIBUTORS

Dirigiste bargaining environments provide regimes with highly favorable economic and political advantages, enabling public-sector elites to play development games that place them very much at the center of regime decision making. In well-developed local democracies, planner regimes are organized to play out the conflicts and rivalries in their local political systems. They draw upon their formidable bargaining resources to organize open governing coalitions that are dominated by party, government, and bureaucratic interests and that hold sway over business participants. The regime's strategic relationship with capital is used to extract social benefits from the expanding capital investment stream

Figure 3 Influence of bargaining contexts of regimes in eight cities.

of the city. Empowered by favorable bargaining advantages, these regimes are able to optimize democratic impulses and use the public sector to pursue aggressive social policies.

Distributor regimes are created when organizational enhancement takes precedence over collective benefits and when popular participation is unable to check party dominance. Political elites capitalize on public resources to dominate local politics and are less likely to organize open governing coalitions or to make ambitious demands on business for collective benefits. Instead, these regimes are content to impose on the capital-investment process to extract and disseminate limited social benefits in ways that enhance their organizations.

Paris as a planner

Paris is probably the classic planner regime. Paris holds a pre-eminent position as one of the most vibrant advanced service economies in France and in the European Union. The city also is governed within one of Western Europe's most integrated urban planning systems. For eighteen years, until 1995, Jacques Chirac and his Gaullist Party capitalized on these bargaining resources to create virtually a state within a state, known popularly as *le system Chirac*. The regime functions in the mode of an interlocking directorate by mobilizing popular institutions and linking them to regional and national government. For nearly two decades, Chirac's Gaullists worked to consolidate power, and by 1995, they held 68 percent of the vote on the municipal council. The mayor delegated overall executive power to just four party loyalists who constitute a quartet of executive power.

The key to *le system Chirac* and, indeed, to all local authority in France is the integrated nature of local, regional, and national power. Known as the *cumul des mandate*, officials are entitled to hold two offices at the same time. Chirac was not only mayor of Paris for eighteen years but during that time also served as prime minister of France. The *cumul des mandate* extends to city councilors, and in 1995, twenty-one of these representatives were also members of the regional council, nineteen were also members of the National Assembly, six were also senators, and four were also government ministers. One can only imagine how powerful a city

council in the United States would be if significant portions of it also served in the House of Representatives, the Senate, both state houses, the national Cabinet and, not least, the White House.

This combination of market power and intergovernmental linkages enable a regime to pursue ambitious state-centered development agendas. The regime coordinates bureaucratic activity and channels public scrutiny into planning and policy decisions. For the most part, business is brought into the process after public objectives have been sorted out.

[...]

Significantly, nothing is built and few changes occur without the watchful eye of city hall. The city adheres to strict architectural codes and limitations on building height. Every development zone ... is duly registered and governed by standards set by city hall and the neighborhood *mairie*. These standards governed development at the direct epicenter of the city when Les Halles was conceived in the 1960s; they governed the construction of La Defense, the new Parisian central business district built by the government in the 1970s; and they govern massive development in southeast Paris today.

Milan as a distributor

Like the rest of Italy, Milan is characterized by strong local influence over development and supported by a high degree of center–periphery integration. Italy's local–national fiscal system enables local government to rely upon the state for revenues, limiting local motivation to use urban development as a generator of supplementary revenues. Land-use laws are extensive and sometimes made at the national level. These regulations can be used by local officials, with access to national government, to shape development if they wish. These advantages are reinforced by Milan's powerful market position. Among European cities, Milan occupied the seventeenth position in 1971 and 1988 rankings. Unlike the other major cities in Italy's industrial triangle (Milan, Turin, and Genoa), Milan has managed to avoid their declining fate by consolidating a diversified mix of production and services, including advanced technology, producer services, fashion, and other leading-edge industries.

Milan is dominated by a highly organized party system that combines partisan, ideological rivalry with patronage politics to constitute a formidable power base for local officials. Unlike France, Italy lacks a prestigious bureaucratic class, allowing the parties a monopoly over decision making. Indeed, the reach of major parties is felt in most areas of Italian society. Parties maintain a strong presence in the economy; they routinely appoint managers in large public companies and financial institutions. Until the investigations of political corruption and electoral reforms since 1992, Milanese voters displayed very stable voter loyalties. Since World War II, all of Milan's mayors have come from the Socialist or Christian Democratic parties, and local government has a long tradition of multiparty coalition building. All of these factors obstruct easy capture of local government programs by business.

Despite these resources, Milan's political classes have not been inclined to seize a very directive role in Milan's development process. The fragmented and patronage-driven party machine limits cooperation, making it difficult to initiate or sustain many long-term economic development programs. Milanese politicians from virtually all parties are absorbed in reinforcing particularistic interests to enhance individual political careers. Party and factional leaders promote plans that are heavy in symbol but modest in substance and use projects to trade favors. As a result, public–private bargaining is difficult, limiting the influence of public officials as well as business interests in development planning.

[. . .]

DEPENDENT PRIVATE BARGAINING: RADICAL AND VENDOR REGIMES

A possible response to hopeless dependency is to organize a radical regime. Leadership gives up attempting to compete economically and focuses on the local political process in hopes of lifting public morale and securing mass support. This response is associated with communities that also have programmatic political parties and high popular participation. Rather than persevere in discouraging capital markets leaders form open governing coalitions that pursue symbolic politics directed at a mass audience.

Another possible response to dependency is to organize a vendor regime. Under these circumstances, regimes are inclined to sell the city's wares at bargain prices. Regimes do this, hoping the city can keep its declining capital from sliding still further. In a period of scarcity, favors to private capital can still yield side payments and selective benefits. Commonly, vendor regimes have low voter turnout, weak popular control, and limited public accountability. The regime is characterized by closed politics, centered around business and apt to grant generous conditions to attract capital.

Liverpool as a radical regime

Liverpool long had a tradition of radical politics that remained at the periphery of power in the city. Immigrant (mostly Irish) workers flocked to the city and had little confidence in a system that they felt was discriminatory. More than this, the city's middle class is weak and has grown smaller with each decade of economic decline. Port traffic in Liverpool dipped sharply through the 1960s and 1970s. The city's manufacturing base shrank and joblessness also rose to depression-like proportions. Absentee ownership left few sources of indigenous leadership, sparse political connections by which to tie classes together, and shrink business commitments to the city.

Until the 1980s, Liverpool lived with these tensions, which were largely ameliorated by central government grants and public employment. Its electorate had been split between Protestant workers (who voted Tory), Catholic workers (who voted Labor), and a smaller middle class (who voted Liberal). Most often government was managed by a "hung council" or loose center–left coalition. The bitter class antagonism was muted by political ambiguities and by a lack of direction.

The Thatcher revolution changed much of this. Upon coming to power in 1979, the Tories took draconian steps toward budget reduction and demanded that localities raise their own revenues. Cuts from central government were especially severe, amounting to nearly 50 percent, with more threatened.

The electorate responded by turning leftward, and by 1982, the Labour party had taken control of the council. This was no mainstream Labour party but one heavily influenced by a Trotskyist faction called the Militant Tendency. While still a minority faction within Labour, Militant Tendency used its organizational coherence to take control of the party and, ultimately, the council.

The new leadership argued that private investment was not responding to inducements and was in exodus from the city. In its view, only the public sector could restore the local economy, and public employment had to be the engine of growth. This view was combined with a strategy that was grounded more in ideology and symbolism than realpolitik. The radicals intended to force Thatcher's government to provide more funds by bankrupting the city and encouraging mass resistance.

The new leadership turned to creative accounting and foreign borrowing to finance major construction programs in housing, leisure facilities, and environmental improvements. Little attention was paid to business or market oriented development. The Labour party refused to cut its spending for public projects and ran large deficits year after year. Most of these initiatives centered on housing construction within the depressed neighborhoods. Militant rule finally ended in 1987, when the Tories took legal action, inflicting heavy fines on the councilors and removing them from office.

Although the Labour party's strategy alienated most of the private sector, it proved successful as a means of building support for the Militant Tendency within party circles, from the city's unions, from council employees, and from the unemployed. Militant Tendency also drew others who were attracted by public-service jobs, social-enhancement programs, and confrontational politics. Labour radicals gained national publicity for their cause, won an enlarged following in the city, and, for a time, managed to undermine alternative leadership.

In effect, Liverpool's strategic development was driven by the internal struggles of the Labour party and by efforts to score ideological victories rather than by an effort to recruit private investment. . . .

Detroit as vendor regime

Like Liverpool, Detroit has also experienced steady economic decline and abandonment. Today the city is financially and politically isolated. Although "golden corridors" have sprung up in affluent outskirts, the suburbs strongly resist the central city. Income disparities between city and suburb are among the highest in the nation and are growing with each decennial census. By 1990, per capita income in Detroit was less than half that of its surrounding suburbs. Over the years, federal aid has shrunk and now accounts for less than 6 percent of the city's budget. Attempts at creating metropolitan mechanisms to share tax bases or undertake planning have failed.

Regime building to cope with Detroit's decline has been colored by relatively weak democratic constraints. Detroit voters took a major step when they elected Coleman Young as mayor. Young's rule was not seriously challenged, and he commanded the city for 20 years [until 1995]. The mayor's power was largely personal, and he presided over a nine-member city council, elected at large in nonpartisan balloting. Voter turnout is less than half the eligible electorate, and Detroit affords scant opportunity for neighborhood expression. The city's singular racial composition became coupled to a politics of black symbolism that muted political opposition to Young's leadership. This was reinforced by Young's success in using jobs, contracts, and city hall patronage to build his political base.

As a new mayor, Young understood that Detroit's business community had a great deal invested in the city. Young tried to capitalize on this influence by forging close links with downtown elites and building a relationship with William Milliken, the former Republican governor. The mayor's efforts met with mixed success but have since been overshadowed by a huge middle-class exodus to the suburbs, declining assistance from other governments, and increasing racial tensions. Distrust between the residents of Detroit and those of its surrounding townships was exacerbated by Young's denunciatory style, his references to "hostile suburbs," and his use of racial rhetoric.

Facing much weaker democratic constraints than Liverpool, Detroit's leaders have played the option of coping with dependency by tailoring their regime to facilitate business privilege. They do this in hopes of extracting limited concessions that can be parlayed to the advantage of public-sector elites who seek to maintain their personal power.

By composition and operation, Detroit's development is lodged with business, facilitated by politicians, and conducted at the initiative of the automobile, banking, and insurance executives. Most of these companies have national or international interests, and personal ties to the city are fast diminishing. Business leaders lay out the sites for development, generate the investment, and determine the terms of that investment. For their part, city leaders (through the mayor and development offices) help locate potential sites and search for inducements to lure private investment.

Renaissance Center exemplifies this politics. It was conceived in the wake of the 1967 riots by 23 business executives. Led by the Ford Motor Company and high-ranking executives from finance, banking, and retail enterprises, the strategy involved a massive, self-contained business complex built along the waterfront and connected to the suburbs by a series of auto routes. By the early 1970s, motor magnate Henry Ford II, financier Max Fisher, and master developer Alfred Taubman put together a "RenCen" partnership worth over $500 million. Other investors included America's insurance industry, some of the nation's largest banks, and General Motors (GM). The city played a peripheral role and was limited to drawing up technical plans for land clearance and preparation. Since then, the Renaissance Center has had a dubious history. The project's backers have defaulted on its mortgage, and it has been sold for a fraction of its investment.

[. . .]

DEPENDENT PUBLIC BARGAINING: GRANTSMAN AND CLIENTELIST REGIMES

Dependent-public contexts provide limited support for cities, and regimes pursue that support in different ways. In cities with a high level of popular participation, grantsman regimes are formed that use public aid for business revival and collective social benefits. This type of bargaining emphasizes intergovernmental contacts and assistance, which serve as substitutes for private investment. Grantsman regimes are then able to limit the influence of business in ways not available to other types of regimes.

In cities in which popular participation is not as strong, clientelist regimes can be produced. These regimes are organized to preserve the privileges of political elites by exploiting public-sector programs

and are used to sustain exclusive, patron–client relationships between politicians and developers (Banfield and Wilson 1963). Collective benefits are negligible, and rewards are heavily skewed toward side payments. The upshot is that projects are not just frustrated but are caught in a cycle of rampant corruption and frequently never get off the ground.

Glasgow as grantsman

Glasgow is among Britain's poorest cities. During the 1970s, the city had rising unemployment (nearly 20 percent) and low per capita income (less than $6,000 per annum). The city has remained a center of unemployment and social decay in a rapidly transforming economic region despite economic improvement of its central business district during recent years. Left-leaning politics has led Glasgow's leaders to try to link conversion of its deteriorating industry with social programs for housing and unemployment. As this strategy was pursued, regime politics became geared toward state-assisted development – for the most part driven by intergovernmental cooperation. The thrust of Glasgow's development has depended on public grants. The city's high level of popular participation and its civic-minded political culture follow the mainstream of British national politics, despite pronounced regional features. Local politicians have cast their lot with regional and national officials, who are able to form strong public-sector alliances to generate economic assistance from the national government. Bargaining agendas focus mostly on chasing grants-in-aid that can be used to leverage private-sector cooperation in planning, redevelopment, and job-creation projects.

Unlike other parts of Great Britain, Scotland has a history of regional policy support. Since the 1970s, Glasgow's leaders have sought assistance from the regional council (Strathclyde) and from a powerful regional development authority, the Scottish Development Agency (SDA). Over the years, the SDA took an increasingly interventionist role, and in 1991, it merged with a job-training agency, renaming itself Scottish Enterprise (SE). The original agency acquired broad confiscatory powers, established special development areas, cleared land, subsidized new investment, and worked on housing and job creation. SDA- and

SE-led projects have been largely state-centered attempts to coordinate local authorities and channel the infusion of money into Glasgow.

Unlike U.S. public-development corporations, SE is not strongly tied to business interests nor is it dependent on private capital or bond markets (despite business-dominated membership on SE's local enterprise boards). The development agencies have drawn their funding from public coffers and profits from their own investments. As a public body, SE is largely tied to regional and national officials, and it is expected to consult widely to mobilize regional and district support.

During the 1980s, the agency's strategies shifted toward greater reliance on the private sector and accorded a lower priority to social investments, such as housing, in preference to downtown business development projects. This shift was not due to business influence in the regime but, rather, largely resulted from national governmental pressures, fiscal constraints, and a more narrow development agenda adopted by SDA planners.

[. . .]

Recently, increased competition and a paucity of public funds have influenced a greater emphasis on attracting private capital. Tighter limits have also been set on the regional agency, limits that have hampered its ability to accommodate regional and district needs. Should the constriction of public funding continue and should private investment pay off, Glasgow could well take on characteristics of a commercial regime.

Naples as a clientelist regime

Naples is one of the poorest cities in Western Europe. Although industrial development in Italy's Mezzogiorno has been disproportionately concentrated in the Neapolitan area, Naples has undergone devastating deindustrialization. Closure or downsizing of state-subsidized heavy industries and the city's failure to attract many new industrial or service jobs have left an economy dominated by the construction industry, small family firms, and artisan shops. These changes have also precipitated a high unemployment rate that exceeds 20 percent. Its population, which is marked by high rates of poverty and disease, has stagnated, and residents search for opportunities elsewhere.

The city's weak economy is heavily dependent on massive state aid. For decades, Naples has been a major recipient of national governmental programs to develop the Mezzogiorno. Along with substantial income-transfer programs, the national government has provided huge amounts of assistance for infrastructure and business development. This assistance is delivered through state enterprises and special public authorities.

Yet neither massive state assistance nor the pressures of economic decline have driven many local politicians to organize grantsman-style regimes because of the city's clientelist political system. Although Italian popular-control institutions offer a wide range of formal possibilities for participation, they afford only a low level of real political accountability in Naples.

Since the 1970s, Naples has had highly unstable governing coalitions of Socialists, Communists, Christian Democrats, and right-wing parties. Since 1993, the Democratic Party of the Left (the reformed Communist party) has taken over. Despite party competition, Neapolitan politics is characterized by poorly developed democratic traditions, a situation that probably is linked to the community's mix of limited social capital, poverty, and weak economic organization. The city's political culture and chronic political corruption beget an inclination to treat government mainly as a source of jobs, favors, and patronage.

Given these weak democratic resources, Naples supports a closed governing regime that exploits its few bargaining advantages – particularly access to big state-aid programs – for only the most limited political purposes. Political elites view economic development as an opportunity for extracting selective benefits that will solidify their power base. They use their connection to national leaders to have access to national financial resources that they distribute at the local level. They actually discourage private investment from sizable and reputable firms not linked to the party system and, instead, encourage activities by marginal enterprises seeking an edge. In this environment, there is also room for organized crime . . . with which politicians strike deals in exchange for votes and payoffs. The presence of illegal organizations dissuades private investment, which, in turn, increases the dependency of the city on state aid – and enhances the position of political elites controlling it.

Rampant corruption has bred consequences for city development. The administration has been unable to provide basic public services. Water shortages continue throughout each summer, pollution plagues the city, traffic lights frequently do not work, and main streets are untended and full of potholes. Naples does not have an urban plan. Market-oriented development is rare and often fruitless. During the 1990s, the city became insolvent, and Naples became the first major Italian city to go broke – although its financial problems and debts were eventually absorbed by Rome.

Public policies in response to the 1980 earthquake illustrate the workings of this clientelist regime. The earthquake's damage and the failure of earlier efforts at regeneration prompted a major shift in state programs, away from investment in large industry and in favor of public works. By 1990, Naples received a 16,000 billion lira (U.S. $10.67 billion) share of the government's massive 50,000 billion lira (U.S. $33.3 billion) reconstruction program. The Piano Speciale per la Ricostruzione provided for an emergency housing program and public works projects.

[. . .]

Public officials adopted a laissez-faire approach to infrastructure development, which allowed developers a large role in planning and carrying out the specific projects. However, the private market was quickly infected by old habits because politicians forged patron–client alliances with both private- and public-sector companies and allocated projects to insiders. Engineering companies, developers, and builders (especially those controlled by the state) grabbed the lion's share of the projects. Later, they successfully worked to increase public funding for more infrastructure spending rather than to pursue market-rate projects that were less profitable.

In the end, these efforts did not spur economic regeneration. Instead, they wasted public funds on incomplete roadways, amputated viaducts, and useless public facilities – accomplished at cost overruns that exceeded 60 percent. To make matters worse, massive public spending diverted private resources from private ventures. The building boom also fed organized crime; underworld families formed construction companies and resorted to violence to claim a share of the public windfall. The results have been disastrous for the city. Naples still lacks affordable housing, many new jobs, and open public space.

MERCANTILE BARGAINING: COMMERCIAL FREE-ENTERPRISE REGIMES

Commercial regimes are cross pressured to incorporate business demands as well as demands for participation by neighborhood and civic interests. These regimes are constrained to be moderately open and supportive of policy agendas that call for collective benefits. Commercial regimes adapt through compromise (negotiations between developers and citizens) or by subterfuge (use of public corporations to shield development).

When popular participation is more feeble, cross pressures diminish and make it easy for free-enterprise regimes to emerge. These regimes are largely driven by economic competition and are manifestly probusiness. They arise in relatively prosperous cities and are supported by progrowth, limited-government cultures. Tensions between business and government or between economic development and political democracy may not always be apparent because political institutions are designed to downplay conflict.

New York City's commercial regime

New York City has a powerful market position. Its reputation as a global capital puts it at the top of the urban hierarchy. The city houses leading banks, international businesses, corporate headquarters, and national media and has a vital downtown. Despite its deteriorating slums, the city maintains a strong civic culture. Since the 1970s, New York City has undergone charter reforms that have created greater access for neighborhood groups. The city's organized, if fragmented, political system is open and highly competitive. The city council is vibrant and is a powerful check on the mayor. Neighborhood councils blanket the city and influence land use, development, and housing.

Regimes in New York City must deal with political and economic cross pressures without substantial policy support from higher-level governments. The city is forever engaged in "place wars" and business pirating from surrounding localities. Located in one of the most fragmented regions in North America, New York City faces a struggle with its suburbs to maintain its job base and has been losing ground steadily during recent decades. Neither the national government nor the state government compensates the city for these losses or replaces revenue losses despite mounting social service burdens.

The need to manage economic competition while contending with activist civic groups has governed regime formation. Government, business, party, and civic groups must somehow get along or exist in contradictory ways. Instances abound in which popular demands are circumvented because governing elites are unable to resist business demands. Public development corporations are used to finance, plan, and implement major development projects. Some of New York City's largest projects have been initiated by the state Urban Development Corporation or the bi-state Port Authority of New York and New Jersey. Both are able to override local government and citizen participatory mechanisms.

New York City's public corporations act in concert with major developers, investment banks, and large businesses. Urban development unites business executives, the heads of public corporations, and the politicians who dominate the regime. This unity depends on the willingness of project sponsors to make strategic concessions or even to abandon projects when public opposition mounts. Coalition building among these interests is commonplace, and they often collaborate to bring projects to fruition.

Collaboration between developers and public authorities marked the reconstruction of the city's epicenter at Times Square, it characterized the creation of a massive complex in Lower Manhattan called Battery Park City, and it governed the development of a high-technology center in downtown Brooklyn. Each of these projects engaged public funding to support private development and was guided by elaborate public–private partnerships.

It would, however, be a mistake to assume development takes place without some popular intervention. Plans to renew Times Square were altered repeatedly in response to protests from neighborhood associations, arts groups, and the like. Battery Park City was affected by successful protests against a massive highway that was designed to connect the project to mid-Manhattan. Renewal in downtown Brooklyn was taken up by community boards and residents. Together, they were able to secure some provision for jobs and housing for moderate-income residents. Thus projects can be stopped, stalled, and, sometimes, shaped by grassroots action. Combined with the uncertainties of finance and unsure investors, popular participation can affect development.

Houston's free-enterprise regime

With fewer pressures from elsewhere and only modest opposition, Houston's business community reigns supreme. Development agendas reflect the progrowth priorities of business leaders, who see local government as an extension of entrepreneurship. Close overlap between government and business characterize social production.

By virtue of its port and its oil and gas industries, Houston holds a strong economic position. More recently, its growth as a major regional center for service and corporate headquarters has made it one of the premier urban centers in the United States. Business interests – particularly developers, real estate interests, bankers, consultants, and lawyers – play powerful formal and informal roles in running city government.

Business-style government is all the more possible because competing interests are weakly organized or lack political access to governing circles. Houston's politics is marked by weak civic traditions. Unions are discouraged by law and practice. There is little collective involvement in development issues. Voter turnout is low, and there is a widely shared sense of individualism and respect for private enterprise – even among blue-collar workers. Houston is one of the few U.S. cities without a general zoning law. In principle, land can be used for anything and is only protected by covenants between sellers and buyers. Indeed, the major purpose of civic organization is to enforce deed restrictions so that neighborhoods can enjoy a modicum of security against zealous development.

Houston's popular-control system is not well organized. Its political parties tend to ignore the city's nonpartisan elections, in which contests are likely to focus on personalities more than programs. Candidates supported by the chamber of commerce, conservative newspapers, developers, and business organizations usually win elections. Mayors are elected for limited terms, making it difficult to build neighborhood or popular constituencies.

This context enables the formation of regimes in which business and governmental elites interlock to support progrowth strategies, even at the cost of sacrificing other public services such as environmental regulation, education, and social services. For decades, city officials pursued low tax policies to attract business and stimulate development. These progrowth strategies were fueled by annexation of suburban districts.

Houston residents are proud of their city's image as a high-growth, low-tax, unregulated city, which gives free rein to business. At the same time, the local government takes laissez faire just so far and has not hesitated to pour money into infrastructure, highways, and commercial services. . . .

STRUCTURE, REGIMES, AND URBAN COMPARISON

Our analysis suggests that regime politics is influenced in systematic ways by particular kinds of bargaining environments. To better understand this,

we step outside the boundaries of regime theory and examine the interplay of local democracy, economic conditions, and intergovernmental institutions on regime dynamics. We suggest these forces have a profound influence on social production by setting the rules, limitations, and opportunities under which bargaining takes place. Using this perspective, we offer an explanation of how regimes vary in their governing coalitions, the manner by which social production is carried out, and their development agendas. Figure 4 summarizes these patterns.

We analyze eight types of regimes, which are constrained by particular bargaining contexts and engage the private sector in different ways. As Figure 4 shows, the vigor of popular participation and democratic institutions is a key factor in determining the particular regimes that arise in a given bargaining context. Planners are likely to emanate in dirigiste contexts, radical regimes are linked to dependent-private circumstances, grantsman types take root in dependent-public settings, and commercial regimes emerge in mercantile contexts.

Structural environment			Regime characteristics		Regime
Bargaining contexts	Popular participation	Coalition partners	Mode of public–private cooperation	Policy agenda for collective benefits	Ideal type
Dirigiste	High	Open (government, bureaucracy, civic)	Government led	High	Planner
	Low	Limited (party elites) patron–client	Party led	Moderate-symbolic	Distributor
Dependent public	High	Open (party, government, bureaucracy)	Government-bureaucracy led	Moderate	Grantsman
	Low	Limited (party elites, patron–client)	Party led	Low (selective benefits, side payments, manifest corruption)	Clientelist
Dependent private	High	Open (party, union, mass)	Party and social movement led	Symbolic	Radical
	Low	Closed (government–business)	Business led	Low (selective benefits, side payments)	Vendor
Mercantile	High	Sporadically open (government, party, civic)	Public–private bargaining	Moderate	Commercial
	Low	Limited (business–government)	Business led	Low	Free enterprise

Figure 4 Bargaining contexts and regimes.

When popular-control conditions are weaker, however, the various bargaining contexts favor distributor, vendor, clientelist and free-enterprise regimes, respectively. This framework can hardly account for all possible types of regimes because it relates regimes to only three critical variables. Nevertheless, it helps to put regimes in their place, as it were. It suggests why several real-world kinds of regimes are not simply by-products of local political dynamics.

Consequently, this approach has considerable comparative utility. The framework draws attention to the politicoeconomic parameters that empower different regimes. Highlighting the interaction of markets and political institutions, it is suggestive of how, when, and why particular kinds of regimes are likely to arise. Understanding structure and context also allows one to explain why particular regimes are able to do things that may not be possible to do in other settings.

Planner regimes, like that in Paris, can impose substantial limitations on business interests and pursue ambitious social agendas. But this type of regime is unlikely except in the most fortuitous political and economic circumstances. By contrast, Detroit's regime operates under an entirely different set of structures and institutions. Detroit's social agenda is far more limited, and no mayor can easily change that circumstance. Do regimes have leeway? Indeed they do, but they must use that discretion within the realities of their bargaining circumstances. Liverpool provides an illustration of how a city was pushed beyond its structural parameters by national government, how a radical regime arose in those circumstances, and how such a regime became a victim of its own actions.

This comparative approach also suggests the likely pathways of regime change. Although the choice of regime depends on the efforts of local politicians to mobilize political cooperation, our framework suggests that these efforts will be easier for some kinds of cities in particular historical contexts than for others. For example, it may be possible for a distributor city, like Milan, to acquire more of the characteristics of a planner city if reforms are made toward citizen accountability and party democracy. Alternatively, Milan could shift toward a commercial-style regime in a mercantile setting if Italy diminishes intergovernmental support programs or if Milan's position in world markets weakens.

In contrast, it is unlikely that vendor-city Detroit will soon acquire many of the characteristics of a planner or even a mercantile city. Its hollowed-out economy portends a weak market position for as far as anyone can see. The provision of federal and state aid programs for the city on the order of its poor European counterparts is remote. However, radical-style regimes are a possibility in Detroit if populist sentiments grow, reform leaders emerge, and democratic politics is enlivened. Similarly, it is unlikely that Naples can acquire the regime characteristics of Houston. Naples' dominant political structures and economic environment would obstruct attempts to convert it into a free-enterprise city and probably would produce entirely different results – perhaps a rise in political protest coupled to new underworld influence.

Of course, these conclusions are necessarily speculative. Indeed, our framework suggests caution in attempting to transfer policies to cities with very different structural soils. It offers only a beginning for more systematic comparative research on regime politics.

REFERENCES FROM THE READING

Banfield, E.C., and J.Q. Wilson. 1963. *City politics*, New York: Vintage.

Lindblom, C.E. 1977. *Politics and markets.* New York: Basic.

Page, E.C., and M.J. Goldsmith, eds. 1987. *Central and local governmental relations.* Beverly Hills, CA: Sage.

Stone, C.N. 1989. *Regime politics: Governing Atlanta, 1946–1988.* Lawrence: University Press of Kansas.

—— 1993. Urban regimes and the capacity to govern: A political economy approach. *Journal of Urban Affairs* 15: 1–29.

"Congress Threatens to Leave D.C. Unless New Capitol Is Built"

from *The Onion* (2002)

Editors' Introduction

In the last reading Savitch, Kantor, and Vicari showed that different sorts of urban regimes are in better or worse positions to bargain with investors who might wish to locate in their city. Quite often, efforts to attract investment appear to be cut-throat, zero-sum games, in which one city's gain is another's loss. Indeed, in the 1980s, a television ad paid for by New York City's economic development office featured the city's mayor Ed Koch boarding up the entrance to the Holland Tunnel, which connected New York to its New Jersey suburbs, symbolizing his willingness to do *anything* to keep businesses in town. The ad made clear that New York City and its regional rivals were at war.

Well, if Boeing can leave Seattle for a better deal in Chicago, and the football Baltimore Colts can move to Indianapolis to get a better stadium, and the New York Stock Exchange can threaten to move to New Jersey, can our most prestigious public institutions be far behind? "Congress Threatens to Leave D.C. Unless New Capitol Is Built," printed in the satirical periodical *The Onion*, seems on the one hand absurd (few national governing bodies are either constitutionally, politically, or ethically able to simply pick up and move), but on the other hand strangely familiar. The offers from cities eager to add to their prestige; the use of private-sector brand names on public facilities, have become common practices. At least to some eyes this story seemed plausible: the *Beijing Evening News*, Beijing's largest circulation newspaper, picked up and reprinted this report, apparently believing it to be true.

The Onion started up as an alternative satirical newspaper in 1988, and has since grown, with print and online editions. The publication may be found at www.theonion.com.

WASHINGTON, DC – Calling the current U.S. Capitol "inadequate and obsolete," Congress will relocate to Charlotte or Memphis if its demands for a new, state-of-the-art facility are not met, leaders announced Monday.

"Don't get us wrong: We love the drafty old building," Speaker of the House Dennis Hastert (R-IL) said. "But the hard reality is, it's no longer suitable for a world-class legislative branch. The sight lines are bad, there aren't enough concession stands or bathrooms, and the parking is miserable. It hurts to say, but the capitol's time has come and gone."

"If we want to stay competitive, we need to up-grade," said House Minority Leader Dick Gephardt (D-MO), who has proposed a new $3.5 billion capitol on the site of the current edifice. "Look at British Parliament. Look at the Vatican. Respected institutions in their markets. But without modern facilities, they've been having big problems attracting top talent."

Its cornerstone laid in 1793 by President Washington, the capitol has been built, rebuilt, extended, and restored countless times over the past 209 years. Legislators say another multimillion-dollar renovation is not an acceptable alternative to a new building.

"How many times can you put a fresh coat of paint over an old, broken-down horse?" asked Sen. Rick Santorum (R-PA), co-chair of the Senate Relocation Subcommittee. "We need a building that befits our status as the nation's number-one democratically elected legislative body. And if D.C. isn't willing to provide that, I can think of plenty of other cities that would be more than happy to."

The leading candidates for a possible congressional relocation are Charlotte and Memphis, both of which have long sought a major organization to raise their national profile. San Francisco civic leaders have also lobbied hard, offering to finance a $4 billion Pac Bell Capitol Building using a combination of private corporate funds (40 percent), a county sales tax (35 percent), and a local cigarette tax (25 percent). Dallas, Seattle, and Toronto have also been mentioned as long shots.

Demonstrating its commitment to "stay in Washington if at all possible," Congress has invited more than a dozen architectural firms to submit proposals for a new D.C. capitol. Among the early favorites is the ambitiously titled "Halls Of Power," a retro-futuristic design by the Kansas City architectural firm of Hellmuth, Obata, and Kassabaum. The Halls Of Power would feature a retractable rotunda for daytime sessions, a Dancing Waters fountain in the front courtyard, and 55 more luxury boxes than the current building.

"This is just the kind of thing we need to stay competitive in today's lawmaking environment," said agent Barry Halperin, who represents many prominent government officials, including Sen. Jim Jeffords (I-VT) and Secretary of Defense Donald Rumsfeld. "Washington can no longer afford to ignore the fact that visitor attendance has dropped every year since 1989. Our elected officials don't like coming to this building and, clearly, neither do their constituents."

Experts attribute the decline in congressional attendance to a number of factors, including increased home viewership of legislative activities on C-SPAN, with which Congress signed an exclusive 20-year, $360 million broadcast pact in 1984. It is not known how a new capitol building would affect the terms of that soon-to-expire contract, but Congress is expected to restructure the deal to increase its share of revenues and secure possible advertising rights, regardless of whether it opts for rebuilding or relocation.

According to the lawmakers' constituents, the capitol is not the problem.

"Sure, the capitol's a little beat-up, but it's got its charms," said Geoff Lapointe, a Glendale, CA, voter. "The real problem is the legislators. Back in the old days, you had big stars like John Kennedy and Richard Nixon. Who've they got today? Evan Bayh? Paul Sarbanes? Who's gonna get excited about those guys?"

Lapointe said he is "fed up" with the legislators and their demands.

"Those guys are all just a bunch of spoiled, overpaid crybabies," Lapointe said. "All they want is money – they don't care about all the hardworking people who pay their salaries. Look at 'em: When's the last time you saw them acting like a team? They can take their capitol and shove it."

"Community Empowerment Strategies: The Limits and Potential of Organizing in Low Income Neighborhoods"

from *Cityscape* (1996)

Peter Dreier

Editors' Introduction

The emergence of community-based organizations as significant political actors is a major development in recent urban political history. To some extent, the rising number and importance of these grass-roots, nongovernmental actors is symptomatic of the decline of other institutions. In North American and Europe, political parties had long taken the lead in mobilizing voters, translating demands into government action, and even offering direct services to needy constituents. The blossoming of community-based movements may be seen in some cases as a sign of the decline of parties as mobilizers, and in particular as evidence of their failure to mobilize new constituents (e.g., youth, members of racial minorities) around new issues (e.g., equity, environmental concerns).

In some European cities, neighborhood-based "self-help" initiatives are a direct response to cut-backs in public services. And of course throughout the industrialized world, economic restructuring has resulted in shifting labor markets, new geographies of production, and declining labor unions. Grass-roots organizations, with political, economic, and social service goals, have arisen to address these deficits.

However, community-based organizing does not just fill in spaces left vacant by traditional political institutions: it represents a new dimension in city politics. In the U.S., community-based organizations have fought locally for civil rights advancements, have resisted urban renewal plans, and have battled for funds to provide services to and build housing for disadvantaged populations. In Europe, such movements have been further influenced by environmental and anti-nuclear concerns, as well as the desire to provide services to marginal groups and to solicit resident input into planning and housing construction. As Dreier notes, such groups are no longer oddities at the margins of urban politics; rather, in some cities, they are key political actors and service providers.

Dreier is careful to note that "community development" is an umbrella term that encompasses (at least) three distinct strategies. "Community organizing" involves mobilizing neighbors to have a voice in the things that affect them; often these begin as protest movements (see Mollenkopf's reading in Part 4 for examples). The European literature on "urban social movements" gives greater theoretical weight to such movements, and indeed in many cases the protest movements spawned in European cities have been more deeply ideological. For further readings on this topic, see Manuel Castells, *The City and the Grassroots* (London: Edward Arnold, 1983), and Margit Mayer, "Urban Movements and Urban Theory in the Late Twentieth Century," in Sophie Body-Gendrot and Bob Beauregard (eds), *The Urban Moment* (Thousand Oaks: Sage, 1999, pp. 209–239).

Community-based development, in contrast, involves local residents in self-improvement activities. In the U.S., these have largely been confined to housing renovation and construction, but some organizations have also sought to create profit-making businesses in an effort to address the economic deficits in their communities. To learn more about work of American community development corporations, see Avis Vidal's *Rebuilding Communities: A National Study of Urban Community Development* (1992), and *From Neighborhood to Community: Evidence on the Social Effects of Community Development Corporations*, (1996) by Xavier DeSouza Briggs and Elizabeth J. Mueller, both available from the Community Development Research Center of the New School's Milano School for Management and Urban Policy. Finally, organizations focused on community-based service provision attempt to provide services, ranging from child care to job training to legal advising, often under contract to the government. Located in and controlled by local residents, such organizations hope to be better positioned to respond to the needs of their neighbors than would state bureaucracies.

These three strategies can co-exist within a single organization. Many groups mobilized originally to mount a protest, and evolved into community developers and/or service providers, but just how groups should balance these different roles, and how central community-based organizations should be to the larger project of urban development, has been controversial.

Some critics, including those on the Right, complain that community-based enterprises almost never become profitable, and that local economic development is better achieved by inducing the private sector to invest in poor communities. But critics on the Left have concerns about the role of community-based efforts as well. For example, Adolph Reed's reading in Part 5 maintains that the work of community-based organizations sends the message that poor people can, and should, fend for themselves, leading quickly to the conclusion that poor communities pressing for outside help simply are not trying hard enough. Randy Stoecker, reprising some of the concerns raised by Frances Fox Piven and Richard Cloward in *Poor Peoples' Movements* (New York: Vintage Books, 1978) and *Regulating the Poor* (New York: Vintage Books, 1993), expresses concern that community development organizations strain themselves to undertake the myriad organizational tasks associated with running an organization, raising funds, and achieving economic development goals, and thus become unable to take on the more promising role of community organizer ("The CDC Model of Urban Development," *Journal of Urban Affairs* 19(1), 1997: 1–22). Finally, journalist Nicholas Lemann, writing in the *New York Times*, notes that community development organizations in U.S. cities are seldom able to foster real economic development, but fears they remain politically popular because, at least in the racially polarized U.S. context, they help avoid broader discussions of racial and economic integration.

But community-based organizations are by now a permanent part of the urban environment, and Peter Dreier's article is concerned with what they do, and how they can be helped to do it better. He notes the importance of intermediary groups that have arisen at the regional and national level, providing technical assistance to smaller organizations and helping to link them to funding sources. Government policy makers, he writes, can target their efforts more effectively by helping groups achieve higher levels of technological and management competence.

Peter Dreier is the E.P. Clapp Distinguished Professor of Politics at Occidental College, where he directs the Public Policy Program and is a senior fellow at the International and Public Affairs Center (IPAC). Previously he served for nine years as the director of housing at the Boston Redevelopment Authority and housing policy advisor to Boston Mayor Ray Flynn. He is the co-author of *Place Matters: Metropolitics for the Twenty-first Century* (Lawrence: University Press of Kansas, 2001), *Regions That Work: How Cities and Suburbs Can Grow Together* (Minneapolis: University of Minnesota Press, 2000), *The Next Los Angeles: A Century of Struggle for a Livable City* (Berkeley: University of California Press, 2004), and *Up Against the Sprawl: Public Policy and the Re-making of Southern California* (Minneapolis: University of Minnesota Press, 2004).

Since the late 1970s, the Nation has witnessed a remarkable resurgence of citizen activism. Residents of America's urban neighborhoods have ignited what Harry Boyte called a "backyard revolution" of community activism (Boyte, 1980, 1989). Most American cities, and many inner-ring suburbs, have at least some level of grassroots neighborhood participation. Today, tens of thousands of neighborhood organizations are involved in a wide range of community improvement efforts.

It is important to distinguish among three strategies for promoting what is often called community empowerment. Community organizing involves mobilizing people to combat common problems and to increase their voice in institutions and decisions that affect their lives and communities. Community-based development involves neighborhood-based efforts to improve an area's physical and economic condition, such as the construction or rehabilitation of housing and the creation of jobs and business enterprises. Community-based service provision involves neighborhood-level efforts to deliver social services (such as job training, child care, parenting skills, housing counseling, immunization, and literacy) that will improve people's lives and opportunities (often called "human capital") within a neighborhood.

The heart of the new community empowerment movement is grassroots organizing to solve social problems and improve economic conditions in distressed urban neighborhoods. Community organizations that engage in successful mobilization efforts sometimes branch out into community development and/or the provision of social services. Although efforts to balance these components are not without tension, this is a logical step toward a comprehensive community empowerment agenda. Community groups that focus primarily on service delivery or community development often lose the energy and momentum required to do effective community organizing. Service delivery and community development are more effective when they are part of a community organizing strategy, especially when the tasks are clearly delineated within the organization.

[. . .]

Community organizations vary widely in size, scope, and competence, but the range of issues and concerns that have been addressed is remarkable. They include public safety, crime, and drugs; tenants'

rights, abandoned housing, and housing discrimination; environmental and public health issues, such as toxic waste dumping, smoking, lead paint, and pollution; community reinvestment, redlining, and related matters; economic development, job training, and plant closings; youth, education, and recreation; and municipal services delivery. Some community organizations focus on a single issue, while others tackle a variety of issues under a single organizational umbrella. Some groups focus solely on problems on their block or in their neighborhood, while others tackle issues across neighborhoods, either by expanding their own "turf" or by forging alliances and coalitions with counterparts in other neighborhoods.

The experiences and activities of the Nation's community-based empowerment organizations provide ample evidence that the American self-help tradition is alive and well. Although many Americans engage in some aspect of community organizing, the public is not well informed about this phenomenon. The mainstream media typically report on the activities of these groups only when they disrupt business as usual. Few newspapers or television stations routinely cover the efforts of community-based organizations. Although funding organizations and some scholars have examined specific groups, there has been relatively little analysis of the experiences of these groups or of the factors that account for their success. Still, there is a sufficient body of knowledge about this growing sector of American urban life to provide a brief overview of recent trends and an evaluation of the factors that contribute to their success or failure.

THE FEDERAL ROLE IN COMMUNITY EMPOWERMENT

In recent years, many government officials, civic leaders, and academics have embraced the notion of community empowerment as a component of a strategy for revitalizing and strengthening America's urban communities. To do so makes sense from both a moral and an administrative perspective. In a democracy, self-government rests on two foundations: citizen participation and reciprocal responsibility. Community empowerment reflects the longstanding American values of promoting strong families in healthy neighborhoods, self-help and volunteerism, and the balancing of rights and responsibilities.

Moreover, if government community development programs are to succeed, social institutions in America's neighborhoods must be strengthened. Community empowerment is consistent with the concept, endorsed by both conservatives and liberals, of using voluntary intermediary community institutions to help rebuild the social fabric – or social capital – of troubled neighborhoods. Neither the public nor the private sector alone can address the problems of America's urban areas; community organizations must play a key role. In recent years, some American business leaders have recognized the benefits of restructuring enterprises to increase the voice of workers, midlevel managers, and even consumers in decision making. . . . Rather than viewing neighborhood residents as passive consumers or clients of government services, it is more appropriate, as well as more efficient and effective, to view them as citizens and partners who can help shape, promote, and even deliver services. In order for America's urban neighborhoods to be healthy, their residents must gain a stronger voice in shaping the physical, economic, and social conditions in their communities.

Government support for community organizing involves a healthy and creative tension. Government's institutional culture encourages lawmakers and bureaucrats to view policymaking and program implementation as their prerogatives, but policies and programs are a two-way street. Citizen participation can sometimes be messy and even conflicting, but it often results in better public policy, more cost-effective programs, and a healthier democracy.

[. . .]

[Government] can build on an existing track record of Federal funding for community organizing, from the Great Society antipoverty programs to current initiatives. Too many journalistic accounts have painted these efforts with the same brush. This is unfortunate, because there is much we can learn from both the successes and the failures of these efforts. Over the years, the U.S. Department of Housing and Urban Development (HUD) [and the Department of Justice (DOJ)] have sponsored a wide range of community self-help and mobilization efforts. These include the Model Cities program, the Tenant Management demon-

stration program, the Neighborhood Development Demonstration program (now called the Heinz Neighborhood Development Program); community-based fair housing monitoring and homeownership counseling; and support for public and HUD-assisted housing tenants who mobilize to fight crime, improve management, attain a stronger role in management, and help tenant associations purchase their homes or negotiate with private owners and nonprofit organizations to assume ownership. . . . Many of these [federally]-funded groups weathered Federal cutbacks of the 1980s and remain rooted in their neighborhoods, working on issues of community improvement.

[. . .]

LIMITS AND POTENTIAL OF NEIGHBORHOOD ORGANIZING

Observers of urban neighborhood problems recognize that sources of urban decay reside primarily outside of neighborhood boundaries. Symptoms of urban decay – poverty, unemployment, homelessness, violent crime, racial segregation, and high infant mortality rates – have their roots in large-scale economic forces and Federal Government policy. The forces and policies include economic restructuring toward a low-wage service economy; corporate disinvestment (encouraged by Federal tax laws); bidding wars among cities and States to attract businesses that undermine local fiscal health; redlining by banks and insurance companies; Federal housing, transportation, tax, and defense spending policies that have subsidized the migration of people and businesses to the suburbs (exacerbating urban fiscal traumas); and Federal cutbacks of various financial assistance, housing, social service, economic development, and other programs. These large-scale forces can undermine the economic and social fabric of urban neighborhoods.

In the face of such realities, neighborhood empowerment organizations face enormous obstacles to repairing the social and economic fabric of their communities. What influence can neighborhood self-help organizations have on policies made in State capitals or in Washington, D.C., and on decisions made in corporate boardrooms? Some would argue that neighborhood crime watches, tenant organizations, community reinvestment coalitions, and similar groups can have only a marginal impact, in light of these major trends and forces. Although there is some truth to this notion, it is ultimately misguided. Community-based organizations cannot, on their own, solve the major problems in their neighborhoods, but they provide the essential building blocks for doing so.

This is a very important point. Most neighborhood and community organizations that operate on their own have only limited success. They can win some victories, but they often have difficulty sustaining their accomplishments. This limitation is due in part to organizations' inability to develop strategies for strengthening their base and moving on to new issues. But, it is also due to the fact that the resources or authority needed to address a neighborhood's problems are not available at the neighborhood level, and often not even at the city level.

Community organizations have won many neighborhood-level victories. Some organizing networks have built statewide coalitions to address State-level issues and change laws, regulations, and priorities. But the hard truth is that despite the tens of thousands of grassroots community organizations that have emerged in America's urban neighborhoods, the whole of the community organizing movement is smaller than the sum of its parts. For every group that succeeds, there are many that do not. With some important exceptions, described below, community groups that *do* win important local victories are not always capable of building on their success and moving on to other issues and larger problems. For the most part, despite local success and growth, community-based organizing has been unable to affect the national agenda – or, in most cases, even the State agenda. As a result, they often improve only marginally the conditions of life in many urban neighborhoods.

SUCCESSFUL COMMUNITY ORGANIZING REQUIRES LEADERSHIP TRAINING AND CAPACITY BUILDING

There is a considerable body of social science knowledge related to the various types, quality, and social structures of urban neighborhoods. Thanks to the decennial census and other data, we know a great deal about the changing demographic,

economic, racial, and family composition of America's urban neighborhoods. We have much evidence about the causes and incidence of residential segregation and concentrated poverty. Many studies have explored indicators of social problems, including rates of poverty, unemployment, crime, juvenile delinquency, and substandard housing. Some studies also examine levels of social cohesion, including such indicators as voluntary association membership, residential turnover, homeownership, and psychological affiliation with the community.

Despite all the research, we know little about so-called *neighborhood effects:* the impact of neighborhood-level factors on individual, household, or group behavior. In terms of our understanding of neighborhood empowerment, social science has paid inadequate attention to the factors that contribute to community mobilization and to the impact of those efforts on neighborhood quality. We know, for example, that community members' involvement in neighborhoods and residents' trust in police – their willingness to report crime, press for arrest, and cooperate with police and prosecutors – help in apprehension and prosecution. However, we know little about the reasons why some neighborhoods mobilize while others do not. Neighborhoods are not simply statistical aggregates; they are *social* places. Macrolevel factors alone cannot account for how, or how well, neighborhood residents organize themselves. Human will and volition are also involved. We know, for example, that (controlling for income) homeowners are more likely than renters to participate in community organizations and to vote, but experience also shows that when renters are mobilized, their neighborhood involvement and voting participation increase.

There is no easy formula to explain when and why residents of a neighborhood, particularly a low-income neighborhood, will join together to address a common problem, or whether or not their efforts will be successful. The situation cannot be explained simply by looking at macroeconomic forces, because community organizing has occurred in good times and bad, when conditions were improving and when they were getting worse. Nor can it be fully explained by looking only at neighborhood-level conditions. Two neighborhoods with similar social and economic conditions – the same levels of poverty, racial composition, church membership, crime, and housing conditions – may manifest two very different levels of community mobilization.

It is here that decisions by organizers and leaders can play a key role. Although a community may be ripe for grassroots mobilization, there is no guarantee that it will occur. Community residents, local institutions (such as religious congregations), or external organizations must make a conscious decision to invest time and resources in mobilizing people around common concerns. Any careful, honest examination of community mobilization must also recognize that there are many false starts on the road to community empowerment. In fact, because we rarely hear about the efforts that went nowhere, we fail to note that many grassroots initiatives never get far beyond the first living-room complaint session, the first church basement meeting, the first phone call that fell on deaf ears, or the first leaflet that appeared in neighborhood mailboxes and went unacknowledged.

But success is not simply about winning victories on specific issues. It is also about changing attitudes. It is about overcoming hopelessness and the sense of futility that infect America's inner cities – that which some have called the *quiet riots* of drug and alcohol abuse, violence, and suicide. It is about giving young people a vision of a different, and better, future. It is about giving everyone more self confidence and self esteem. It is, in other words, as much about winning "hearts and minds" as it is about winning better police protection, a new stoplight on the corner, or a new bank branch in the neighborhood. It is these changes in attitude that give people and neighborhoods the inner strength to organize around issues and to develop a vision that things can be different. Religious institutions often play a key role in community organizing, in part because they provide the moral solidarity that adds an important dimension to self help efforts that transcend narrow concepts of self interest.

The process of developing strong leaders and community organizations is not simply a matter of expanding the self confidence and skills of certain individuals. It is about building solid organizations to change economic conditions, strengthen families and communities, and improve the social fabric of urban neighborhoods. Moreover, transforming social conditions in urban areas has important

ripple effects for the entire society because . . . their destinies are interwoven.

Many community organizations are extremely fragile entities. Although staff members and leaders may have enormous commitment and energy, these attributes alone cannot create a strong organization. Funding is clearly a major problem. Many grassroots organizations lack sufficient resources to maintain adequate staff, office space, equipment, and other basics, and most have little financial stability or continuity. In the past decade, many foundations concerned with low-income neighborhoods have shifted priorities, putting more resources into community-based development than into community organizing. Few community organizations are adept at grassroots fundraising, which includes such activities as collecting membership dues, canvassing for donations, and mounting the proverbial bake sale.

[. . .]

But not even additional funding will, on its own, guarantee effectiveness; leaders and staff must be skilled in building organizations. Too many community groups rely on a small number of leaders and, in most cases, a few staff members. When these people leave or "burn out," the organization often collapses, because there has been no plan for developing or recruiting new leadership and staff. In some situations, charismatic or dominant leaders resist recruitment of new leaders and members. They may feel threatened by perceived competition, or they may not realize that delegating tasks and giving more people a stake in the organization strengthens a group's effectiveness. Success also depends on the ability of poor people's movements and community groups to mobilize resources and generate external support for their activities from various members of the public (the "conscience constituency"), government officials, the media, and funding groups, including religious institutions, philanthropic organizations, businesses, and government.

Successful community empowerment requires a number of factors, such as strong, skilled, indigenous leadership; a stable organization in terms of membership and funding; a clear sense of mission that includes a long-term stake in the community; and an overall strategy that allows an organization to build on its defeats and its victories. These attributes do not emerge overnight. . . . Successful community empowerment efforts depend a great deal on indigenous leadership development and organizational capacity building – the important "how-to" matters that encompass such skills as chairing meetings, dealing with the media, negotiating with government and business institutions, fundraising, and handling budgets. . . . Individuals must develop the skills, stamina, and willpower to succeed as community activists, organization builders, and problem solvers. The popular notion that most leaders and movements emerge spontaneously is misleading – the stuff of folklore.

Many Americans believe, for example, that the 1955 Montgomery, Alabama, bus boycott and the subsequent civil rights movement were triggered spontaneously by Rosa Parks' sudden refusal to move to the back of the bus. In fact, Mrs. Parks and her husband were longtime civil rights activists involved with the National Association for the Advancement of Colored People (NAACP) and other organizations. She had attended the Highlander Folk School, a training center for citizenship education, and was part of a network of African-American community leaders that included E.D. Nixon of the Brotherhood of Sleeping Car Porters. This network had the capacity to mobilize resources quickly and efficiently. It arranged meeting sites (particularly in churches), had access to mimeograph machines and telephone lists, raised funds, organized a complex alternative transportation system, and identified candidates for a variety of leadership roles, including Dr. Martin Luther King and a number of less-heralded individuals.

The example of Rosa Parks illustrates the important point that people interested in successful community mobilization need not reinvent the wheel. Groups can draw on recent experiences in leadership development and organization building that have been informally codified through a variety of training centers, organizational networks, and other vehicles. Strong grassroots community leaders, as well as strong grassroots community organizations, are born *and* made.

LESSONS FROM THE COMMUNITY DEVELOPMENT SECTOR

This scenario may sound familiar to those who have closely observed community-based development during the past two decades. Many CDCs of the late 1960s and 1970s – with roots in well-intentioned community organizations, churches, and social service agencies – tripped over their own inexperience. With funding from foundations and the Federal Government, this generation of CDCs struggled to undertake physical redevelopment projects. But many of them lacked the financial, developmental, and management experience needed to construct and manage low-income rental housing competently. Although a few of these early groups managed to survive, grow, and prosper, many fell on hard times and ultimately went out of business. Some of their housing projects were mismanaged; some fell into foreclosure.

In the early 1980s, as the Government began cutting assistance for low-income housing sharply, few observers would have predicted that the decade would witness something of a renaissance for the nonprofit community development sector. As the decade began, only a handful of organizations had the capacity to undertake complex projects that required multiple sources of funding. Even fewer had the capacity to manage rental housing occupied by populations with many social and economic problems. Although the Nation's community-based development sector is still relatively small and its track record varies from region to region, observers acknowledge . . . it has made significant headway against overwhelming odds, which include an unsympathetic Federal administration . . . patchwork financing, high-risk development projects, and undercapitalization. This sector is moving increasingly from the margins to the mainstream of the Nation's community revitalization efforts.

[. . .]

The key ingredient in the numerical growth and improved capacity of the community development sector has been the creation and expansion of . . . intermediary institutions over the past decade. These include organizations such as the Local Initiatives Support Corporation (LISC), Enterprise Foundation, Neighborhood Reinvestment Corporation . . . and McAuley Institute. These organizations provide technical assistance to help existing organizations improve their skills and to help new organizations learn the basics of community development. They help channel private, philanthropic, and government funding – including

Federal HOME/Community Housing Partnership funds and Low Income Housing Tax Credits – to community-based development groups to help them undertake projects successfully.

Thanks in part to the work of these intermediary institutions, community-based development organizations have become increasingly sophisticated in terms of finance, construction, management, and other key functions. This has been accomplished not simply by targeting technical assistance and funds to individual groups but by enabling groups to learn from one another, build on one another's successes, and form partnerships and coalitions. The Metropolitan Boston Housing Partnership, the Chicago Rehab Network, the Coalition of Neighborhood Developers in Los Angeles, and other citywide umbrella organizations – many of them public–private–community partnerships – have expanded exponentially the capacity of CDCs in their cities. These collaborative efforts have, in turn, provided community development groups with the resources to become key players in their neighborhoods, not only in housing and economic development but also as sponsors or facilitators of improved human services, public safety, and other components of vibrant, healthy neighborhoods.

Although in some parts of the country the community-based development sector is still barely noticed, it has become a highly visible and important part of community rebuilding efforts in many areas. As it becomes more sophisticated, its success triggers other successes in a cumulative process. Communities gain hope when they see buildings being repaired and new businesses opening. Other neighborhoods recognize that they can do the same thing. Neighborhoods that once objected to subsidized housing projects are more likely to welcome developments sponsored by community-based groups that can demonstrate success in design, construction, management, and local hiring.

The intermediary organizations have expanded their activities during the past decade. Initially they were funded primarily by corporate and private philanthropy, but the Federal HOME/Community Housing Partnership program provides a specific set-aside that enables intermediaries to provide technical assistance and training to community-based development organizations.

[. . .]

COMMUNITY REINVESTMENT

Perhaps the most successful community-based organizing in the past decade has been around the issue of redlining and community reinvestment. It is worth looking closely at this movement in order to understand its success. In the mid-1970s, small groups of community activists in cities across the country recognized that the invisible hand of market forces wrote with a red pen. In Baltimore, Boston, Chicago, Cleveland, New York, and other cities, neighborhood residents and small business owners began to recognize a pattern in bank lending decisions. Banks were refusing to make loans to homes and businesses in certain neighborhoods, creating a self-fulfilling prophecy of neglect and deterioration.

Local activists concluded that their neighborhoods were experiencing systematic disinvestment, not isolated lending decisions by misguided loan officers, and they began efforts to convince banks to revise their perceptions and lending practices. Some were simply educational campaigns to change the way bankers – often suburban residents with stereotyped images of city neighborhoods – viewed the areas. Other efforts involved consumer boycotts – "greenlining" campaigns – of neighborhood banks that refused to reinvest local depositors' money in their own backyards. Most of the efforts ended in frustration, with little impact on the banks' practices. But some neighborhood groups achieved small victories, including agreements between banks and community organizations to provide loans or maintain branches in their neighborhoods. Eventually, activists across the country who were working on similar issues discovered one another and recognized their common agendas. From such localized efforts grew a national "community reinvestment" movement to address the problem of bank redlining.

In response to grassroots pressure from the emerging neighborhood movement, Congress sponsored a number of initiatives to promote community self-help efforts against redlining. These included two key pieces of legislation, the Home Mortgage Disclosure Act (HMDA) of 1975 and the Community Reinvestment Act (CRA) of 1977.

In combination, HMDA and CRA provided an effective tool that enabled local groups to pressure banks to invest in low-income and minority neighborhoods. HMDA provided the data needed to analyze banks' lending patterns systematically (for housing loans but not commercial loans). HMDA gave many community groups and university-based scholars – and some newspapers, local governments, and other agencies – the data with which to investigate geographic and racial bias in lending. By requiring banks to meet community needs as a prerequisite for obtaining various approvals from Federal bank regulators, and by giving consumer and community groups the right to challenge these approvals, CRA provided the groups with leverage to bring banks to the negotiating table.

From 1977 through the late 1980s, Federal regulators failed to monitor and enforce CRA. As a result, community reinvestment activities primarily involved bottom-up enforcement: local campaigns by community organizations or coalitions against local banks. In the late 1980s, these local activities coalesced into a significant national presence. Thanks to the work of three national community organizing networks – ACORN, CCC, and NPA – these local efforts became building blocks for a truly national effort that has produced dramatic results in the past few years. Locally crafted CRA agreements alone have catalyzed more than $60 billion in bank lending and services. But even more important is the fact that many banks are now much more proactive in working with community organizations to form successful neighborhood rebuilding partnerships.

Training centers and organizing networks have helped local organizations significantly expand their capacity to identify redlining, work with local media, negotiate with lenders, persuade State and local governments to support their efforts through linked deposit policies and public–private lending partnerships, and work with CDCs to take advantage of new lending products. With funding support from several foundations and technical advice from these national networks and training centers, community groups have been able to hire experts to help interpret HMDA data, publish reports, and expose systematic bank discrimination. Whereas in the past most HMDA studies focused only on one bank or one city, groups such as ACORN that have a base in neighborhoods in many cities were able to demonstrate that the problem is not confined to just a few places. In 1989 the Federal Reserve began to respond with several studies of its own.

Community groups and organizing networks have gained the respect of the Nation's mainstream media, which began to report the redlining issue with some regularity. In fact, the *Atlanta Journal and Constitution* won a Pulitzer Prize for its 1988 series "The Color of Money" on this subject. Through these networks, acting on their own or in concert, grass-roots groups pressured Congress to strengthen both CRA and HMDA several times in the late 1980s. These were dramatic legislative victories against overwhelming political odds. In the early 1990s, the national networks, along with community development intermediaries such as LISC, the National Congress for Community Economic Development, and The Enterprise Foundation, formed the National Community Reinvestment Coalition to strengthen the community reinvestment agenda.

Indeed, the entire community reinvestment climate has changed dramatically in the past few years. Banks are now much more proactive in working with community organizations to identify credit needs and create partnerships to meet them. Government regulators are much more active in evaluating lenders' CRA performance and using regulatory incentives to ensure compliance. Fulfilling its campaign pledge, the Clinton administration has made the issue of redlining and community reinvestment by banks and insurance companies, as well as support for community-based development, a centerpiece of its urban policy agenda.

What were the *key ingredients for success* in community reinvestment?

- First, it was an issue that affected many people and was clearly linked to economic and social conditions in urban neighborhoods.
- Second, the HMDA law provided community groups with usable tools to identify the problem, illustrating the importance of community organizations having access to key information.

[...]

- [Third], local groups working on the same issue were able to learn from one another through several national organizing networks and training centers, such as ACORN, NPA, and CCC, which helped expand the capacity of local community groups to use CRA and HMDA to rebuild and revitalize neighborhoods. The networks provided groups with training and linked them together to make the Federal

Government legislators and regulators alike more responsive to neighborhood credit needs.

- [Fourth], local groups had access to training and leadership development that empowered them to stabilize the membership and fundraising of their organizations; to form coalitions with a variety of groups (including church-based organizations, civil rights groups, nonprofit developers, and social service agencies) that often crossed boundaries of race, income, and neighborhood; to learn how to develop strategies for working on several issues simultaneously and building on small victories; to develop a strategy for negotiating with lenders and government; and to deal with the media. . . .

- Sixth, local groups had access to expertise and technology that enabled them to take advantage of HMDA and CRA. To make such Federal laws work, community groups must learn how to use them, and that usually involves having money to hire experts or to train staff in the computer skills needed to analyze complex data and translate them into reports understandable to the general public and the media.

TENANT ORGANIZING IN PUBLIC HOUSING DEVELOPMENTS

The strength and success of the grassroots community reinvestment movement stands in contrast to organizing efforts in both public and HUD-assisted housing developments. Without doubt, there has been a great deal of grassroots organizing among the Nation public housing and HUD-assisted housing tenants. Although we can point to important success stories in developments across the country, the cumulative impact of tenant-led efforts has been marginal at best, in terms of building strong, stable community organizations and making a significant impact on economic and social conditions in the developments and their neighborhoods.

Public housing tenants have organized to improve the local housing authority management, especially in making repairs and improving the physical condition of developments; to deal with questions of security and public safety, including the epidemic of drugs and gangs in the developments; to start or expand job-training, child-care,

counseling, and other human service programs in their developments; and to address such environmental and public health hazards in and near the developments as lead-based paint, toxic dumps, and asbestos. Some public housing tenants have used direct action and litigation tactics to save their homes from the wrecking ball. Some of the struggles have led tenant organizations to demand a stronger tenant voice in the day-to-day management of their housing, including the creation of resident management corporations and even tenant ownership.

Although a growing number of tenant associations, resident councils, and tenant management corporations exist in public housing today, they represent only a handful of the Nation's public housing developments. Many tenant groups are relatively weak in terms of leadership and organizational capacity. Few enjoy widespread participation by residents, several tiers of leadership and subcommittees, or regular elections. A number could be categorized as company unions, lacking the level of independence from the housing authority management that makes tenant groups effective. A framework for accountability between residents and tenant leaders is often problematic. In only a few cities including Milwaukee, Boston, Los Angeles, Baton Rouge, and Kansas City, Missouri have residents formed citywide tenant councils that bring together leaders from various developments to negotiate their common concerns with the housing authority. Most citywide groups are quite fragile.

Thus resident organizing and participation in the Nation's public housing is still extremely thin. In part this has to do with the overwhelming problems confronting the low-income residents of public housing. But there are enough examples of successful organizing by public housing tenants to demonstrate that effective organizing is possible, as it was even during the 1980s when public housing had few friends in high places – not in the Federal Government, the media, or foundations.

What has been missing is the ability to disseminate the lessons of success (and failure) from one development to another in a given city and to share the lessons among cities in order to build a national infrastructure of public housing residents who can become effective advocates in Washington for public housing, especially in dealing with Congress. Organizing in the Nation's public housing developments is ad hoc and unfocused. Few of the tenant groups are linked to broader organizations or networks, and few have the resources to undertake leadership training or capacity building. There is no effort to create a new empowerment movement among public housing tenants. Even when local housing authorities recognize the importance of tenants organizing, public housing authority (PHA) staffs rarely have the mandate or training to build an effective grassroots organization among residents, and most tenant organizations lack the resources to hire staff.

[. . .]

POLICY RECOMMENDATIONS: PROGRAM CRITERIA

What can we learn from these experiences in order to forge a partnership with community organizing groups to strengthen inner-city neighborhoods, cities, and metropolitan areas and our Nation's well-being and productivity? Strategies to expand community-based organizing efforts must be viewed as part of the larger agenda for improving economic and social conditions in urban areas. Three areas in which the Federal Government can be most helpful in promoting community empowerment are organizing and training, access to information, and leverage points.

Organizing and training

Community organizations need multiyear funding for organizing and training. It is critical for these groups to get effective technical assistance in leadership development and organizational capacity building. Options include providing funds to community organizing training centers and networks (intermediaries) and/or providing categorical grants to community groups for ongoing training.

Focus on organizing groups

Community organizing groups are a special type of community institution, and it is important to ensure that only bona fide organizations are eligible. Although a group may engage in development and/or service delivery, funding should be restricted to groups whose primary activity is the mobilization and empowerment of neighborhood residents. It should be a nonprofit organization that is not part of a local government or a government-controlled entity. Its governing board and leadership should be democratically elected by its membership, and the board should hold regular meetings and use accountability mechanisms. Although community organizations may work in economically diverse neighborhoods, low-income people should be well represented on an organization's governing board.

Make categorical grants to community groups and intermediaries

Federal initiatives to help community organizing should be administered as competitively awarded categorical programs under Federal agency supervision, instead of being directed through local governmental jurisdictions. Funds should be allocated in two ways: (1) through national, regional, and local intermediaries (training centers and organizing networks) with good track records in community organizing; and (2) to groups that work with these intermediaries. Federal funds should focus on two types of activities: (1) support for community organizations' day-to-day operations and (2) technical assistance to help train community organization leadership and expand their capacity in such areas as fundraising, budgeting, and membership recruitment.

Build alliances across income and race

Recent discussions of urban conditions have focused attention on the social, economic, and political isolation of the Nation's inner-city poor. Low-income people need to develop strong organizations and leadership to help overcome this isolation, but they also need to build alliances with moderate-income people who share common concerns about the condition of their neighborhoods, families, schools, and the economy. It is often difficult to find issues and develop strategies that cut across the boundaries of income and race, but some of the most successful community organizations have done so. Federal support for community-based organizing should recognize the importance of both empowering the poor and building alliances with those only a step or two above the poverty level.

Target distressed urban and suburban neighborhoods

Secretary Cisneros has spoken of the nterwoven destinies of America's cities and suburbs. A growing body of research has shown that suburbs cannot remain healthy if their central cities are decaying. Equally important, many so-called suburbs have social, economic, and demographic conditions similar to those in inner-city neighborhoods. The artificial boundaries between cities and suburbs, particularly the inner-ring suburbs, must be broken down. One way to do that is to encourage residents of distressed suburban communities to organize and find common ground with their counterparts in the inner cities. This does not mean providing funds for affluent suburban neighborhood associations to promote NIMBY (not in my back yard) attitudes; it means identifying troubled low-income neighborhoods in such places as Compton, California; Harvey, Illinois; Somerville, Massachusetts; and in other communities. Funding formulas and targeting should not focus exclusively on low-income neighborhoods in central cities but should be flexible enough to identify areas outside inner cities.

Access to information

The Federal Government should provide easy access to information, such as HMDA data, modernization estimates for public and HUD-assisted housing developments, Superfund inventories, crime statistics, and community right-to-know laws about chemicals, and should help community organizations gain access to the expertise and technology necessary to interpret and work with these data. Additional measures that would be helpful include putting HMDA data online so that community groups have easy access to it and adding commercial lending data to the HMDA law.

Promote community access to technology and expertise

If community organizations are to be effective problem solvers, they must have access to expertise and technology. The access should not be an afterthought, but instead should be a key component of the community organization operating budget. The organizations need funds with which to hire experts who can help them evaluate environmental impact statements, HMDA data, housing rehabilitation and financing estimates, architectural design and zoning guidelines, utility company documents involving rate structures, and similar matters. They also need access to computers for desktop publishing of newsletters and other forms of communication; for research using such data as the census, HMDA, and crime incidence reports; and for compiling membership lists.

To promote community access to such expertise, the Federal Government might encourage community groups and local colleges and universities to form partnerships based on existing models, such as the Center for Community and Environmental Development at Pratt Institute in New York, the Policy Research and Action Group in Chicago, the Public Research Institute at San Francisco State University, and the Center for Neighborhood Development at Cleveland State University. At these centers, academic researchers work closely with community groups, not only to provide technical and scientific expertise but also to train community organizations to use these tools.

[...]

Improve media coverage of community initiatives

The media play an important role in either enhancing or thwarting community-based problem solving. For the most part, the Nation's mainstream media treat urban neighborhoods as magnets for social problems. In so doing, they distort reality; exaggerate urban ills; undermine the public will to address these problems; and inadvertently sabotage efforts by government, community organizations, and the private sector to forge solutions. With some important exceptions, the media generally ignore or trivialize the community-building efforts of neighborhood groups and the policy-making efforts of government. Community groups can help improve the media coverage of the urban condition and the community-based efforts to solve urban problems. Training programs for community organizations should include the topic of dealing with the local media. Equally important, the Federal Government should help community groups forge partnerships with local journalism schools. Together, they could sponsor workshops for journalists on urban issues and community-based problem solving and analyze the content of print and broadcast news coverage to help identify institutional blind spots. They could sponsor walking tours of neighborhoods for reporters and editors and point out problem-solving activities that could become topics for news stories.

Leverage points

Community organizations need to have a regular and legitimate role in shaping public policy and enforcing laws and regulations. Organizers call such activities *handles*; that is, points of leverage or access to the policy process.

Promote laws and regulations that give communities a voice

Certain laws and regulations offer community organizations opportunities to voice their concerns and become part of the public process. CRA, the Superfund law, and community right-to-know laws have been helpful in giving communities a voice in the public policy process and catalyzing effective grassroots organizing. HUD efforts to give tenants in public and assisted housing a voice in management have been helpful but have put too much emphasis on the goal of resident management or ownership rather than resident mobilization. A revised crime bill that would give community organizations a role in all community policing initiatives could do the same thing.

[. . .]

REFERENCES FROM THE READING

Boyte, Harry. 1980. *The Backyard Revolution: Understanding the New Citizen Movement.* Philadelphia: Temple University Press.
—— 1989. *Commonwealth: A Return to Citizen Politics.* New York: Free Press.

F
O
U
R

"Gender, Development and Urban Social Change: Women's Community Action in Global Cities"

from *World Development* (1997)

Amy Lind

Editors' Introduction

Across North America and Europe, community development and self-help organizations have arisen to respond to the dislocations of economic restructuring. Facing geographically targeted disadvantages and, in many cases, unresponsive national policy makers, urban residents have sought ways to use collective action to address their needs and seek access to the policy process.

But even disadvantaged people living in American and European cities are privileged relative to the poor of the less developed world. Amy Lind is interested in how Latin American women have organized to provide services and effect political change. Nowhere are the connections between the global and the local clearer than they are here: a global economic agency like the International Monetary Fund (IMF) demands economic reforms in debtor nations such as Peru; national leaders impose austerity measures and cutbacks in their meager welfare states; poor families struggle to survive. In some cases, survival strategies include organizing to provide for basic needs such as food and water. Lind's case describes the development of communal kitchens in Lima, Peru, offering families a way to pool their resources and feed their families more efficiently. She makes clear that such strategies are integrally linked to national and international economic policies: each time Peru's leadership has adopted a new "structural adjustment package" to win IMF approval, the number of communal kitchens has risen.

However, Lind adds a further dimension to her analysis of urban social movements in South America and the U.S. by focusing on the importance of gender. The organizations she studies have been initiated by women, and have almost entirely female memberships. In this they are not alone: many community-based organizations have been shaped by women. In many cases, they have grown out of shared concern about what have traditionally been viewed as women's affairs (education, welfare, consumption), and women have organized to defend their ability to work effectively as mothers and guardians of the domestic sphere. Quite often, such projects move into more clearly political arenas, as women fighting for affordable food or better educational opportunities for their children confront resistance and expand their battles to confront new targets. Because economic restructuring pressures exact such a heavy toll on the consumption and reproduction needs of the urban poor, women have become increasingly important initiators of local self-help and resistance efforts.

Lind's work draws from, and contributes to, a growing debate about the role of gender in conditioning grassroots activism in developing and fully industrialized urban areas. Some scholars have argued that women's activism requires entirely new models of community mobilization and leadership, as the approaches advocated by Saul Alinsky in his classic works *Reveille for Radicals* (first published in 1946), and *Rules for Radicals* (1969), emphasizing confrontation with authority, privileged a male model of political activism. Women-led movements, in contrast, pay more attention to internal maintenance and development matters, and do not privilege the public sphere over the private sphere. Thus women's activism springs from concerns for family and community, growing to encompass the polity when such forms of activity seem most fruitful. Other scholars have been less inclined to create entirely new categories for women activists. Some believe that race, ethnicity, and class remain more important lines of cleavage in urban political struggles. Others are concerned that emphasis on a distinctive women's activism can serve to reinforce gender stereotypes and bar women from participating in male-dominated spheres of political life.

These debates are reflected in the work of Nancy Naples, author of *Grassroots Warriors: Activist Mothers, Community Work and the War on Poverty* (New York: Routledge, 1998), and co-editor, with Manisha Desai, of *Women's Activism and Globalization: Linking Local Struggles and Transnational Politics* (New York: Routledge, 2002). Temma Kaplan provides a different perspective, along with profiles of individual women activists in a number of settings in her *Crazy for Democracy: Women in Grassroots Movements* (New York: Routledge, 1997). Susan Stall has written on women's community activism in the U.S.: see her 1998 article with Randy Stoeker, "Community Organizing or Organizing Community? Gender and the Crafts of Empowerment" (*Gender and Society* 12: 729–756), and her book, co-authored with Roberta Feldman on organizing in the Chicago public housing system, *The Dignity of Resistance* (New York: Cambridge University Press, 2001). Amy Lind is the author, most recently, of *Gendered Paradoxes: Women's Movements, State Restructuring, and Global Development in Ecuador*, published in 2005 by the Penn State University Press. She is a lecturer at the Studies in Women and Gender Program at the University of Virginia.

1. INTRODUCTION

[. . .]

This paper addresses the gender dimensions of women's community action in the context of economic restructuring and urban poverty. As I will show, the gender dimensions of urban structural change are far-reaching, and while initial research documents the potential gender effects of macroeconomic frameworks with "male biases" less has been done to understand the nature of women's collective participation in the reorganization of social reproduction at the community level, and to draw out the implications of this type of collective action for neo-liberal development frameworks and practices. In some cases, women have acted based on their traditional gender roles – as in the case of "mothers' movements" against violence and human rights abuses, and in the process have transformed public understandings of women's participation in the development and political process. These organizations have made public a set of issues about violence and its impacts on family structures as well as about the (historically invisible) roles of women in local development and urban social change.

In other cases, women's organizations have participated actively in local initiatives, which stem from policies of decentralization and state downsizing – through municipal planning structures, nongovernmental organizations, and grassroots movements. Supporters of these policies often herald increases in local power on the basis that "more people are participating – more effectively, and more democratically." However, as I argue in this article, increases in local power may not automatically transfer into power for women if and when the "hidden" transfer of welfare responsibilities to community organizations and households is left unexamined, and for as long as women's community participation is perceived as "outside" the planning and development process.

Local women's organizations are at the forefront of challenging these biases as they affect the participants' everyday lives and their surrounding urban and policy environments. This holds true comparatively among the cases in this study, despite important cultural, political and historical differences among the organizations and cities. The similarities and differences in local movements within a global context depend upon a number of related factors

F
O
U
R

that reflect both the organizations' and the participants' locations within community and urban structures; their institutional networks; and their roles in the household and family, and by extension, in their communities – roles which are often unacknowledged because they are not accounted for in the market and have been undervalued culturally. Whether it be through participating in one of the more than 2,000 communal kitchens in Lima, in neighborhood women's organizations in Quito, or in mothers' anti-violence movements in U.S. cities, women's organizations have played key roles in generating women's involvement in community decision-making and addressing the daily impacts of economic restructuring: in this sense, both their community involvement and the implications of their action for policy frameworks merit further attention if we are to promote more equitable national and urban policies.

[. . .]

2. STRUGGLES FOR LIVELIHOOD: WOMEN'S ORGANIZATIONS IN COMPARATIVE PERSPECTIVE

[. . .]

Much research has been conducted on women's organizations and movements in Western, industrialized countries; until recently, much less so has been conducted on developing regions. The literature on Western women's movements has tended to focus on middle-class movements, although many studies document the important contributions of poor and working class women's organizations and movements in the United States, Canada and Western Europe. These and other studies point out that poor women's and other urban social movements have arisen in response to de-industrialization, massive unemployment, and struggles for decent living spaces in deeply segregated and economically overburdened cities. For many, this is coupled by a lack of citizenship rights, insufficient health and educational systems, and growing rates of political and racial violence in urban areas. In contrast to other urban social movements, women's organizations and movements address a set of gender-specific issues including: violence against women, their roles as mothers and as working women, gender-based discrimination in the workplace and/or in the

informal sector, the gender impacts of social policies, and children's rights. To my knowledge, many studies, which focus on community development and local power often overlook women's protagonistic roles in these processes. Studies that focus specifically on women's organizations provide an important basis for explaining why women choose to create their own organizations and movements, and how gender-specific forms of community action might inform local government, urban and national policy.

[. . .]

In Latin America, a large number of poor women's organizations emerged during the period of economic crisis and democratic transition in the late 1970s and 1980s. These movements have been explored extensively. While original research focused on women's struggles against military authoritarianism and participation in processes of democratization, more recent scholarship has focused on the dynamics of power and structural inequalities which emerge and become consolidated under formal democracy. In the 1990s, further attention has been placed on women's collective responses to economic crisis and structural adjustment policies. This literature discusses the political potential and limitations of contemporary women's organizations, including their long-term impact on institutional and social change. In many countries where structural adjustment and/or neo-liberal social and economic policies have been introduced, scholars and policy makers have begun to analyze women's collective survival strategies in this context. . . .

In Latin America, a proliferation of research has documented women's collective responses to urban poverty and economic restructuring; this is due in part to the rich tradition of women's collective organizing, and to the explicit responses many women's organizations have made to adjustment measures. In this section I draw from this literature and suggest its relevance for other regions undergoing similar processes. An initial question is the extent to which grassroots women's organizations are likely to sustain themselves through crisis periods, and influence state policy agendas in the long run. Alvarez (1996) argues that neo-liberal development policies have served to institutionalize what were once viewed as spontaneous strategies to cope with a momentary crisis. Alvarez contends that women's organizations are increasingly placed in a paradoxical position under neo-liberal development policies. On one hand, many community-based women's organizations were initiated in the late 1970s/early 1980s to confront the economic crisis and the negative impacts of structural adjustment policies. Their struggles therefore emerged out of economic necessity, although many developed more complicated critiques of power and structural inequalities through the process of organizing, and their political demands, similar to those of traditional party politics and class-based movements, were directed at the state. Under neo-liberal reform, as the welfare state is dismantled, poor women's organizations have lost crucial state funding – as well as access to state welfare services – and have become even more dependent on an "undependable" state. Women's organizations are therefore left with little recourse but to continue their efforts on their own or to seek funding elsewhere, primarily from international NGOs, bilateral and multilateral agencies. Alvarez's paradox, therefore, refers to the institutional crisis that organizations face as they are no longer certain where to direct their demands.

Indeed, many organizations have had to develop new strategies in order to secure funding and maintain their institutional structure. In the context of state retrenchment, as international donors increasingly channel funds to local NGOs and grassroots organizations in their new roles as service providers, Alvarez's paradox holds true. Women's organizations may in fact benefit from donors' emphasis on NGO participation if they are incorporated into the new local structures. Most development frameworks however, either do not account for gender or assume that women have indefinite time to participate in volunteer-based community groups. This being the case, despite donors' intentions to promote the participation of NGOs, strengthen civil society, and build democratic practices, women's organizations are likely to lose out entirely or continue serving in their roles as unpaid managers of social reproduction. This depends, to a large extent, on country-specific neo-liberal measures as well as on the policies which preceded them and the effects of the shifts in welfare provision on local communities.

A related problem is the extent to which people will seek collective answers or retreat to the private realm of the family and other informal networks for survival. Beneria (1992) found that there has been a "privatization of the struggle" for daily survival along with the broader process of privatizations taking place in Mexico. In her study of fifty-five households in Mexico City, Beneria concludes that poor households are increasingly responsible for social reproduction, with little or no help from the state, or even from private organizations or informal networks. Thus Beneria's work suggests that, in the case of Mexico City, rather than becoming more dependent on the state, poor women and their families have become more reliant on direct family networks than on any other form of welfare provision and/or social support. Beneria observes a lack of collective action; Alvarez observes that women are acting collectively and have no alternative but to continue doing so, even if funding is scarce. Both analyses reflect accurately the dilemmas faced by neighborhood and other local women's organizations (and individual households) in cities throughout the world.

Studies of household survival strategies indicate that women's motivations for participating in organizations depends not only upon their poor economic situations but also upon the particular relationships they develop with public institutions and social movements. This observation merits further attention as it remains unclear why some women choose to participate and others do not. Research which focuses on the political, institutional context within which community-based women's organizations develop their strategies is one way to address why certain groups of women initiate organizations and what relationships they develop, over time, with public and community organizations. It also sheds light on why some groups deeply influence policy agendas while others do not. Barrig (1996) argues that the communal kitchen movement in Lima, Peru is a "needs-based" movement. She . . . argues that members of communal kitchens in Lima position themselves politically as "consumers" and/or as "clients" of the state, rather than as a political class pushing for more fundamental institutional change. Specifically, Barrig argues that as kitchen members struggle for their rights and needs as poor women, mothers, and members of collective kitchens, they compete with other kitchens for scarce state resources. This has led to a situation in which the kitchens position themselves, hierarchically, as "clients" of the state and fail to make strategic connections between their own struggles and those of other kitchens, organizations and movements.

Barrig suggests that these organizations are increasingly isolated from other movements, and often do not build coalitions nor envision a broader transformation of society (a situation, she suggests, which results both from institutional constraints and from the organizations' inability to conceptualize new political strategies). Thus, communal kitchens are largely reactive and unable to influence state policy in meaningful ways.

Other research concludes more positively, although cautiously, about the potential of women's organizations to influence policy agendas and negotiate power in their local communities – through interactions with neighborhood associations and/or cooperatives, political parties, municipalities, religious institutions, and NGOs. Schild (1991) analyzes women's roles in local (mixed) organizations in authoritarian Santiago, Chile (1973–1980). She contends that women's participation in these organizations – and their struggles to establish their own, gender-specific organizations – deeply engendered the traditional arena of class politics and human rights struggles in Santiago. Their participation, therefore, must be viewed not only in terms of their actual involvement, but also in terms of how they negotiate gender and class relations, and political ideologies, in their daily lives and the consequences of these actions for changing consciousness. She contends that much of what is decided politically within the organizations depends upon their relationships to, and interpretations of, state development practices in Chile. Thus women's struggles for the seemingly most basic needs – such as the basic right to life and the right to a decent living space – are often ideological struggles over gender (and other) inequalities deeply engrained in state policy frameworks, in the law, and in community structures.

Schild's analysis allows us to make connections between women's political identities and state development policies. Poor women's massive participation in (both mixed and gender-specific) organizations in Santiago, and social solidarity networks established between middle-class and poor women's organizations, contributed a great deal to the incorporation of gender issues into the state policies developed by the transition government of President Patricio Alwyn (1990–1994). This assertion is confirmed by other researchers . . . who argue that the wide production of knowledge about Chilean poor women through grassroots, anti-authoritarian political activism in the 1970s provided the basis upon which policy makers could make a case for incorporating gender issues into the public agendas of the 1980s and 1990s. While important criticisms have been made of studies which attribute the emergence of urban social movements directly to international solidarity and development policies, these historical factors in Chile exemplify the complicated and intertwined relationships that women's organizations in poor countries have had with public institutions since the inception of their collective strategizing.

All of these studies provide important examples of the transformative power and limitations of local women's organizations in the 1980s and 1990s. They also point out that women's struggles for livelihood are often determined as much by their dire economic needs as by their positions, roles and relationships in family and political structures. In the remainder of this paper I examine some regional cases and draw out their gender implications for state policy and economic restructuring.

3. SHARING THE COSTS OF SOCIAL REPRODUCTION: COMMUNAL KITCHENS IN LIMA, PERU

In Peru, the state has undergone a series of measures to liberalize the economy, decrease state spending and transfer the responsibility of social reproduction to the private realm of the family, economy and civil society. Unlike most other Latin American countries they were originally carried out in a highly unorthodox fashion – one which differed from the recommendations of the International Monetary Fund (IMF). Breaking somewhat with Peru's heterodox policy tradition, the administration of President Alberto Fujimori applied an IMF-inspired structural adjustment program which . . . escalated the cost of living and doubled poverty rates according to important indicators of social crisis. [For] twelve years, the Peruvian state's project [was] implemented during a period of intense civil war, promulgated by Shining Path, which . . . cost over 25,000 lives.

One result of these measures has been that poor neighborhoods have organized to collectivize costs and confront the economic and political crisis in

Lima, the capital city of seven million inhabitants. Lima's *pueblos jovenes* (literally "young towns," or poor neighborhoods) have increased in population from 1.5 million in 1981 to approximately 3.5 million in 1993. In this context of increased urban poverty, communal kitchens are one example of how women have developed a strong activist network to address the problem of poverty – in particular, food consumption and distribution. Every morning, some 40,000 low-income women belonging to the *Federacion de Comedores Populares Autogestionarios* (FCPA, or Federation of Self-Managed Popular Kitchens) gather at 2,000 sites throughout Lima's poor neighborhoods, pooling their human and material resources to feed their families, some 200,000 persons. Twenty to 30 female friends, relatives, church mates and neighbors participate in each *comedor*. Women are joined by shared concerns and are welcomed, in theory, regardless of political positions or religious affiliation. The women rotate in positions of leadership and all take turns collecting dues, buying foodstuffs, and preparing the meals, usually in one of the member's kitchens. The kitchens accept donations, and continue to be dependent on them to varying degrees.

The first kitchens were organized during 1979–1986 as a response to the impact of structural adjustment programs which drastically cut – or eliminated – both real income and public food subsidies. Until 1990, participants were mainly middle aged women migrants seeking to escape rural poverty and violence. Rather than directly demand social benefits from the state, the women designed autonomous, self-help solutions based on their own resources.

Between September 1988 and March 1989 alone, the number of kitchens jumped from 700 to more than 1,000 in response to implementation of a structural adjustment package implemented by the administration of President Alan Garcia (1985–1990).

A similar surge took place in 1990, this time a result of a particularly drastic structural adjustment package commonly referred to as President Alberto Fujimori's *paquetazo*, incorporating thousands of younger, newly impoverished Lima residents into the kitchens. A combination of intense economic reforms, coupled with historically unprecedented levels of violence associated with Peru's internal war, led to the growth of new forms of poverty among

Lima's settlers and to the emergence of the "new poor." Many women in this group sought refuge in the communal kitchens that already existed in their neighborhoods, or followed the example of other communities to create their own. Organizational support for some of the kitchens has come from the Catholic Church, political parties, or the state; other "autonomous" kitchens receive support elsewhere. There are many differences between these kitchens, including their levels of democratic structure and participation, reliance on external funding, self-sustainability, and quantity and quality of food served. Much debate has taken place about the relative autonomy of the different types of kitchens, and two major studies concluded that those kitchens organized in a "top down" fashion by the church or state were less likely to transform gender roles and consciousness than were autonomous organizations. In autonomous kitchens, members participated more actively in decision-making and became more active in broader community planning processes.

Today the communal kitchens are organized into federations and confederations, including the FCPA, which represents the movement to government officials, purchases wholesale inputs, organizes micro-enterprise activities, and elaborates and transmits a broader view of gender and women's community participation. In December 1988, the FCPA's predecessor, the *Comision Nacional de Comedores* (CNC), achieved passage of a law, which called for the creation of a fund to support the kitchens, and a new legal status for them (as a "social base group"). This achievement resulted from the efforts of the CNC leadership, the base group membership, professionals from local support organizations, and sympathetic government officials.

The communal kitchens serve as a powerful example of women's community action. Fifteen years of successful results, despite the political and economic odds, demonstrate the kitchens' sustainability; their expansion also demonstrates their replicability. Members of the organizations have gained a new awareness of their roles not only in social reproduction, but also in community and civic action. The kitchens constitute an important part of the broader "popular" women's movement in the country. Their perspectives and demands have generated a great deal of attention from NGOs, political parties, feminist activists, church groups and other base groups who work in poor neighborhoods. This was particularly evident when Maria Elena Moyano, ex-Vice Mayor of the municipality of Villa El Salvador, and ex-President of the *Federacion Popular de Mujeres de Villa El Salvador* (FEPOMUVES), a women's federation which has organized several communal kitchens in Villa El Salvador, was assassinated by Shining Path in front of a local fundraising event on February 15, 1992. For many neighborhood and feminist activists, her death represented the severe contradictions that local communities had faced since the inception of heavy political violence and economic crisis. In this context, communal kitchens in this slum area and in others came to represent much more than a struggle to put "bread and butter" on the table. Rather, members of the kitchens were forced to deal with the infiltration of violence, and particularly death threats from Shining Path, were they to take an explicit stance against Shining Path's presence in their community. This led to a situation in which members of communal kitchens were forced to strategize even the most seemingly mundane or "basic" aspects of their daily lives – the provision of food – under highly adverse, difficult and often dangerous conditions.

[. . .]

The fact that younger, Lima-born women are currently participating in the kitchens may . . . point toward a broader understanding of women's community action. Their participation may be purely out of economic necessity, but it also may be due to the fact that the kitchens have become an accepted practice in daily life for young Limenans or, most positively, that they provide a sense of empowerment for many of their members. It is clear . . . that while the struggle for survival is a key motivating factor, the organizations have become an accepted means to raise awareness about other community issues. The perseverance of members to keep the kitchens active, despite the political and economic odds, testifies to the fact that there continues to be a broad need for more adequate distribution of social welfare. In addition, the participation of many members has transformed their own understandings of domestic labor as well as the broader community's understandings of the shared costs of reproduction, something which has proved invaluable for the efforts of poor women in Lima.

The weaknesses in this form of organizing lie in the fact that women members may become "burnt out" from participating for so many years. After fifteen years, the original excitement about the kitchens has worn off; many women are tired and would prefer to seek employment opportunities elsewhere rather than participate as volunteers in the kitchens. Their volunteer participation, furthermore, is above and beyond their already strenuous (unpaid) domestic workloads. While in general the kitchens remain a successful survival strategy, there are nevertheless many constraints that the women members face. To begin with, women are the primary, if not exclusive, members of the kitchens. While awareness of the shared responsibility of reproduction has increased, it is women who remain responsible for food preparation and distribution in the kitchens. In this regard, the fact that communal kitchens have become accepted practices may not mean that they are desired, but rather necessary for survival.

Policies and projects which support the kitchens often exacerbate this problem by leaving unexamined women's unequal burdens in community food provision and by assuming that women have expendable time and energy to participate in social reproduction. Economic restructuring and social policies deepen gender inequalities when and if the hidden transfer of reproduction to families is left unexamined. Communal kitchens exemplify contradictions in women's entry into community decision-making: on one hand, women have organized a massive movement which has transformed how central governments, local municipalities, and NGOs understand women's traditional, "private" role in food provision. It has politicized the women around issues of class and gender inequalities as well as political violence – a public awareness which cannot be removed from the historical record, neither for the participants themselves nor for the communities in which they live.

This, however, is not enough to change important policy and political approaches which continue to reinforce gender inequalities by excluding gender as a variable in their frameworks, although initial efforts have been made to engender development frameworks by feminists and other social scientists. Nor is it enough to improve the lives of the participants and their families in significant ways. The . . . economic and social policies of the

Fujimori administration offer little hope for extremely poor communities in Lima who have little access to the benefits of the new state policies. For communal kitchens, this seems to imply a long-term struggle to provide for their families and seek basic levels of dignity in an urban system characterized by deep-rooted structural inequalities.

[. . .]

URBAN POVERTY AND VIOLENCE: MOTHERS' MOVEMENTS

In other cases, women have not mobilized directly to confront the economic crisis, but their efforts reflect related issues of urban violence and poverty. Mothers' movements, or movements of women to combat violence and human rights abuses against their families, have emerged throughout Latin America and other regions in the past twenty years. These movements must be understood in the context of authoritarianism – both under military and democratic states – as well as in the context of rising rates of random violence in urban areas. The *Maes de Acari*, a mothers' movement in the Acari slum neighborhood of Rio de Janeiro, highlights one example of this type of mobilization. This organization began when Mariene de Souza and five other mothers learned that their children had been abducted and "disappeared" in July 1990. . . . [T]he mothers began a long process of investigation and protest of the "Acari Eleven" case. They carried banners before the Police Secretariat building, and made appointments for interviews with the Secretaries of Justice, Public Safety, and the Civil Police. With the participation of local government officials, they organized public rallies. These efforts led them to coalesce with the Center for the Articulation of Marginalized Populations (CEAP). CEAP mobilized support for the mothers from local and international human rights organizations such as Amnesty International, which continues to pressure the Brazilian government on this case.

[Although] the mothers have not learned what happened to their abducted and missing children, the fact that two lawsuits are in process for such a politically loaded case involving indigent slum-dwellers is an impressive victory. In the press, the Acari Eleven are no longer considered criminal

youngsters, but eleven citizens – even if second-class citizens. The murders are now appropriately treated as a political event, rather than a common crime, by activists, the media, and even by some state officials.

Most positively, the struggle of the Mothers of Acari has contributed to greater solidarity among the members of the Acari community, encouraging them to fight more strongly for their social, economic and political rights. The Acari slum has become a symbol of resistance to arbitrary government abuses and associated forms of systemic violence, and the Mothers of Acari have become a reference and a motivator for the broader human rights movement in Rio de Janeiro.

An interesting factor is that the mothers attribute the violence and the disappearances of their children to the conditions of poverty that they face as poor people, and not explicitly to their gender roles and their conditions as Afro-Brazilians. They perceive themselves as constituting an economic class, rather than a class of women and/or a racial class. Nevertheless, they play increasingly strong public roles as women and continue to organize for a less violent society. For the Mothers of Acari, this includes a fundamental transformation in class relations.

[. . .]

Increasingly, mothers' anti-violence movements have emerged in U.S. cities. In the United States with a current uneven income distribution which only compares to the Great Depression of the 1930s, poor communities have been faced with growing sets of problems related to poverty, housing, homelessness, racial violence, domestic violence and lack of affordable healthcare. In the midst of deep racial tensions in urban areas, some women have acted in their roles as mothers to combat violence and reclaim their urban spaces. New York City-based Mothers Against Violence (MAV) is one such organization, which was created to address neighborhood violence at a city-wide level. Mothers Against Violence was created in 1991 to focus attention on the multiple dimensions of the problem of violence, its impact on individuals, families and communities, and to develop practical community responses to this complex problem. In New York City alone, violence is the leading cause of death for young people ages 15–19, and the fourth ranked cause of death for children ages 5–14. MAV is unusual in that it was initiated by City Hall staff, but immediately became an independent nonprofit organization in which the power and decision-making emanate from the neighborhoods. The Deputy Mayor and a number of city commissioners supported the creation of a new nonprofit because they did not believe that government policy was the right vehicle for addressing racial violence.

Members of MAV are victims of violence from diverse cultural, ethnic, and socioeconomic backgrounds. They share the understanding however, that the criminal justice mind set with which the problem was usually addressed had to be changed to a public health mind set, seeing violence as a disease rather than a crime. MAV seeks to increase community-based violence prevention programs, to advocate for better services for victims and survivors, to raise awareness of the extent of the problem, and to reduce the presence of violence in the media. Pain, trauma, fear and apathy have been transformed into effective local prevention and advocacy strategies for program and policy reforms at the city-wide level. MAV's proactive role in mobilizing the community against violence is key in healing the mothers themselves, and also enables young people, parents, and others affected by violence to have similar opportunities, promoting community activism in the process.

MAV activities include public advocacy, providing safe havens for youth, memorial events, and a youth leadership project that involves peer counseling and youth employment initiatives. MAV develops gun violence elimination strategies, convenes annual conferences and publishes conference proceedings and other reports to disseminate information about the problem and what communities and individuals can do to prevent it. The youth in the program have articulated that they need a safe place to come together with their peers, so MAV facilitates recreational programs that keep youth off the streets. MAV has reached over 2,000 people through their chapter activities in seven neighborhoods, and thousands more through the media.

Like the Mothers of Acari and other mothers' movements, MAV members originally acted out of sheer anger, frustration and pain at the disappearance of their children; through the process, however, they have become important public voices in neighborhood and city-wide decision-making processes. From their inception, MAV has addressed an intersection of issues ranging from violence to economic poverty to the stigmatization of working class and minority communities. This perspective on violence has led MAV through a transformative process, from a purely reactive movement to a proactive one where they lead citizen actions to promote positive institutional and policy reforms such as youth leadership training and public advocacy for underrepresented groups.

5. ENGENDERING COMMUNITY ACTION

Women's responses to urban poverty and their entry as a visible political class into community decision-making often occurs indirectly, through their struggles for seemingly non-gender specific needs. Mothers' movements against violence and neighborhood organizations are two such catalysts for women's active involvement in local development. Basu describes poor women's movements as processes of "shared oppression," in which women mobilize around certain issues, only to later discover the gender dimensions of their actions and/or be described as "women's movements." She argues that it is less important whether a particular movement defines itself as a "women's movement," and more important if it responds to the women's concerns and to those of external actors.

[. . .]

These types of women's organizations – neighborhood and anti-violence – reflect the diverse ways in which women coalesce around a shared sense of oppression to meet their needs and those of their communities. They, along with other types of movements such as housing, provide a starting point from which women have entered community decision-making and in some cases, influenced state development policies.

[. . .]

The examples discussed in this paper suggest that what catalyzes women's collective action (or conversely, what prevents women from participating at the community level) stems from the ways in which gender is structured into families and communities, development frameworks, and political and ideological movements (among others), and from how specific groups of women perceive and respond to these structures and practices. One way to understand why community action is (or is not) important to women is by analyzing local power in terms of gender. This includes an analysis of power relations within households, an approach developed in the feminist literature on household survival strategies. Despite the fact that this literature has focused largely on intra-household relations and has not theorized the community or larger public sphere, it has nevertheless positioned the household as a central analytical category and has analyzed the links between households and broader institutions and structural changes. Disaggregating the household on the basis of gender and identifying it as a site of conflict and cooperation, as socially constructed and as an essential part of the economy, rather than as a non-market, natural unit characterized by altruism, provides a different – and indeed complementary starting point for understanding why and how women develop strategies and approaches to community development that broaden our understandings of economic and political participation. As opposed to studies which posit households *vis-à-vis* the economy or civil society without analyzing power relations within them (as in conventional neoclassical and Marxist analysis), the feminist critiques suggest that households, like broader societal institutions, are not neutral, safe or cohesive with respect to gender. Women in particular observe this in their daily lives, and in cities where there is severe economic crisis or restructuring measures, gender inequalities in family structure, job opportunities, household maintenance and childcare tend to increase, often catalyzing women's collective action.

These examples . . . suggest . . . that men and women experience and interpret urban poverty differently, according to their roles in everyday life: their perspectives on parenting, violence, safe living environments, the provisioning of food, schooling, health care, etc. Beneria's (1992) Mexico study, for example, demonstrates how the Mexican debt crisis and subsequent structural adjustment policies also led to the "restructuring of everyday life," including an intensification of domestic work; changes in purchasing habits; and changes in social life. The particular ways in which women experience these effects contribute to the survival strategies they develop – whether they be individual (i.e. family/household-based) or collective (i.e. through community participation). The UNRISD studies suggest a similar process in other cities and demonstrate that community participation can empower women, yet also increase their reproductive workloads if and when community participation is not analyzed in terms of gender. This is especially evident in the communal kitchens, where women complain of being "burnt out" and of their unequal burden in food provision, despite increased community awareness of the shared costs of social reproduction, and in the case of the Mothers of Acari, where members have had great difficulties in balancing their activism with their family responsibilities. Thus the restructuring of everyday life is both an effect of broader restructuring measures and structural inequalities, as well as of communities' locations in that process and the distributive and decision-making mechanisms they develop, both formally and informally.

[. . .]

[T]the successful incorporation of women into the new structures and the acceptance of gender-aware planning agendas would most likely transform the outcome. The literature on urban social movements tends to overlook the important questions of how gender inequalities are reproduced in community structures, why more men are in community leadership roles than women, and how this determines policy agendas in general and women's participation in particular. Urban planners may applaud increases in local power – such as in the role of community development corporations (CDCs), local governments and social movements – without considering how local power is structured along gender lines, and what differential effects the restructuring of communities have for women and men. In other words, increases in local power may not automatically translate into power for women; at the very least, this needs to be explored in future research.

An analysis of gender in this context would include the assertion that (i) gender is an analytical category affecting the allocation of political, social and economic resources; (ii) perceived sex differences help determine patterns of social, political, and economic organization; and (iii) the concept of gender is used to assign men and women to different areas of the economy and society and thus contributes to the distribution of power in both public and private spheres. The gender dimensions of decentralization measures and shifts in local power depend largely upon the structures of local production, local government, civic organizations, and family networks (among others), and on the nature of the shifts that take place and their resulting gender impacts. In terms of economic development, what policy makers may regard as a more productive local economy may instead be a shifting of costs from the paid to the unpaid economy, much of which falls upon women. In terms of local power and community action, it is important to engender analyses of local power structures as well as broaden the scope of the question: for women and women's organizations, empowerment begins with addressing inequalities within their families as well as in society at large. . . . One way to understand the gender aspects of this process is to broaden working definitions of community development and planning to encompass both formal and informal, or both institutionalized and grassroots, planning practices. . . .

6. CONCLUSION

This paper shows that shifts resulting from economic restructuring, combined with the demands generated by increased urban poverty, have catalyzed women to organize collectively, ultimately contributing to new gender-based understandings and community practices in global cities. Women's organizations have responded to the local effects of globalization by creating their own organizations which reflect their gender locations in family structures and in the broader political economy. Throughout the world, women have played protagonistic roles in anti-violence and neighborhood movements (among others) and have engendered the landscape of urban social movements and change. Despite this, the roles of grassroots women's organizations in community development and planning processes remain largely undocumented, and both their collective work and the private household work of women in general have yet to be more fully incorporated into analyses of restructuring, decentralization and other measures associated with the new neo-liberal policy frameworks. One way to overcome these conceptual biases and the

resulting gender impacts of development policies is by integrating feminist approaches to development and to urban social movements.

[. . .]

It is clear that the move toward integrating nation-states into the global economy and shifting welfare responsibilities to local (public and private) levels are not likely to subside or be reversed. Given these circumstances, it is especially important that researchers and policy makers begin to address the gender dimensions of this process in order to prevent further structural constraints and burdens for women and to integrate them into community development.

[. . .]

REFERENCES FROM THE READING

Alvarez, S. (1996) Concluding reflections: 'Redrawing' the parameters of gender struggle. In *Emergences: Women's Struggles for Livelihood in Latin America*, ed. J. Friedmann, R. Abers, and L. Autler, pp. 137–151. UCLA Latin American Studies Center, Los Angeles.

Barrig, M. (1996) Women, collective kitchens and the crisis of the state in Peru. In *Emergences: Women's Struggles for Livelihood in Latin America*, ed. J. Friedmann, R. Abets and L. Autler, pp. 59–77. UCLA Latin American Studies Center, Los Angeles.

Basu, A. (1995a) Introduction. In *The Challenge of Local Feminisms: Women's Movements In Global Perspective*, ed. Amrita Basu, pp. 1–24. Westview Press, Boulder.

Basu, A. ed. (1995b) *The Challenge of Local Feminisms: Women's Movements In Global Perspective*. Westview Press, Boulder.

Beneria, L. (1992) The Mexican debt crisis: Restructuring the household and the economy. In *Unequal Burden: Economic Crisis, Persistent Poverty, and Women's Work*, ed. L. Beneria and S. Feldman, pp. 83–104. Westview Press, Boulder.

Beneria, L. (1995) Toward a greater integration of gender in economics. *World Development 23*(11), 1839–1850.

Beneria, L. and Feldman, S. (eds) (1992) *Unequal Burden: Economic Crisis, Persistent Poverty and Women's Work*. Westview Press, Boulder.

Schild, V. (1991) Gender, Class and Politics: Poor Neighborhood Organizing in Authoritarian Chile. Ph.D. Dissertation, University of Toronto.

The politics of race, ethnicity, and gender

Plate 5 Anti-Spanish graffiti, Barcelona.

INTRODUCTION TO PART FIVE

As the introduction to Part 4 indicated, most large cities in the United States – and for that matter Europe – experienced a massive shift in their populations and economies in the six decades since the end of World War II. Beginning as places where blue-collar, white ethnic, native-born populations were a predominant feature of the social and political landscape, they have ended as cities where members of minority communities, working in the service sectors, increasingly from immigrant backgrounds, now predominate. Cities in different regions and national settings have followed different trajectories; the old cities of the American industrial heartland, and the largest cities of the South, followed a white to black trajectory; elsewhere, in port, trading, and commercial cities such as New York, Los Angeles, Miami, and San Francisco, the minority population is much more diverse. The post-industrial transformation of European cities has been marked by the economic stranding of the guest-workers recruited from Turkey and North Africa, along with people from former colonies who migrated to the metropolitan core in the wake of decolonization.

Whatever their specific form, these population transitions made the political class of elected officials, which tends to change much more slowly than the underlying population, increasingly less representative of the resident populations of their cities. Like all political elites, these elected officials have been loath to give up power simply because their constituencies are declining in number and new ones are appearing. In democracies, however, this situation can be a powder-keg, especially when the new, growing, but under-represented populations have access to the political process. (In European cities, this is much less likely; in American cities, birthright citizenship and relatively easy naturalization provide new immigrant ethnic groups with a somewhat wider set of political opportunities.)

The first and most explosive stage of urban racial and ethnic succession took place in American cities in the 1960s. In city after city, patterns of racial exclusion, discrimination, and subordination angered growing black and Hispanic populations who had been drawn to cities by the promise of economic opportunity. The use of urban renewal, public housing, and urban freeway construction by white urban political elites to manage and reshape patterns of black settlement was deeply disturbing to many. Not only was the need for growing minority groups to be active in urban politics painfully obvious, but the expansion of the urban public sector offered an important avenue for community advancement, especially for African-Americans. For a time, federal policy sought to enhance this involvement through the programs and civil rights legislation of the Great Society.

In city after city, grass-roots community activism, the Civil Rights movement, and urban protest gave way to the election of black and Hispanic city council members, state legislators, and mayors. This began in the late 1960s and culminated in the late 1980s, as cities with large black populations that had not yet elected a black mayor, like New York and Baltimore, finally did so. Because growing Hispanic populations contained more immigrant non-citizens, were not as clearly politicized by a history of formal, legal, racial subordination, and did not live in as high a level of segregation as blacks, they were a bit slower to achieve the same levels of political success. By the late 1980s, however, many cities with large Hispanic populations, like San Antonio and Miami, had also elected Hispanic mayors. (In 2005, Hispanics achieved a political breakthrough with the victory of Antonio Villaraigosa in Los Angeles, the nation's second largest city.) Gradually, between the 1960s and 1980s, the growth of native-born minority populations in U.S. cities began to subside, and in some cases even decline, while immigrant minority populations surged. This development has set the stage for an entirely new kind of political disempowerment – of immigrant minority non-citizens by native-born minorities as well as native-born whites. With the slowing of the racial transition from native-born whites to native-born blacks or Hispanics, and in the wake of many successes for the Civil Rights movement, one can begin to speak of a "post-Civil Rights era" in urban politics.

The readings in Part 5 describe, explain, and assess the process of racial empowerment in urban politics that has occurred over the past three decades. It opens with a summation from Rufus P. Browning, Dale Rogers Marshall, and David H. Tabb of their years of research into the struggle for minority empowerment in American cities. Beginning with a comparative study of ten California cities begun in the late 1970s, published as *Protest is Not Enough*, they edited a volume with essays from leading urban politics scholars who used their analytic framework to explore the nation's other major cities. This volume, *Racial Politics in American Cities*, has gone through three editions as the urban political situation has evolved and matured. The reading concluded here summarizes the fruits of all this work. They find that while the movement for minority political empowerment has accomplished much, many of its original objectives have not been met.

One of the principal achievements of the Civil Rights movement was to propel African-American and Hispanic challengers into office as chief executives of large American cities. Perhaps it was naive to believe that these new leaders would be able to transform the economic, social, and political situations of the communities from which they came, but they often won office on the basis of claims – and hopes – that they could and would do so. In reality, as the readings in Parts 3 and 4 make clear, the powers of mayors and urban governments are highly constrained both by the larger political and economic context in which cities operate and the specific dynamics of forging a governing coalition. Political scientist Adolph Reed, both an acerbic critic of the shortcomings of the political establishment and a seasoned participant in the administration of a black mayor, presents a penetrating analysis of how and why minority mayors have failed to deliver on their initial promises.

The subsequent readings follow some of the developments in urban politics in the "post-Civil Rights" era. (We use quotation marks because, although the era of formal segregation is behind us, racial inequities are still highly characteristic of life in American cities.) Michael Jones-Correa presents a path-breaking analysis of how the new Latin American immigrant groups have attempted to organize themselves in the face of a highly partisan regular Democratic organization in New York City. Journalist Rob Gurwitt provides an overview of how black mayors are seeking to broaden their appeal and move beyond the rhetoric of black exclusion and empowerment. Scott Bollens provides a comparative analysis of the challenge of governing racially and ethnically divided cities that reaches far beyond the American experience to such difficult terrain as Belfast and Johannesburg. Clearly, this challenge, in many different forms, will remain central to urban politics for decades to come.

Can People of Color Achieve Equality in City Government? The Setting and the Issues

from *Racial Politics in American Cities* (2003)

Rufus P. Browning, Dale Rogers Marshall, and David H. Tabb

Editors' Introduction

Latino and especially African-American populations grew substantially in many American cities in the 1960s, but their representation among the urban political leadership lagged. In cities like Newark and Chicago, white-led political machines continued to dominate local politics, alternately ignoring or co-opting black demands.

This lack of political representation was problematic for a variety of reasons. First, it led to material disadvantage, as people of color were largely excluded from the benefits that accrued to dominant groups, including municipal jobs and contracts. Second, it could translate into poorer and less responsive service delivery, as administrators would favor their own (white) neighborhoods, and city employees such as police officers or social workers would operate with less sensitivity to the needs of unrepresented groups. Finally, it had a symbolic resonance: "city hall" was yet another place to which African-Americans could not aspire.

This situation changed dramatically after the 1970s, shaped by shifting demographics (the rising black and Latino populations in many cities), the successes of the Civil Rights movement, and the legacies of some of the Great Society programs (see O'Connor in Part 6) that proved to be a stepping stone for many black and Latino community activists to pursue electoral politics. In many cities, people of color gained power by forming strategic alliances with other groups. These developments prompted a new urban politics literature that sought to document how groups mobilize and seek common ground in cities like New York (see John Mollenkopf, *Phoenix in the Ashes*, excerpted in Part 3), Los Angeles (see Raphael Sonenshein, *Politics in Black and White: Race and Power in Los Angeles* (Princeton: Princeton University Press, 1993)) among others (see James Jennings (ed.), *Blacks, Latinos and Asians in Urban America* (Westport, CT: Praeger 1994)).

Browning, Marshall, and Tabb's 1984 book, *Protest is Not Enough* (Berkeley: University of California Press) made a unique and significant contribution to this literature. They studied the electoral politics and policy responsiveness in ten racially mixed northern California cities over a twenty-year period. They asked three central research questions. First, how open are American urban political systems to the demands posed by newly mobilized groups? Does the optimistic pluralist view (that virtually all groups can mobilize and press demands) or the less sanguine structural analysis (that the less privileged face important institutional obstacles to participation) best describe American urban politics? They found considerable variation between cities: some accommodated demands from newly active black and Latino groups with ease, while others were resistant.

Second, they ask whether mobilized groups achieved what they call "political incorporation." The representation of minority groups, they posit, can be easily measured by counting the number of minority group members in positions of authority. However, some of these elected and appointed officials may be marginal to real decision-making processes, so mere numerical representation may not signal a group's access to power. In their analysis, "political incorporation" is a step beyond representation; they measure a group's incorporation by studying its representation within the coalition that actually dominates the city's decision making. Finally, they ask whether achieving incorporation has enabled blacks and Hispanics to win policy concessions from local governments. They found that it has made a difference in such matters as the appointment of civilian police review boards (often of great interest to minority communities that have had uneasy relationships with the local police), public contracting, and city employment.

The significance of Browning, Marshall, and Tabb's work extends well beyond their insight into the politics of their ten California case study cities. By looking systematically at the ways in which people of color sought local power, and how their achievement of power affected policy outcomes, they provided a framework and a vocabulary for the study of urban electoral politics that has shaped this area of research since their original book's publication. Their focus on coalition building around electoral victories and local policy conflicts builds on the tradition of pluralist analysis, with its emphasis on the ballot-box as the key point of participation (see Dahl and Mollenkopf in Part 3), although they also acknowledge the structural factors that can limit a group's participation, and thwart efforts to effect significant change even once a coalition has achieved political power. As such, their work is not inconsistent with regime analysis (see Stone in Part 3), and many subsequent authors have adopted their concepts while writing more explicitly in the regime tradition.

Since their book appeared, many case studies of race, ethnicity, and local politics have used their framework to analyze urban coalition building. The edited volume containing this introductory chapter includes case studies of how race shaped electoral contests in more than a dozen cities. The case studies in Michael Jones-Correa's *Governing American Cities: Inter-ethnic Coalitions, Competition and Conflict* (Thousand Oaks, CA: Russell Sage Foundation, 2001) offer further examples of political conflict and cooperation between urban racial and ethnic groups, adding a further layer of sociological analysis to considerations of electoral coalition building.

Further reflections on the promise – and limitations – presented by the election of African-American mayors are found in Adolph Reed's reading in Part 5. H. Paul Friesema's "Black Control of Central Cities: The Hollow Prize" (*American Institute of Planners Journal* 35 (March 1969): 75–79) provided a prescient introduction to the topic. More recently, Neil Kraus and Todd Swanstrom have tested Friesema's prediction that white flight and economic disinvestment would undermine the progressive agendas of black-led cities. Their findings are reported in "Minority Mayors and the Hollow-Prize Problem" (*PS* (March 2001): 99–105).

Rufus P. Browning is Director Emeritus and Senior Faculty Researcher, Public Research Institute at San Francisco State University. Dale Rogers Marshall recently stepped down as President of Wheaton College. David H. Tabb is Professor of Political Science at San Francisco State University.

■ ■ ■ ■ ■ ■

[. . .]

THE CIVIL RIGHTS MOVEMENT AND BLACK PROTEST

Waves of political mobilization, demand, and protest – sometimes peaceful, but often violent – swept across the United States from the late 1950s to the mid-1970s. African Americans and their allies mounted assaults on the institutionalized structures of racial exclusion and domination nationwide and in all cities with significant black populations. Latinos, who have a long history of engagement with civil rights issues, accelerated their mobilization too.

First came the civil rights movement, challenging the exclusion of African Americans from politics, government, and education, etching scenes that will forever mark the American consciousness: National Guardsmen escorting black children into school through mobs of enraged whites, lunch counter sit-ins, Governor George Wallace – "segregation today, segregation tomorrow, segregation forever" – blocking the doorway of the University of Alabama to federal officials, Martin Luther King,

Jr.'s impassioned plea for equality from the steps of the Lincoln Memorial, marches in Selma and Birmingham in Alabama and the attacks on them, the murder of civil rights workers, burnings of black churches.

Mass violence erupted in the mid-1960s. Riots in Los Angeles, Detroit, Newark, and dozens of other cities both expressed and aroused fear, anger, and hatred. Leaders struggled to control events and prevent cities from burning. The riots were followed by recriminations, investigations, and heightened demands.

The federal government initiated programs aimed at poverty, racial inequality, and discrimination – and at defusing protest. President Lyndon Johnson pushed aggressively for passage of the Civil Rights Act of 1964, the Voting Rights Act of 1965, and the "war on poverty" created by the Economic Opportunity Act of 1964. These were followed by Model Cities and a tidal wave of other programs in employment, housing, education, and health, many of which changed the activities and resources of city governments but also the prospects and resources of blacks, Latinos, the unemployed, low-income workers, and inner-city residents. During the first Nixon administration (1969–1973), the federal system of grants to cities was reorganized but continued to expand with the institution of general revenue sharing and block grants.

Since the 1970s, the great passion and commitment of the civil rights movement has been defused by its achievements, both real and symbolic. The support of whites for fundamental civil rights, evoked with the deeply moral and religious voice of Martin Luther King, Jr., could not be sustained and transformed into support for the economic agenda that beckoned after federal power had been applied to voting registration and the integration of schools and universities. The movements for civil rights and black power were also suppressed by assassination of their strongest and most charismatic leaders – King and Malcolm X – and eclipsed by other issues, in particular the war in Vietnam. They suffered the attrition of exhaustion, fear, and generational change.

With a series of Republican and moderate Democratic presidencies beginning in 1969, the organizations that carry the mantle of the black civil rights movement, such as the National Association for the Advancement of Colored People, the Southern Christian Leadership Conference, and the Urban League, lost visibility and access to the federal government and became less active. Electoral organization and office holding at all levels, by African Americans especially but also by Latinos and Asians, have grown nationwide, while the dramatic protests that so gripped public attention in the 1960s and 1970s virtually ceased, though with notable exceptions: violent civil unrest in Los Angeles in 1992 and Louis Farrakhan's Million Black Men march in 1995.

In many cities, biracial or multiethnic coalitions formed, and African Americans and Latinos rose from exclusion to positions of authority as mayors, council members, and top managers and administrators. Where this happened, the politics of mobilization and mass action were replaced by the politics of administration, implementation, planning, and economic development – and sometimes by crises of competence and corruption, as in governments generally. Open conflict both within and between minority groups now represented in city governments has sometimes replaced the unity that was once attained when city government and its white power holders were the common enemy.

THE STRUGGLE FOR DEMOCRACY IN CITY POLITICS

Much of the denial of civil rights and of rights to equality in employment, education, housing, and government services occurred at the local level, where people lived, worked, voted, and were subject to the imposition of police power and other local regulation. Accordingly, much of the civil rights movement and of local mobilization by African Americans, Latinos, and other groups aimed to force city governments to end their massive, blatant, common, and virtually complete discrimination and exclusion, and to engage the power of city governments on the side of reducing discrimination in private employment and housing. These historic efforts became tests not only of the ability of groups to sustain a high level of political activity and achieve their goals, but tests as well of the American polity itself, a running experiment on the proposition that excluded groups in a racially obsessed society could realize the democratic promise of the American political ideal.

The continuing efforts of African Americans and Latinos – and of Asians, in some cities – to achieve access to government and responsive policies from cities, and the response to their efforts, are the subjects of this [reading]. As we shall see, these struggles have achieved changes that are striking in their scope and significance. Standing in 1960 and looking forward from the near-total exclusion of African, Latino, and Asian-American people from government in the United States at that time, it would have seemed incredible that an African American could become a general in the U.S. Army, Chairman of the Joint Chiefs of Staff, and Secretary of State or the powerful Speaker of the California Assembly; or that blacks would be mayors of New York, Los Angeles, Chicago, Philadelphia, Washington, DC, Seattle, San Francisco, and many other cities.

Such achievements stemmed from rapid social and economic gains, including the breakdown of barriers to higher education and the professions, a change in attitudes and practices that had kept minority people out of many trades and jobs, and the extraordinarily rapid growth of an African-American middle class from the 1940s on. On the other hand, while "the American condition is overall dramatically improved" in racial matters, it remains "in important respects, continually depressing," and "tenacious ills remain," including intractable gaps in education, employment, and income; continued segregation and racial isolation; and "the vastly disproportionate involvement of black males in the criminal justice system" (Foreman 1999, 5). Even the astonishing long boom of the 1990s, though it reduced unemployment and poverty among many groups, fell considerably short of prosperity for all.

In city politics, the value of the benefits gained through mobilization and participation is questioned by some. The momentum of the movement has slowed, its successes have been uneven over time and from city to city, and its gains have been subject to attack and reversal: A long tide of reaction to racial and other social changes of the 1960s strengthened forces at all levels of government that dismantled programs intended to undo or counterbalance discrimination and cut funding that many people of color (and many whites) as well as their organizations and leaders believe is necessary for continued progress toward equality and

regard as rightful compensation for the barriers and deprivations of a racialized society. In the courts, decisions were entered against affirmative action in government contracting and in both admissions and financial aid in higher education.

In California, four successful ballot initiatives that had their greatest impact on people of color were sponsored by conservative interests and the California Republican Party. The cleverly misnamed "California Civil Rights Initiative" (Proposition 209, 1996) prohibited by constitutional amendment a wide range of state affirmative action programs that provided selective educational and employment support to people of color. . . . And in the largest cities in the country – Los Angeles, New York, Chicago, Philadelphia – African-American mayors were replaced by significantly more conservative whites. It was not unreasonable to fear that the great expansion of office holding by African Americans was only temporary or would fail to sustain long-term equality in law and policy and be followed by re-establishment of white rule at the local level and reversal of legislative gains at all levels.

Even with the expansion of office holding by blacks, Latinos, and Asians in some cities, many with substantial black, Latino, or Asian populations still have no, or very little, minority representation in city council and mayoral offices. Even where they hold office, how much power do they have? Can black, Latino, and Asian office holders really make city governments responsive to the interests of their groups? Clearly, in some cities, people of color can control local policies on some issues and at least some departments of city government. Control of police departments in particular remains especially difficult and onerous in many settings, as in Los Angeles. Even where people of color hold office, can they make any headway against unemployment and poverty, which remain painful and intractable problems in a racialized society, especially for people of color? If they try to reallocate resources to their people, can they still attract the investors and financial institutions on which cities depend for investment and economic growth? Will economic and demographic forces – recession, globalization, high rates of immigration, and reaction to immigration – hinder or support political equality? These questions ask us to look beyond the achievement of local office to the problematic nature of local office in a racialized society, in which by far the greater power remains in the hands of the dominant group in the economy and at the higher levels of government.

In short, even with the growing number of black, Latino, and Asian officials, it may be that the limited powers of cities in a federal system and in a racialized, capitalist society render that gain more symbolic than real. When forces succeed in reversing policies of the 1960s and 1970s, as with the dismantling of race-targeted affirmative action in California, the future of progress for African Americans and Latinos especially is thrown into doubt. (Whether the end of race-targeted affirmative action will actually improve the status of those groups in the long run is, of course, a point of contention.)

African Americans and Latinos are the two largest minority groups in the United States. Together, they comprised 21 percent of the population in 1990 and 25 percent in 2000, about equally split between the two groups, and much larger proportions in many states and cities. Include Asians, and we have 29 percent. The quality of their mobilization and their capacity to sustain political power in cities are crucial to their ability to gain continuing access at the national level of government as well as a voice in the governance of the cities where most of them live. And because many contenders for state and national office first hold local office and learn from their formative experience in city politics, it is important to understand local office holding and the diverse lessons that experience teaches.

Most important, this book offers a current report on the efforts of racially subordinated, excluded groups to gain equality by election-continuing experiments in democracy. Their efforts often arise out of frustration over persistent, racially determined inequality. We know from experience in Los Angeles and elsewhere – for example, the videotaped police beating of Rodney King in 1991 and the racially charged violence of April 1992 – that racialized official violence and explosions of rage among inner-city black and Latino populations can still occur. We know that racially motivated crimes continue to be a problem. We know that racial profiling and differential treatment of suspects based on race remain intensely hated practices of state and local police in many parts of the country. We know that educational opportunities are typically much worse for low-income groups, especially people of color, than for middle- and upper-income groups. Whether governments are really able to respond effectively to demands for equal treatment and social justice remains an open question to be decided in the ebb and flow of social and political contention.

THE CITIES

This book addresses these questions by bringing together thirteen chapters on the political mobilization and political power of African, Latino, and Asian Americans in twenty cities. Table 1 shows the fundamental demographics of these cities. In the top half of the table . . . ten [large] U.S. cities [analyzed in the book] are presented; in the bottom half, the ten cities in Northern California that are discussed in [that were the subject of an earlier study]. In each half, cities are shown in order of increasing percentage of white, non-Hispanic residents.

These twenty cities include the four largest cities in the country – New York, Los Angeles, Chicago, and Philadelphia – and other major cities in diverse regions. Black mayors hold office in six of the cities (30 percent) as of early 2002 – Philadelphia, Atlanta, San Francisco, Birmingham, New Orleans, and Denver – a net gain of one since 1995. Oakland and Baltimore lost black mayors; Philadelphia, San Francisco, and Denver added them. Latinos hold the mayoral office in Miami and San Jose (10 percent of the cities), with no net change since 1995: San Jose added, Denver lost. One city has an Asian (Filipino) mayor: Daly City, CA, where the Asian population, mainly Filipino, spurted to slightly more than 50 percent by the 2000 Census.

The ten U.S. cities studied are larger and blacker on average than the ten California cities. The California cities have much larger Asian populations, in percentage terms, than the ten U.S. cities.

To locate these twenty cities in the universe of U.S. cities, Figure 1 places them in a scatterplot with the other 230 U.S. cities with populations of 100,000 or more in 2000. The vertical axis is the relative size of the white, non-Hispanic population; the horizontal axis is total population. As you can see, most cities with at least 100,000 people are relatively small and mostly white compared

Table 1 Populations of the twenty cities studied in this book

City	Total population in 2000 (1,000s)	Percentage			
		White	Black	Latino	Asian
Ten U.S. Cities					
Miami	363	11.8	19.9	65.8	0.6
Birmingham	243	23.5	73.2	1.6	0.8
New Orleans	485	26.6	66.7	3.1	2.3
Los Angeles	3,695	29.7	10.9	46.5	9.9
Baltimore	651	31.0	64.0	1.7	1.5
Chicago	2,896	31.3	36.4	26.0	4.3
Atlanta	417	31.3	61.0	4.5	1.9
New York City	8,008	35.0	24.5	27.0	9.7
Philadelphia	1,518	42.5	42.6	8.5	4.4
Denver	555	51.9	10.8	31.7	2.7
Mean	1,883	31.5	41.0	21.6	3.8
Ten Northern California Cities					
Daly City	104	17.7	4.3	22.3	50.3
Richmond	99	21.2	35.6	26.5	12.2
Oakland	400	23.5	35.1	21.9	15.1
Hayward	140	29.2	10.6	34.2	18.7
Vallejo	117	30.4	23.3	15.9	23.8
Stockton	244	32.2	10.8	32.5	19.3
San Jose	895	36.0	3.3	30.2	26.6
Sacramento	407	40.5	15.0	21.6	16.4
San Francisco	777	43.6	7.6	14.1	30.7
Berkeley	103	55.2	13.3	9.7	16.3
Mean	328	33.0	15.9	22.9	22.9

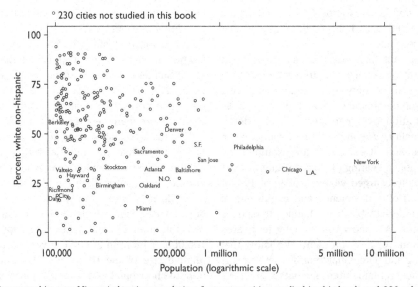

Figure 1 Percent white non-Hispanic by city population for twenty cities studied in this book and 230 other cities with 100,000 or more population in 2000.

to the cities studied in this book. The cities of this book include the largest cities and other cities of various sizes, and eighteen of them are in the range of 20 to 55 percent white, non-Hispanic population. In this middle range, the non-white and Hispanic population is large enough to win or influence elections if they can be united and mobilized but, typically, not so large that the role of whites is negligible. (Even where whites comprise only 20 percent of the total population, they may be a majority of the voting population for many reasons, including more adults, fewer noncitizens, and higher levels of education among whites.) In short, this is a range in which the political opportunities for ethnoracial minorities are good, but coalitions will typically be necessary if control over city government is the goal.

Race and ethnicity

Measurement is social and political. The measurement method of Table 1 maximizes the reported sizes of Latino populations by including in them all persons who identify themselves as Hispanic or Latino, regardless of their racial self-identification on U.S. Census forms. Why is this a problem? As Orlando Patterson has pointed out, "Although all reports routinely note that 'Hispanics can be of any race,' they almost always go on to neglect this critical fact, treating Hispanics as if they were, in fact, a sociological race comparable to 'whites' and 'blacks'." In fact, however, "48 percent of so-called Hispanics classified themselves as solely white" in the 2000 Census (Patterson 2001). As Patterson argues, repeatedly naming and classifying a group under a single category, in the context of an explicitly racial classification, is part of the process by which a group is racialized in a racially obsessed society.

We do not wish to aid the process of racialization. On the other hand, we do want to use indicators of group presence that approximate the size of the group if it were politically mobilized as a group. Because bad things do happen to people who can be identified as Hispanic/Latino and do lead to political mobilization, a measure of the maximum size of that group is relevant. The real racial and ethnic diversity of Latinos is also politically relevant in some settings, as the studies in this book point

out. This point applies also to Asians, who are wholly a socially constructed category and even more diverse in terms of ethnicity and language than Latinos.

WHAT IS POLITICAL POWER IN CITIES?

All the chapters in this book use a framework that we developed studying cities in northern California (Browning, Marshall, and Tabb 1984). Here we identify its main outlines in order to frame the fundamental political problem faced by excluded minorities. Look at the question – What is political power? – from the perspective of people who have long been excluded from holding office or from any significant influence over city government. Suppose they decide to contest their exclusion from city politics, and suppose power over city government is the target of their efforts. They know they have achieved power when they wrest concessions from an unwilling city hall, when they win office against determined opposition, when they succeed in forming a coalition that defeats an incumbent group, when their coalition is able to change the policies and personnel of city government, and when they are able, over a period of years, to institutionalize the changes they sought.

Examples of demands that African Americans and other groups have won in this way include representation in elective offices, access to employment in city government, appointments to head city agencies, application of the enforcement powers of city government to reduce or punish discrimination, and equitable allocation of funds for city services. These are fundamental and legitimate interests clearly within the authority of city governments.

The goal of representation has been at the center of the struggle for political equality. Considering their virtual exclusion from city governments in the late 1950s, it is an astonishing achievement that many people of color now hold office as mayors, council members, and other officials. As important as representation is, however, it is not enough. The presence of people of color in office is essential, but it does not ensure that they will pursue and realize the substantive demands.

To change the direction of city government in the face of opposition requires control of local legislation, programs, spending, governmental structure, and governmental personnel over a period of years. This means that blacks, Latinos, and Asians, if they are to achieve their political goals in any city, must secure control of city council and the mayor's office and hold it for years in the face of opposition. A group must either constitute a majority on its own or participate in a *governing coalition* that can dominate city council on issues of greatest concern to it and secure re-election. Such a coalition does not have to consist entirely of people of color, but it does need a strong commitment to their interests if they are to obtain the changes in city government that they want.

The key is not just representation but a coalition that controls city government. Even substantial minority-group representation – 30 to 40 percent of a city council – will have no effect on policy if they are opposed at every turn by an entrenched and intransigent governing coalition. Where minority groups fall short of 50 percent of the voting-age population, as they usually do, biracial or multiracial coalitions between groups and with liberal whites are necessary to replace, a resistant governing coalition if these groups are to gain control of city government.

Political incorporation

We use the term political incorporation to refer to the extent to which group interests are effectively represented in policy making. We measure the political incorporation of a group by the extent to which it is represented in a coalition that dominates city policy making on issues of greatest concern to that group. This measure involves a supposition – that coalition control of city government really does, at least on key issues, "effectively represent" a group in policy making. This is not an assumption; it is a hypothesis to be tested against the results of such control.

Political incorporation, as a measure, thus refers to a range of possibilities for group presence in city government. At the lowest level, a group is not represented at all – there are no officials from the group, and the group does not participate in a coalition that controls city government on the issues of greatest concern to it. At the next level, there is some representation, but on a council dominated by a coalition that is resistant to minority interests. Finally – the strongest form of incorporation – a group has an equal or leading role in a dominant coalition that is strongly committed to minority interests. The highest levels of political incorporation may afford substantial influence over policy.

What did blacks, then Latinos and Asians, want of city governments?

First, they wanted to end their exclusion from government and the political process. In the twentieth century before the 1960s, they were almost universally and totally excluded. And exclusion remains an issue in some cities: Weak mobilization and lack of political incorporation is common, especially in smaller cities.

They wanted respect from government, access to it, and real influence over policies and programs of special interest to them. They wanted to have their concerns taken seriously, to hold office, and to shape city policies and spending priorities. Real influence is very much an issue, or is nonexistent, in many cities.

They wanted a share of the benefits of government and an end to discrimination. Starting from nearly zero in many cities, they wanted – and often still want – increased minority employment in city government. They want to see minority administrators in top city jobs. They want minority businesses to get some of the city's contracts and purchases. If economic development funds are considered, they want minority business districts to get a share. They want police to stop shooting and beating minority suspects and to end racial profiling, which imposes real risk and unending humiliation. They want low-income housing, parks and recreation programs, police protection, libraries, and health and other services in minority neighborhoods.

They wanted, in short, government that includes them, that is fair, and that is responsive to a broad range of deeply felt demands. However, because in 1960 racism was pervasive, not rare; because whites controlled all the functions of city governments and discrimination was accepted practice; and because people in power do not willingly give it up, a prolonged struggle was necessary if minority demands were to be met, even in part.

Figure 2 Protest strategy for a newly mobilizing group.

What forms did the struggle take?

Groups pursue political objectives in several ways. Aside from terrorism, groups may petition or pressure government from outside – the interest-group and protest strategy – or they may achieve representation and a position of influence inside the electoral strategy. These are not mutually exclusive approaches, and large groups typically pursue both. The strategy that is dominant in a given setting and the intensity of protest will shape the political struggle that ensues.

In 1960, African Americans, Latinos, and Asians had almost no representation in city governments and not even a serious audience for their concerns. In the 1950s, African Americans commonly met steadfast resistance to even the most basic requests for an end to discrimination. The great force of the national civil rights and black power movements pushed these encounters from request to demand to protest. Mobilization of Latinos came later, and of Asians later still. Latinos and Asians also face real civil rights issues and significant barriers to political mobilization and incorporation, but their mobilization in cities has been less intense than the urban mobilization of African Americans was in the 1960s and 1970s – less enraged, less violent, less disruptive, and more consistently oriented toward electoral mobilization.

We can portray a successful protest strategy for a newly mobilizing group as follows [Figure 2].

The group mobilizes and applies pressure to city government with demands backed by protest-marches, pickets, displays of anger and determination, disruptions of public meetings, and the like. A co-optative or a responsive governing coalition in city government may respond by appointing one or more minority representatives to vacancies as

they occur on city council or to managerial positions in city government; in any case, the coalition responds positively to some of the group's demands. The short-term success of the strategy depends on the responsiveness of the governing coalition – a quality in short supply during the 1950s and 1960s.

Where a group is sufficiently large or can find allies for a coalition, an electoral strategy might be feasible. A successful electoral strategy looks like this [Figure 3].

In this scenario, the focus is on electoral mobilization. If successful, it leads to representation and some level of group incorporation into city government; in turn, the extent of incorporation determines the extent to which city government is responsive to group interests. Protests may be carried out, but their primary function is to arouse minority populations and their potential supporters, to raise the level of anger and create the possibility and the determination to act. The position of power in city government achieved by minority-group office holders, rather than solely group pressure on city government from the outside, leads to changes in city government that make it more responsive.

There is nothing inevitable to these scenarios: a coalition in power may be utterly unresponsive to minority demands, or electoral effort may lead to no victories and no representation. Still, they constitute possible ways of influencing city governments. Both strategies were and are employed by mobilizing groups.

URBAN POLITICS AND ECONOMIC POWER

The interest-group and electoral strategies are commonplace for all of us, because they are the

Figure 3 Electoral strategy for a newly mobilizing group.

stuff of public discourse about politics and political power. The formation of coalitions, the importance of political leaders in framing issues and building coalitions, mobilization for protest and for elections, public disputes about policy and public funds – these elements of news and talk about politics are both familiar and real, but it is important to understand that the emphasis on them – the assumption that they are important – is a matter of perspective. In political science, this perspective is called the *pluralist* perspective. It is a view of politics as contention among many groups for control of political institutions, with the presumption that those institutions are significantly autonomous, possess real authority, and control important resources.

A *class* or *structural* perspective, in contrast, sees politics from a different vantage point, looking not at the decisions of a given city government but at the relationship between government and the economic structure of society. Through the structural lens, we observe that government is not autonomous, that it is fundamentally constrained by the structure of business interests [Mollenkopf 1992, p. 27]:

> Private property, market competition, wealth and income inequality, the corporate system, and the stage of capitalist development pervasively shape the terrain on which political competition occurs.

In the structural view, city politics reveal deep and lasting inequalities that government can do little about, and the forces that produce and maintain inequality are likely to be more important than the limited autonomy of local government and the limited benefits that governmental action might secure. The structure and dynamic of capitalism, the institutional power of local corporations, and the cumulative inequalities embedded in capitalist institutions exert profound effects on urban politics and government.

Consider how the transformative dynamic of capitalism can undermine the prosperity of a group. African Americans migrate from the South to escape the oppression of sharecropping and seek employment in the great industries of Detroit, Chicago, and Cleveland. Decades of struggle to overcome the racism of both employers and white workers

eventually yield great gains in employment and income. Then, beginning in the 1970s, these gains are steadily undermined as global competition drains market share and employment from steel, automobiles, machinery, and other manufacturing industries. Unemployment and poverty among African Americans increase again, and a portion of the African-American population is mired without hope or opportunity in big-city ghettos. The wealth those industries created is no longer available to finance local governments and local needs; it disappears or moves elsewhere, leaving behind the people who depended on it. Even if people of color have gained control of city governments in such a setting, they will be under constant pressure to compete with other cities for outside investment to reduce taxes and channel resources toward infrastructure that is attractive to investors, thus reducing resources for new programs.

In addition to the overarching structure and transformative dynamic of the global capitalist system, organized and powerful local economic elites may dominate city government. To express the likelihood that strong economic institutions mold city government to their interests, we must refer to a broader governing coalition – a network of political and economic leaders that these institutions seek to form and control according to their economic objectives and that, in turn, is able to control city government. To differentiate this broader coalition from a governing coalition located entirely within the political/governmental sphere, it is called a regime (Stone 1989, 2). While electoral coalitions are typically public because they depend on public support, regimes are partly hidden – they thrive on secrecy about the extent of their power and their ability to produce governmental actions that enhance their power and profits.

Seen from the structural perspective, elections and voting are of limited interest, because they cannot explain why city governments behave as they do. Instead, the extent of structural inequality and the power and objectives of local economic elites in a city constitute the fundamental explanation: "the unequal distribution of economic, organizational, and cultural resources has a substantial bearing on the character of actual governing coalitions" in cities (Stone 1989, 9).

Though often presented as being opposed, the pluralist and structural perspectives are, we believe, complementary. They are both necessary. They offer different truths and allow us to see different possibilities; but neither has a viable claim to be the truth. Their different central claims are both at least sometimes true, and must employ both to understand urban politics fully.

Between them, the two theoretical perspectives identify five fundamental elements of urban political systems. This list does not exhaust the institutions with power over city governments – for example, it does not include state and federal governments – but it does identify the key actors in the immediate vicinity of local institutions:

1. Political entrepreneurs, leaders who are typically coalition builders and who seek to lead city government.
2. Public sector producer, interests within government – administrators, employees, unions.
3. Popular or constituency interests, who may express their demands through elections and interest-group activity.
4. Private market interests, especially corporations with discretion over local capital investment.
5. A dominant political coalition – "a working alliance among different interests that can win elections for executive office and secure the co-operation it needs from other public and private power centers in order to govern" (Mollenkopf 1992, 38). This governing coalition is the product of the efforts of political entrepreneurs who build support from the other key actors.

This governing-coalition approach borrows from the pluralist view the proposition that a key interaction takes place between popular interests (3) and political entrepreneurs (1) as the latter seek to develop viable coalitions (5). From the structural approach, the governing-coalition perspective borrows the proposition, that an exercise of power occurs in the broader sphere between the political coalition that is able to dominate city government (5) and private market interests (4), with market interests almost always influential but with a wide range of possibilities for dominance, collaboration, or stalemate. It is also possible that public employees (2) play significant roles in the allocation of public money and privilege.

While we accord private market interests a prominent place in any account of the capabilities and inclinations of city government, we also look for the possibility that political entrepreneurs and the alliances they may build may create some autonomy for the political sphere and obtain the co-operation, support, or forbearance of market interests.

Thus, the achievement of political power by a hitherto excluded group in or over city government must overcome two potential obstacles – the presence of a dominant political coalition that will not give up power without a struggle, and the presence of market interests that may organize to dominate and shape any political coalition.

[. . .]

REFERENCES FROM THE READING

Alford, Robert R. and Roger Friedland. 1985. *Powers of Theory Capitalism, the State, and Democracy.* Cambridge, UK: Cambridge University Press.

Browning, Rufus P., Dale Rogers Marshall, and David H. Tabb. 1984. *Protest Is Not Enough: The Struggle of Blacks and Hispanics for Equality in Urban Politics.* Berkeley: University of California Press.

Foreman, Christopher H., Jr. 1999. The Rough Road to Racial Uplift. In Foreman, ed. *The African-American Predicament.* Washington, DC: Brookings Institution.

Mollenkopf, John H. 1992. *A Phoenix in the Ashes. The Rise and Fall of the Koch Coalition in New York City Politics.* Princeton, NJ: Princeton University Press.

Patterson, Orlando. 2001. Race by the Numbers. *New York Times*, May 8, A31.

Stone, Clarence N. 1989. *Regime Politics: Governing Atlanta 1946–1988.* Lawrence: University Press of Kansas.

F
I
V
E

"Demobilization in the New Black Political Regime: Ideological Capitulation and Radical Failure in the Post-segregation Era"

from Michael P. Smith and Joe R. Feagin (eds),
The Bubbling Cauldron (1995)

Adolph Reed Jr.

Editors' Introduction

The election of record numbers of people of color to positions of authority in recent years has been a striking victory for the Civil Rights movement. The ever-growing numbers of African-American and Hispanic mayors, council members, and administrators is testament to successful mobilizations along many fronts. Voting rights struggles have reduced legal barriers to the fair exercise of the franchise; equal opportunity efforts have enabled previously excluded groups to benefit from higher education. As a result, voters have more opportunities to choose candidates of color, and there is a growing pool of blacks and Hispanics eager and able to take on complex tasks of governance. In sum, these previously marginalized groups have now achieved, in the words of Browning, Marshall, and Tabb, a high degree of political incorporation, and nowhere is this more evident than in the arena of city politics.

Adolph Reed applauds these developments. He understands that political representation translates into more jobs and better allocation of public goods, which are not trivial achievements. But these gains have certainly fallen short of the promises of the Civil Rights movement and of progressivism more generally – indeed, in some ways political incorporation has itself thwarted further, farther-reaching gains for those seeking truly equal opportunity and equal justice. As a result, he claims, as the number of African-American office holders has increased, the material circumstances for most African-Americans have deteriorated. This can be explained, he believes, by looking at the factors behind the "demobilization" of black politics in the 1980s and 1990s.

First, he notes, incorporation itself tends to have a demobilizing effect on any group. Once elected to positions of power, black leaders cannot just protest any more – they need to govern, and they need to govern within the limits placed by a system that, as Paul Peterson notes in Part 3 and Martin Shefter underscores in Part 4, forces all leaders to pursue policies that conform to fiscal imperatives. Once they are in authority, regardless of how they got there, political leaders tend to share a desire to maintain the status quo. Mobilizing new voters does not seem so appealing to those who have managed to get elected by the old group of voters. So black political officials, like their white counterparts, become more interested in protecting what *is* than in fostering what *could be*.

Second, the mobilization of people of color has been hampered by discourses on the Left and the Right that focus on individual rather than institutional behavior as an explanation for persistent inequality. Reed takes issue with those (academics, journalists and politicians) who have popularized the concept of the "underclass" – a group of ultra-poor, mostly black city-dwellers, who are defined not just by their poverty, but by their dysfunctional behaviors: out-of-wedlock births, high divorce rates, unemployment, substance abuse, and criminal activity most prominently (Ken Auletta's *Underclass* (New York: Random House, 1982) was one of the first books to put forward this argument). Critics of the "underclass" concept have noted that many of these behaviors can easily be found, and indeed have become increasingly common among other social classes. When those with wealth get divorced, drink to excess, and cheat on their taxes, however, journalists and research institutes don't rush to label them as deviant. By framing the discussion of poverty around these behavioral paradigms, Reed maintains, social critics and policy makers can blame the poor for their own predicament. Thus, instead of policy solutions that address structural economic inequalities, we get welfare reforms that punish the poor for their allegedly problematic behavior (for an excellent review of this area of inquiry we recommend Michael Katz's edited volume, *The "Underclass" Debate: Views from History*, (Princeton: Princeton University Press, 1993)).

The corollary of a discourse blaming the poor for their poverty is the rhetoric of "self-help." Liberals and conservatives, claims Reed, found consensus in the 1980s around the idea that poor communities can help themselves by forming wealth-generating initiatives, rather than waiting for government policies to deliver economic benefits. Organizations such as community development corporations (see Peter Dreier's article in Part 4), which sought to build housing and start businesses on a not-for-profit basis, appealed to the Left because of their promise of grass-roots activism and eventual empowerment, and to the Right because they suggested that the barriers of poverty could truly be overcome by hard work (and that those failing to escape poverty simply had not worked hard enough). Whatever benefits community-based organizations have managed to generate, Reed argues, they have furthered the demobilization of black politics. Just as political incorporation forces leaders to focus on picking up the garbage and balancing the books, the creation of community-based enterprise draws local leadership into problems of raising funds and cutting payroll checks. In both cases, little energy is left for the pursuit of a broader social change agenda. This dilemma is further analyzed by Frances Fox Piven and Richard Cloward in their classic study, *Poor People's Movements: Why They Succeed, How They Fail* (New York: Vintage Books, 1977, 1979).

Reed raises a number of provocative points, some specific to African-American politics of the 1990s, others broadly applicable to the study of social justice and urban policy at the present time. Although he is clearly writing from the Left, he is quite critical of liberals and progressives who have given up their struggle – progressive political leaders who have too readily become the arm of an inherently unfair state; radical thinkers who have "retreated ever more hermetically into the university." In his conclusion (which is not included in this reading) he calls for a new progressive black politics that focuses more intently on structural economic inequality.

Adolph Reed is a unique voice in the discipline of political science. He has written eloquently and prolific-ally on American political thought, incorporating discussions of contemporary political concerns with a deep understanding of their historical and theoretical underpinnings. His books include: *The Jesse Jackson Phenomenon: The Crisis of Purpose in Afro-American Politics* (New Haven: Yale University Press, 1986), *W.E.B. DuBois and American Political Thought: Fabianism and the Color Line* (New York: Oxford University Press, 1997), and *Stirrings in the Jug: Black Politics in the Post-segregation Era* (Minneapolis: University of Minnesota Press, 1999). He is also a frequent contributor to such periodicals as the *Nation, The Progressive*, and *The Village Voice*, a leading member of the U.S. Labor Party, and an outspoken advocate of other pro-gressive political causes. He is Professor of Political Science at the University of Pennsylvania.

It is ironic that the exponential increases in black public-office holding since the 1970s have been accompanied by a deterioration of the material circumstances of large segments of the black citizenry. Comment on that irony comes both from those on the Left who underscore the insufficiency of capturing public office and from those on the Right who disparage the pursuit of public action on behalf of blacks or push oblique claims about black incompetence. In the middle are liberal social scientists and journalists who construe this inverse association as a puzzling deviation from the orthodox narrative of American interest-group pluralism. The liberal and conservative tendencies especially are often elaborated through a rhetoric that juxtaposes black political power and white economic power, treating them almost as naturalized racial properties, rather than as contingent products of social and political institutions.

At the same time, a different anomaly bedevils those on the Left who presume that oppression breeds political resistance to power relations enforced through the state apparatus. The intensification of oppression over the 1980s – seen, for example, in worsening of material conditions and an expanding regime of social repression – has not produced serious oppositional political mobilization. This is the key problem for articulation of a progressive black urban politics in the 1990s.

Making sense of these anomalies requires examining critical characteristics of post-segregation-era black politics. Although the disparate fortunes of black officialdom and its constituents are not causally linked, their relation sheds light on popular demobilization. This relation connects with each of the three features of the contemporary political landscape that hinder progressive black mobilization: (1) political incorporation and its limits, (2) the hegemony of underclass discourse as a frame for discussing racial inequality and poverty, and (3) the Left's failure to think carefully and critically about black politics and the ways that it connects with the role of race in the American stratification system.

THE LIMITS OF INCORPORATION

Systemic incorporation along four dimensions has been the most significant development in black urban politics since the 1960s. First, enforcement of the Voting Rights Act has increased the efficacy of black electoral participation; invalidation of cruder forms of racial gerrymandering and biased electoral systems; as well as redress against intimidation, have made it easier for black voters to elect candidates.

Second, a corollary of that electoral efficacy, has been the dramatic increase of black elected officials. Their existence has become a fact of life in U.S. politics and has shaped the modalities of race relations management. Black elected officials tend to operate within already existing governing coalitions at the local level and within the imperatives of the Democratic party's internal politics, as well as with an eye to their constituents. The logic of incumbency, moreover, is race-blind and favors reelection above all else. Not surprisingly, black officeholders tend to be disposed to articulate their black constituents' interests in ways that are compatible with those other commitments.

Third, black people have increasingly assumed administrative control of the institutions of urban governance. Housing authorities, welfare departments, school systems, even public safety departments are ever more likely to be run by black officials, and black functionaries are likely to be prominent at all levels within those organizations. Those agencies have their own attentive constituencies, within the black electorate, radiating out into the family and friendship networks of personnel. And a substratum of professional, often geographically mobile public functionaries with commitments to public management ideologies may now constitute a relatively autonomous interest configuration within black politics. This dimension of incorporation short-circuits critiques of those agencies' operations crafted within the racially inflected language most familiar to black insurgency. A critique that pivots on racial legitimacy as a standard for evaluating institutional behavior cannot be effective – as a basis for either organizing opposition or stimulating critical public debate – in a situation in which blacks conspicuously run the institutions. Because they have their own black constituencies and greater access to resources for shaping public opinion, public officials have the advantage in any debate that rests simplistically on determining racial authenticity.

A fourth and related dimension of incorporation is the integration of private civil rights and uplift organizations into a regime of race relations management driven by incrementalist, insider negotiation. The tracings of this process could be seen dramatically at the national level during the Jimmy Carter administration with the inclusion of Jesse Jackson's Operation Push and the National Urban League as line item accounts in Department of Labor budgets. The boundaries between state agendas and elites and those of black non-government organizations may even be more porous at the local level, where personnel commonly move back and forth from one payroll system to another and where close coordination with local interest groupings is woven more seamlessly into the texture of everyday life.

An effect has arguably been further to skew the black politically attentive public toward the new regime of race relations management. On the one hand, generation of a professional world of public/private race relations engineers drawn from politically attentive elements of the black population channels issue-articulation and agenda-formation processes in black politics in ways reflecting the regime's common sense. On the other hand, insofar as the nongovernmental organizations and their elites carry the historical sediment of adversarial, protest politics, their integration into the new regime further ratifies its protocols as the only thinkable politics.

These trajectories of incorporation have yielded real benefits for the black citizenry. They have enhanced income and employment opportunity and have injected a greater measure of fairness into the distribution of public benefits in large and small ways. Black citizens have greater access now to the informal networks through which ordinary people use government to get things done – find summer jobs for their children, obtain zoning variances and building permits, get people out of jail, remove neighborhood nuisances, or site parks and libraries. Objectives that not long ago required storming city council meetings now can be met through routine processes. These accomplishments often are dismissed in some quarters on the Left as trivial and evidence of co-optation. Certainly, such characterizations are true "in the last analysis," but we don't live and can't do effective politics "in the last analysis." For them to function effectively as co-optation, for example, the fruits of incorporation cannot be trivial for those who

partake or expect to be able to partake of them. The inclination to dismiss them reflects instead problematic tendencies within the Left to trivialize and simultaneously to demonize the exercise of public authority.

The new regime of race relations management as realized through the four-pronged dynamic of incorporation has exerted a demobilizing effect on black politics precisely by virtue of its capacities for delivering benefits and for defining what benefits political action can legitimately be used to pursue. Ease in voting and in producing desired electoral outcomes legitimizes that form as the primary means of political participation, which naturally seems attractive compared with others that require more extensive and intensive commitment of attention and effort. A result is to narrow the operative conception of political engagement to one form, and the most passive one at that.

Incumbent public officials generically have an interest in dampening the possibilities for new or widespread mobilization because of its intrinsic volatility. Uncontrolled participation can produce unpleasant electoral surprises and equally can interfere with the reigning protocols through which public agencies discharge their functions. As popular participation narrows, the inertial logic of incumbency operates to constrict the field of political discourse. Incumbents respond to durable interests, and they seek predictability, continuity, and a shared common sense. This translates into a preference for a brokered "politics as usual" that limits the number and range of claims on the policy agenda. Such a politics preserves the thrust of inherited policy regimes and reinforces existing patterns of systemic advantage by limiting the boundaries of the politically reasonable. The same is true for the insider-negotiation processes through which the nongovernmental organizations now define their roles, and those organizations often earn their insider status by providing a convincing alternative to popular political mobilization.

UNDERCLASS RHETORIC AND THE DISAPPEARANCE OF POLITICS

Fueled largely by sensationalist journalism and supposedly tough-minded, policy-oriented social scientists, underclass rhetoric became over the

1980s the main frame within which to discuss inner-city poverty and inequality. The pundits and scholars who created this "underclass" define the stratum's membership in a variety of slightly differing ways; however, they all circle around a basic characterization that roots it among inner-city blacks and Hispanics, and they share a consensual assessment that the underclass makes up about 20 percent of the impoverished population in inner cities.

The underclass notion is a contemporary extrapolation from a Victorian petit bourgeois fantasy world, and it is almost invariably harnessed to arguments for reactionary and punitive social policy. Even at its best – that is, when it is connected with some agenda other than pure stigmatization and denial of public responsibility – this rhetoric is de-politicizing and thus demobilizing in at least three ways.

First, the underclass frame does not direct attention to the political-economic dynamics that produced and reproduced dispossession and its entailments but focuses instead on behavioral characteristics alleged to exist among the victims of those dynamics. The result is to immerse discussion of inequality, poverty, and racial stratification in often overlapping rhetorics of individual or collective pathogenesis and knee-jerk moral evaluation. Conservatives bask in the simplicity of a discourse that revolves around racialized stigmatization of people as good, bad, or defective. Even those versions propounded by liberals, like that offered by William Julius Wilson, which purport to provide structurally grounded accounts of inner-city inequality, describe the "underclass" in primarily behavioral terms.

In both conservative and putatively liberal versions, the underclass rhetoric reinforces tendencies to demobilization by situating debate about poverty and inequality not in the public realm of politics – which would warrant examination of the role of public action in the reproduction of an unequal distribution of material costs and benefits (for example, federal and local housing and redevelopment policies that feed ghettoization and favor suburbs over inner cities, that favor homeowners over renters in the face of widespread and blatant racial discrimination in access to mortgages, and subsidies for urban de-industrialization and dis-investments) – but on the ostensibly private realm of individual values and behavior, pivoting specifically on images of male criminality and female slovenliness and irresponsible sexuality. The specter of drugs and gangs is omnipresent as well, underscoring the composite image of a wanton, depraved Other and automatically justifying any extreme of official repression and brutality. Even when acknowledged as unfounded, invocation of suspicion of the presence of drugs and gangs exculpates arbitrary violation of civil liberties in inner cities and police brutality to the extent of homicide.

[. . .]

Second, the underclass rhetoric reinforces demobilization because of its very nature as a third-person discourse. As a rhetoric of stigmatization, it is deployed about rather than by any real population. No one self-identifies as a member of the underclass. To that extent, as well as because the rhetoric presumes their incompetence, exhortations of the stigmatized population to undertake any concerted political action on their own behalf are unthinkable.

Its association with "self-help" ideology is in fact the third way that the underclass narrative undercuts popular mobilization. Because behavioral pathology appears in that narrative as at least the proximate source of poverty, inequality, and even contemporary racial discrimination, the programmatic responses that arise most naturally within its purview are those geared to correcting the supposed defects of the target population. This biases programmatic discussion toward bootstrap initiatives that claim moral rehabilitation of impoverished individuals and communities as part of their mission.

In this context two apparently different streams of neo Jeffersonian romanticism – those associated respectively with the 1960s' New Left and Reaganism – converge on an orientation that eschews government action on principle in favor of voluntarist, "community-based" initiatives. Particularly when steeped in a language of "empowerment," this anti-statist convergence overlaps current manifestations of a conservative, bootstrap tendency among black elites that stretches back at least through Booker T. Washington at the turn of the century. Indeed, it was the Reagan administration's evil genius to appeal to that tendency by shifting from a first-term tactic that projected combative black voices, like Thomas Sowell and Clarence

Pendleton, to a more conciliatory style exemplified by Glenn Loury . . . and Robert Woodson of the National Center for Neighborhood Enterprise. . . . Although this wave of black Reaganauts could be pugnacious with adversaries, they were far more inclined than their predecessors to make overtures to the entrenched race relations elite. Those overtures disarmed partisan skepticism by emphasizing the black middle class's supposedly special responsibility for correcting the underclass and the problems associated with it.

Underwriting this version of self-help are three interlocked claims: (1) that black inner cities are beset by grave and self-regenerative problems of social breakdown and pathology that have undermined the possibility of normal civic life, (2) that these problems are beyond the reach of positive state action, and (3) that they can be addressed only by private, voluntarist black action led by the middle class. Over the late 1980s and early 1990s these three claims – each dubious enough on its own, all justified at most by appeal to lurid anecdotes, self-righteous prejudices, and crackerbarrel social theory – congealed into hegemonic wisdom. Black public figures supposedly identified with the Left, like Jesse Jackson, Roger Wilkins, and Cornel West, have become as devout proselytizers of this catechistic orthodoxy as are rightists like Woodson, Loury, and Clarence Thomas.

The rise and consolidation of the Democratic Leadership Council and the "New Liberalism" as dominant within the Democratic party no doubt reinforced and were reinforced by black self-help bromides' elevation to the status of conventional wisdom. On the one hand, black self-help rhetoric historically has been associated with presumptions that blacks have no hope for allies in pursuit of justice through public policy, and the successful offensive of Democratic "centrists" and neoliberals . . . certainly lends credence to the impression that the federal government is not a dependable ally of black objectives. Even the celebrated declamations by New Liberal consciences Bill Bradley and John Kerry for racial justice and tolerance were mainly, after brief statements against bigotry, extended characterizations of impoverished inner cities as savage hearts of darkness, saturated in self-destructive violence and pathology; the speeches carried no particular warrant for action addressing inequality and its effects except calls for moral uplift.

Despite its foundation on notions of grassroots activism, the self-help regime is best seen as community mobilization for political demobilization. Each attempt by a neighborhood or church group to scrounge around the philanthropic world and the interstices of the federal system for funds to build low-income housing or day-care or neighborhood centers or to organize programs that compensate for inadequate school funding, public safety, or trash pickup, simultaneously concedes the point that black citizens cannot legitimately pursue those benefits through government. This is a very dangerous concession in an ideological context defined largely by a logic that, like that in the post-Reconstruction era of the last century, could extend to an almost genocidal expulsion of black citizens toward a "Bantustanized" periphery of society.

[. . .]

The problem with self-help ideology is that it reifies community initiative, freighting it with an ideological burden that reduces to political quietism and a programmatic mission it is ill equipped to fulfill. It is absurd to present neighborhood and church initiatives as appropriate responses to the effects of government-supported disinvestment, labor market segmentation, widespread and well-documented patterns of discrimination in employment and housing as well as in the trajectory of direct and indirect public spending, and an all-out corporate assault on the social wage.

Its endorsement by public officials is a particularly ironic aspect of the self-help rhetoric. That endorsement amounts to an admission of failure, an acknowledgment that the problems afflicting their constituents are indeed beyond the scope of the institutional apparatus under their control, that black officials are in fact powerless to provide services to inner-city citizens effectively through those institutions.

A key to overcoming the demobilizing effects of self-help ideology, as well as those of underclass rhetoric more generally, lies in stimulation of strategic debate – grounded in the relation between social conditions affecting the black population and public policy and the larger political-economic tendencies to which it responds – within and about black political activity. This in turn requires attending to the complex dynamics of interest and ideological differentiation that operate within black politics, taking into account the who-gets-what-when-where-how dimension of politics as it appears among black political agents and interest configurations. In principle, the Left should be intimately engaged in this project, which is the stock-in-trade of Left political analysis.

THE LEFT AND BLACK POLITICS

By outlawing official segregation and discriminatory restrictions on political participation, the Voting Rights Act and the 1964 Civil Rights Act rendered obsolete the least common denominator – opposition to Jim Crow that for more than a half-century had given black political activity coherence and a pragmatic agenda plausibly understood to be shared uniformly among the black citizenry. (This effect no doubt is a factor – along with the spread of self-help ideology and the aging of the population that can recall the ancient regime – driving contemporary nostalgia for the sense of community that supposedly flourished under segregation. That perception was always more apparent than real; the coherence and cohesiveness were most of all artifacts of the imperatives of the Jim Crow system and the struggle against it. In black politics as elsewhere, what appears as political cohesiveness has been the assertion of one tendency over others coexisting and competing with it – in this case, first, white elites' successful projection of Booker T. Washington's capitulationist program and then, for the half-century after Washington's death, the primacy of the focus on attacking codified segregation.) The Voting Rights Act, additionally, ensued in opening new possibilities, concrete objectives and incentives for political action, and new, more complex relations with mainstream political institutions, particularly government and the Democratic party at all levels.

In the decade after 1965 black political activity came increasingly to revolve around gaining, enhancing, or maintaining official representation in public institutions and the distribution of associated material and symbolic benefits. The greatest increases in black elective-officeholding occurred during those years. That period also saw the rise of black urban governance, both in black-led municipal regimes and in growing black authority in the urban administrative apparatus.

At the same time this shift exposed a long-standing tension in black political discourse between narrower and broader constructions of the practical agenda for realizing racially democratic interests. The narrower view has focused political objectives on singular pursuit of racial inclusion, either accepting the structure and performance of political and economic institutions as given or presuming that black representation is an adequate basis for correcting what might be unsatisfactory about them. The essence of this view was distilled, appropriately, in two pithy formulations in the late 1960s: the slogan demanding "black faces in previously all-white places" and the proposition that, as an ideal, black Americans should make up 12 percent of corporate executives, 12 percent of the unemployed, and 12 percent of everything in between. The broader tendency is perhaps best seen as an ensemble of views joined by inclination toward structural critique. This tendency sees simple racial inclusion as inadequate and argues for tying political action to insurgent programs that seek either to transform existing institutions or to reject them altogether in favor of race nationalist or social revolutionary alternatives.

The tension between these two views has been a recurring issue in black politics, overlapping and crosscutting – and, arguably, being mistaken for – other fault lines that appear more commonly in the historiography of black political debate (for example, the militant/moderate, protest/accommodationist, and integrationist/nationalist dichotomies). In the 1960s, however, the combination of broad popular mobilization and heightened prospects for victory against legally enforced exclusion made this tension more prominent than at any prior time except during the 1930s and early 1940s, when Ralph Bunche and other Young Turks pushed sharp, Marxist-inspired critiques into the main lines of black debate.

Black accession to responsible positions in the apparatus of public management enabled for the first time – save for fleeting moments in Reconstruction – a discourse focused on the concrete, nuts-and-bolts, incrementalist exercise of public authority. Three factors compel the new pragmatic orientation toward incrementalism. First, the inclusionist program had developed largely as an insider politics, seeking legitimacy in part through emphasis of loyalty, particularly in the cold war context, to prevailing political and economic arrangements except insofar as those were racially exclusionary. To that extent it has been predisposed to take existing systemic and institutional imperatives as given. Second, experience in War on Poverty and Great Society programs socialized the pool of potential black officials into the public management system's entrenched protocols and operating logic, initiating them into existing policy processes. This socialization spurred articulation of a rhetoric exalting realpolitik and keying strategic consideration only to advancement of black representation among beneficiaries within existing institutional regimes. This notion of political pragmatism not only reinforces incrementalism; it also requires a shifting construction of "black" interests to conform to options set in a received policy and issue framework. . . . Finally, inclusionist politics affords no larger vision around which to orient a critical perspective on either the operations and general functions of political institutions or the general thrust of public policy. This characteristic, which might appear as political myopia, is rationalizable as pragmatic; in any event, it further reinforces incrementalism by screening out broader issues and concerns.

The hegemony of incrementalism has facilitated elaboration of a political discourse that sidesteps a critical problem at the core of post-segregation-era black politics: the tension between black officials' institutional legitimation and their popular electoral legitimation. The institutions that black officials administer are driven by the imperatives of managing systemic racial subordination, but the expectations they cultivate among their constituents define the role of black administrative representation in those institutions as a de facto challenge to racial subordination.

So by the 1990s it was commonplace to see black housing authority directors' policy innovations run to advocating lockdowns and random police sweeps, black school superintendents discussing their duties principally through a rhetoric of discipline and calling for punishment of parents of transgressors in their charge, black mayors and legislators locked into a victim-blaming interpretive frame accenting drug abuse and criminality as the only actionable social problems – and all falling back on the bromides about family breakdown and moral crisis among their constituents

to explain the inadequacy of public services. This rhetoric obscures their capitulation to business-led programs of regressive redistribution – tax breaks and other subsidies, as well as general subservience to development interests in planning and policy formulation – that contribute further to fiscal strain, thus justifying still further service cuts, which increase pressure for giving more to development interests to stimulate "growth" that supposedly will build the tax base, and so on. From this perspective, Sharon Pratt Kelly's Washington, D.C., mayoralty [1990–1994] is emblematic; her tenure was distinguished only by repeated service and personnel cuts and her 1993 call for the National Guard to buttress municipal police efforts – even as the District of Columbia already has one of the highest police-to-citizen ratios in the United States. There could hardly be a more striking illustration of the extent to which minority public officials are the equivalent of Bantustan administrators. Incrementalism serves as blinders, sword, and shield. It blocks alternative courses from view, de-legitimizes criticism with incantations of realpolitik, and provides a Pontius Pilate defense of any action by characterizing officials as incapable of acting on their circumstances.

Continued debate with the oppositional tendency in black politics could have mitigated the corrosive effects of incrementalist hegemony. Such debate might have broadened somewhat the perspective from which black officials themselves define pragmatic agendas. It might have stimulated among black citizens a practical, policy-oriented public discourse that would either have supported black officials in the articulation of bold initiatives and/or held them accountable to autonomously generated programmatic agendas and concerns.

Yet few would dispute the argument that radicalism has been routed in post-segregation black politics. Some fit that fact into a naturalistic reading of incorporation: radicalism automatically wanes as avenues open for regular political participation. Others concede incrementalist, petit bourgeois hegemony in electoral politics but claim that radicalism's social base has not been destroyed but only displaced to other domains – dormant mass anger, Louis Farrakhan's apparent popularity, rap music and other extrusions of youth culture, literary production, and the like – suggesting a need to re-conceptualize politics to reflect the significance of such phenomena. Both sorts of response, however, evade giving an account of how the radical tendency was expunged from the black political mainstream, which is critically important for making sense of the limitations of inherited forms of black radicalism and for the task of constructing a progressive black politics in the present.

The oppositional tendency in post-segregation black politics was hampered by an aspect of its origin in black power ideology. Radicals – all along the spectrum, ranging from cultural nationalist to Stalinoid Marxist – began from a stance that took the "black community" as the central configuration of political interest and the source of critical agency. This stance grew from black power rhetoric's emphasis on "community control" and its projection of the "community" as touchstone of legitimacy and insurgent authenticity. This formulation is a presumptive claim for the existence of a racial population that is organically integrated and that operates as a collective subject in pursuit of unitary interests. That claim, which persists as a grounding principle in black strategic discourse, is problematic in two linked ways that bear on elaboration of a critical politics.

First, positing a black collectivity as an organic political agent preempts questions of interest differentiation. If the "community" operates with a single will and a single agenda, then there is neither need nor basis for evaluating political programs or policies with respect to their impact on differing elements of the black population. Any initiative enjoying conspicuous support from any group of black people can be said plausibly to reflect the community's preference or interest; the metaphorical organicism that drives the "black community" formulation presumes that what is good for one is good for all.

Similarly, because the organic black community is construed as naturalistic, the notion precludes discussion of both criteria of political representation and the definition of constituencies. Those issues become matters for concern when the relevant polity is perceived to be made up of diverse and not necessarily compatible interests and/or when the relation between representatives and represented is seen as contingent and mediated rather than cellular or isomorphic. By contrast, in the black community construct those who appear as leaders or spokespersons are not so much representatives as pure embodiments of collective aspirations.

As the stratum of black public officials emerged, black power radicalism's limitations became visible. Blacks' accession to prominence within the institutional apparatus of urban administration did not appreciably alter the mission or official practices of the institutions in their charge. Putting black faces in previously all-white places was not sufficient for those who identified with institutional transformation along populist lines or who otherwise rejected the status quo of race relations management. Yet, because black power's communitarian premises reified group identity and could not accommodate structural differentiation among Afro-Americans, the only critical frame on which radicals could draw consensually was the language of racial authenticity.

By the end of the 1960s, black power's inadequacy as a basis for concrete political judgment had begun to fuel radicals' self-conscious turn to creation and adoption of "ideologies" – global political narratives encompassing alternative vision, norms, and strategic programs – that promised to provide definite standpoints for critical judgment and platforms for political mobilization. This development underwrote a logic of sectarianism that embedded a cleavage between Marxists and cultural nationalists as the pivotal tension in black oppositional politics.

Ironically, the impetus propelling the ideological turn – the need to compensate for the inadequacies of black power's simplistic communitarianism – was thwarted by failure to break with the essential flaw, the stance positing the "black community" as the source of political legitimation and its attendant rhetoric of authenticity. Indeed, the turn to ideology may have reinforced propensities to rely on communitarian mystification because the flight into theoreticism made the need to claim connections with popular action all the more urgent. . . .

[The author here reviews black political movements of the 1970s, and offers a critical analysis of Jesse Jackson's two presidential campaigns.]

In the current situation black (and white) radicalism has retreated ever more hermetically into the university, and the unaddressed tendency to wish fulfillment has reached new extremes, so that oppositional politics becomes little more than a pose livening up the march through the tenure ranks. The context of desperation and utter defeat enveloping activist politics outside the academy has not only reinforced the retreat to the campus; it has also removed practical fetters on the compensatory imagination guiding the creation of intentionally oppositional academic discourses. In this context the notion of radicalism is increasingly removed from critique and substantive action directed toward altering entrenched patterns of subordination and inequality mediated through public policy.

The characteristics of this dynamic are mainly crystallized in the turn to a rhetoric pivoting on an idea of "cultural politics." The discourse of cultural politics does not differentiate between public, collective activity explicitly challenging patterns of political and socioeconomic hierarchy and the typically surreptitious, often privatistic practices of "everyday resistance" – the mechanisms through which subordinates construct moments of dignity and autonomy and enhance their options within relations of oppression without attacking them

head on. The failure to make any such distinction – or making and then eliding it – dramatizes the fate that befalls black radicalism's separation of abstract theorizing from concrete political action when academic hermeticism eliminates the imperative to think about identifying and mobilizing a popular constituency. Participating in youth fads, maintaining fraternal organizations, vesting hopes in prayer or root doctors, and even quilt making thus become indistinguishable from slave revolts, activism in Reconstruction governments, the Montgomery bus boycott, grassroots campaigns for voter registration, and labor union or welfare rights agitation as politically meaningful forms of "resistance."

[In his conclusion, Reed outlines the challenges and opportunities for a revived progressive movement.]

F
I
V
E

"Wanting In: Latin American Immigrant Women and the Turn to Electoral Politics"

from *Between Two Nations* (1998)

Michael Jones-Correa

Editors' Introduction

U.S. cities have long been shaped by immigration. At the start of the twentieth century, one-third or more of the people living in New York and Chicago had been born elsewhere, mostly in Europe. At the start of the twenty-first century, similar patterns may be found in New York, Los Angeles, and Miami, although immigrants today are more likely to come from Latin America and Asia. Increasingly, immigration has also become a central feature of Canadian and European cities. A 2004 U.N. study found that 44 percent of those in Toronto and 37 percent in Vancouver were born outside of Canada. Similar patterns may be found in Europe, where 30 percent of those residing in Frankfurt and 35 percent of those in Amsterdam in the mid-1990s were non-citizens.

The social and cultural impact of new immigrant waves is immediately apparent to all. In the introduction to *Between Two Nations*, Jones-Correa notes that the transformative power of immigration "can be seen in the lettering of signs over the storefronts ... in the faces of store owners, street vendors, waiters, pedestrians, passengers on the bus lines." New languages may be found in the newspapers at the local news-stand; unfamiliar cuisines featured in small, family-run eateries are signs of new immigration waves.

New immigration may not, however, be immediately felt in the political arena. Nearly all countries restrict the franchise to citizens (some exceptions: immigrants from European Union countries can vote in EU elections while living in a different EU member nation; in New York City, immigrants could vote in school board elections until their recent abolition), and in most countries becoming a citizen is a lengthy process. Furthermore, for many immigrants the stresses of surviving in a new land, often at the margins of the labor force, are so consuming that political involvement would appear to be a luxury. Often, immigrants become the *objects* of politics long before they are its *subjects*.

Disagreements over immigration have become key lines of cleavage in many European countries, but these are debates among native-born voters, with little participation from immigrants themselves. Some European countries have very low naturalization rates, which create barriers to political participation; in some, proportional representation systems, in which candidates are elected off a party list, and do not represent a geographic district, might make it more difficult than it is in the U.S. for geographically concentrated minority populations to achieve political incorporation. Immigrant communities throughout Europe have, however, become engaged in civic life and self-help activities. Janice Bockmeyer writes about immigrant mobilization in Germany in "Social Cities and Social Inclusion: Assessing the Role of Turkish Residents in Building the New Berlin," a paper presented at the DAAD-Wayland Collegium Conference "Social Integration in the New Berlin," at the Watson Institute for International Studies, Brown University, Providence, Rhode Island, on

March 14, 2005. Romain Garbaye is one of the few authors to have systematically studied the political incorporation of European immigrants in his *Getting into Local Power: The Politics of Ethnic Minorities in British and French Cities* (Blackwell, 2005).

In an earlier reading (see Part 2), Jones-Correa described the process through which Latinos were ignored by the established party politicians in Queens, New York. This reading looks at gender differences in political participation among Queens Latino immigrants. Some of the earlier scholarship on immigration and political participation neglected the role of women, assuming that immigration patterns were driven primarily by men (in part because in the past male immigration outnumbered female, since only men could immigrate as heads of households, sponsoring their wives). This has changed in recent decades. Moreover, there are indications that at least along some measures, women are more likely than men to participate in the political processes in their new home countries. Jones-Correa cites ABC exit polling data taken in New York City during the 1988 presidential elections that found Latino women greatly outnumbering Latino men among primary voters (43 percent male to 57 percent female).

Jones-Correa's research identifies several factors behind these patterns. Whereas many male immigrants see their sojourn in the U.S. as temporary, with the goal of earning money to return home, women are more likely to want to extend their stay, in part because they may enjoy greater independence in the U.S. than they had in their home country. Their longer term perspective may make them more likely to naturalize, or to become engaged in local concerns. In addition, women are likely to be more connected to government services: they oversee their children's schooling; they apply for public benefits, and the more educated among them may seek employment in social service bureaucracies. These contacts make them more aware of the role of government in their lives, and more motivated to participate.

There are a great many books and articles written on the political incorporation of immigrants. Some of the sources on which Jones-Correa has relied include: Sherri Grasmuck and Patricia Pessar, *Between Two Islands: Dominican International Migration* (Berkeley: UC Press, 1991), and several articles by Douglas Gurak and Mary M. Kritz, including "Social Context, Household Composition, and Employment among Migrants and Nonmigrant Dominican Women" (*International Migration Review* 30 (2): 399–422). Contemporary and historical perspectives on immigrant incorporation are presented in Gary Gerstle and John Mollenkopf's edited volume, *E Pluribus Unum?: Contemporary and Historical Perspectives on Immigrant Political Incorporation* (New York: Russell Sage, 2001).

In their attempts to resolve the dilemma of having to choose between two irreconcilable polities, mainstream immigrant organizations, dominated by men, appeal to their countries of origin to allow dual citizenship. Women and others on the margins of immigrant organizational structures choose instead to appeal directly to arbiters in their present political environment. The choice of strategy, I argue, is gendered.

Not only are women in general more likely than men to shift their cultural orientation toward the United States, but female activists are more inclined to participate in American politics. This chapter explores both phenomena. In particular, Latina immigrant activists' identity as women, together with their political socialization here and in their country of origin, leads them to become more involved than men in American politics, and inclines them to be more likely to look here, rather than to their countries of origin, for solutions to ease the costs of participation within the American political system.

WOMEN AND IMMIGRATION

In the last fifteen years there has been some change in the study and evaluation of immigration patterns. Previously it was assumed that men migrated first, and women and children followed; men were assumed to have made most of the decisions about immigration – when and where it would take place. In the 1970s there was a new recognition in the academic literature that decisions took place within households, and that therefore one had to look at each person's strategy within the context of families. With this revised perspective there is increasing recognition of the extent to which legal immigration to the United States is, by a small but significant margin, female. In the last few decades, the United States has received more immigrant women than men from almost every sending country. This is true for New York City as well. Immigrants from the Dominican Republic and Colombia make up the first and second largest non-Puerto Rican Latino migrant populations in New York. From 1982 to 1989 women made up, respectively, 51 percent and 54 percent of the migrant stream from these countries to the city; from 1976 to 1978 the percentages were as high as 57 percent and 66 percent respectively.

Arriving to the United States as members of households, Latin American women often share with men the idea of returning to their home country. The goal of rapidly accumulating savings in order to return requires women to play a much greater role in contributing to the household's income, and they are much more likely to work in New York City than they are in the home country. For example . . . while 31 percent of Dominican women were employed at some time before migration, 91.5 percent worked for pay at some time in the United States and 51 percent were in the work force at the time of the study. . . . Women's employment goes counter to middle-class expectations in the home country, but is rationalized – perhaps more by their husbands than by the women themselves – by saying that it will only be for a short while. Women's work is justified as long as it is for the good of the family. It is not meant to be an end itself.

Immigrant women's entrance into the work force does not entail economic or social parity with immigrant men. More women work after coming to the United States, but the jobs they take are more likely than men's to be low-skill work in manufacturing or the garment industry. Women stay at such jobs longer than men and are less upwardly mobile. Women generally work fewer hours and less regularly and get paid less for the work they do than do men, in part because work does not relieve them of their traditional family responsibilities. . . . However, these women's role in the family does affect their choice of jobs. Women

are more likely than men to take jobs in smaller businesses in the immediate neighborhood . . . presumably so they can stay closer to their families, and to take more flexible jobs at lower pay. Women are still expected to play certain traditional roles in the family, particularly with respect to children, while engaging in work.

Despite the difficulties, employment gives immigrant women economic resources they did not have in their home country. Employment may even be easier in this country for women than for men, and although they often work fewer hours, they make a significant contribution to the household's total income. As women take pay home, this may lead to a renegotiation of decision-making in households where couples are married. Whereas prior to immigration the senior male usually controlled all household expenditures or allocated an allowance to his spouse, in the United States the most common practice becomes the "pooled household" where all income earners put earnings in a common pot and jointly decide how the money will be spent. Women may keep part of their paychecks for their own expenses and remittances.

As women experience the benefits of working and controlling their earnings, their long-term strategies can begin to diverge from men's. While men hold to the maxim that "five dollars spent here means five more years before returning home," women may begin spending savings in this country. This progressively postpones return to their country of origin, where opportunities for work are limited and social controls are stricter. Women are aware that return will mean, in most cases, going back to the male head-of-household pattern, and they may not be eager to relinquish their new-found decision-making authority. A group of my neighbors, all women, raised such issues in talking about their relationships with men. Elizabeth and Anna said they don't want men to support them; they have no intention of being obligated to men. My next-door neighbor, on the other hand, said that when she married her first husband at age eighteen her husband supported her and she didn't have to work (she is now separated from her husband and lives with her daughter). Elizabeth and Anna both said they left home to get away from that feeling of obligation. Having their own means of income allows women to enter into relationships with men on their own terms. Such independence extends as well to their relationships with family authority in general. A woman who had left her husband in Colombia and come with her daughter to New York told me about her strict upbringing: "My father used to lock the telephone in his closet so that I couldn't talk with any boys. . . . My brothers were given the keys to the house when they were older, but I have never had a key to the house, not even now, when I go to visit. When I want to stay out late I have to ask for a key to the house." She thinks this experience may be why she came to the United States. In short, women have economic and personal incentives for abandoning the original strategy of accumulation and return. They may reevaluate the idea that success entails a return to the home country and staying means failure. Staying, for women, may be a significant improvement in their situation.

[. . .]

WOMEN AND SETTLEMENT

Women on their own have to develop alternative economic strategies, since working full-time at low wages and simultaneously bringing up children is next to impossible. Women who find themselves in this situation are likely to receive assistance from the state. In general immigrant women are more aware than immigrant men of social welfare programs; their coworkers and friends, most of whom are women, keep them apprised of what programs are available. In times of sustained need women are much more likely to turn to government programs than men. A large number of immigrant women receive welfare or aid for their dependent children (AFDC): 56 percent of Dominican women, for example, and 25 percent of Colombians. Note that this is closely correlated with the percentage of women in each group who are single heads of household with children: 63 percent of Colombian and 88 percent of Dominican female heads of household receive welfare benefits, and 45.7 percent of Colombians and 69 percent of Dominicans receive AFDC. Childcare is prohibitively expensive, and networks for childcare (through friends or relatives), which existed in the home country are missing or unreliable. A Colombian woman described how she would get home from work at eleven at night, and her daughter would be

ravenous so she would cook her a meal. She hired a baby-sitter, but the baby-sitter was expensive, and she was trying to work and study at the same time. "Really," she said, "I didn't know anyone here – not my parents, my family, no one." Without the social networks that existed in their home countries women turn for help to city and state agency programs; being alone with the children means that women are more likely than men to have contact with the institutions and programs of the federal and local government.

[. . .]

It is easy to see how men's and women's economic strategies begin to diverge with their experiences in this country. If they stay married or attached, women acquire greater independence and power with their income, which gives them an increased incentive to stay in this country. If they separate, even if they wish to go back, their strategy will likely change perforce; accumulating savings becomes almost impossible with the strain of raising children and working. In either case, if they have children they will come into contact with a wide range of public institutions, giving them a broader experience of governmental structures than immigrant men have. In general, then, immigrant women have both negative and positive incentives to orient their strategies toward this country, and away from their country of origin.

These incentives generally lead to an increased desire to stay in the United States, a desire which, among other things, translates into eventual naturalization. A survey of immigrants from seven Latin American countries suggests that women tend to become citizens much more often than men, even taking into account their greater proportion among legal immigrants to the city. This is particularly true for Dominicans and Colombians, who are the second and third largest Latino populations in the city after the Puerto Ricans, and for Hondurans and Salvadorans.

These differences, however, are not sufficient to explain why women are disproportionately represented as the mediators between governmental institutions and other immigrants. After all, while the majority of women may shift their orientations somewhat to further their interests and those of their families, this doesn't mean that they will necessarily choose a life of political activism. Though immigrant women are more likely to become citizens than men, and perhaps more likely to vote once they are citizens, few will devote their full energies to political involvement. Many women view politics with suspicion, and the number of women so involved is actually quite small. For this select group of women, activism is reinforced by other factors in their lives.

WOMEN AND POLITICS

Like men, women participate actively in immigrant organizations, but the organizations are usually founded and run by men. Male domination of organizations is especially evident in the Dominican community in Queens, which revolves around social clubs. *The Hermanos Unidos* club in Corona, for example, has about 350 male members and twenty-five female members. South American organizations have a more even sex ratio in their membership, but men monopolize leadership posts in almost all of these groups too (except those which are specifically women's groups, or *ramas femininas* – women's branches – of more general [male] groups). Men are presidents and vice-presidents; women are minor functionaries. "A lot of men," said one female Colombian activist, "think that power is something they hold privately and personally for themselves. It's accepted that women will organize. But if a woman pushes for leadership, this creates certain worries, a certain discomfort. They feel a woman can be secretary or serve in public relations but that none really has the ability for leadership." Women are allowed to run the concession stands and prepare events, while men do the public speaking and posturing. At a city-wide public forum, a Colombian organizer sat and took names by the door. I asked her how she had gotten stuck with that job. "This is women's work," she said sarcastically. "You'll never see a man doing this." The handful of organizations specifically organized by and for Latin American women provide some outlet for their initiative and leadership, but these organizations, like those dominated by men, are also generally oriented toward the home country rather than toward American politics. The difference is that the women in these organizations often play a dual role: they are also involved as activists and organizers for Latin American immigrant interests in New York.

Women, like men, draw on their pre-migratory experiences for the knowledge and expertise needed to run organizations. Activist women's initial experiences with politics are usually in their country of origin. Their first mentors are members of their own families who initiate them into political and organizational life – their parents, siblings, aunts, and others, both men and women. One Ecuadorian woman, for instance, told me she had been influenced by her mother: "Well, my mother was always very active. My mother was a teacher, and for her education was fundamental, and she always encouraged my education. She was very active, as well; she was a member of the Press Club, for example, and a member of la Sociedad de Quitenos." A Colombian woman said that her father and sister had been very involved in politics. Her father was most involved, although only in campaign work. She got interested, and now "it's stuck in my blood." Another Colombian said her father had taken her to meetings of the Conservative Party in Cali. Later she joined the Conservative Youth organization. "I learned how to be a leader at home," she said. "It's like a virus. It doesn't leave you later." Activist women often comment on how their political interests are something inherited, almost genetics. This inheritance becomes a model for their political work in the United States.

Latina activists' political socialization, however, is also partly the result of sharing many of the same experiences of other Latin American immigrant women – work, marriage, motherhood. Like many other immigrant women, their initial years in this country are taken up with work and family. "There was a time," one woman said, "when I was only a housewife [*una madre de familia*], until my children were a little older." As women have more time for themselves, husbands have difficulties accepting their new roles and commitments, and often oppose the idea of their wives getting involved in organizing and going off alone to meetings. A Latina block leader told me ruefully: "I saw myself as a politician. [She laughs] I like it a lot. . . . But here my husband has clipped my wings. Because he doesn't think that women should go out and scream and yell. I do my things, but only up to a point, so he doesn't have a problem with it. After all, this is my house, and I have to take care of the house as well. And little by little I get him used to it, I get him used to it. So ... if my husband had backed me, I would be doing politics, yes. Because I like politics."

Activists' full devotion to organizational life often does not begin until such concerns are taken care of in one way or another. Their children grow older, or they are divorced, both situations giving them greater independence. Like other immigrant women, activists' experiences with families and work often lead them to reevaluate their commitments.

It is true that many of these activists are less vulnerable than other immigrant women. Many of them have college educations and are fluent in English, whereas most immigrants must try hard to know enough of the language to defend themselves (*para defenderse*), and women generally know less English than men. Finally, their employment, often initially in the social service bureaucracy, gives them both financial security and an entree into American political life. These advantages are crucial and provide activists with many of the necessary skills to work as intermediaries between immigrants and the local and state governments. As women talk about their work they recognize the special role they play. An Ecuadorian activist describes her work in Queens family court:

I had a certain political interest there too, if you want to call it that, because I realized that the Hispanic woman who went to the court to ask for help with her family problems (and I say Hispanic woman because it was mostly the Hispanic women who sought help) had a lot of difficulties – she didn't know the system, she didn't understand, she was poorly informed. . . . The courts had interpreters and we were the bridge that helped a person make herself understood in the institution, but we also helped a little to orient her in this country, right?

If Latina activists become mediators between immigrants and the governmental bureaucracies, they are also acknowledging and building on their experiences as immigrants and as women. While men are likely to keep a sojourner mentality, and organizations dominated by men will focus on the home country, activist immigrant women are more likely to turn to the problems of the immigrant community in this country. An Ecuadorian woman noted:

It was natural that people who came over in large numbers would want to get together, so they formed civic organizations. And men were always the leaders, because in Latin America men were always the leaders. Women in politics were seen as strange. When I was growing up as a girl in Ecuador, it was not the thing for women to do. Men here are more interested in politics there. They do good things, raise money, but they are not interested in what goes on here. They have status in the community; they are *caciques* [leaders]. But they aren't interested in starting over – to begin with, to have to learn English. If they got involved in politics here they wouldn't be *caciques* anymore. They would only play a small part. So women and Puerto Ricans tend to dominate local politics in Queens. Puerto Ricans because of their experience in politics. Women because they are willing to work with others.

Activist women have both the motive and the opportunity to play the role of political intermediaries. As immigrants, Latinas' primary loyalties remain with their home countries, but as women they find themselves facing new problems in New York. Despite their skills, they are marginalized within immigrant organizations and so turn to alternative forms of participation.

For city and state government agencies looking for people to serve as intermediaries between government and the fast-growing population of immigrants, these activist women are ideal. For the women themselves, these positions offer the chance for leadership unavailable in immigrant organizations. These mediating roles may be frustrating – women remain beholden to their political patrons within New York's political establishment – but the activity also gives women a great deal of visibility among Latin American immigrants in Queens and makes them likely contenders for political office, should any choose to run.

ELECTORAL REDISTRICTING IN QUEENS

Latina activists occupy a peculiar niche in New York City politics. As intermediaries with the broader American political establishment, they have some influence both within and outside the immigrant community, but they remain on the margins of both the American and the immigrant institutional establishments. Their role and their resources (along with their allies, who are often equally as marginalized in the immigrant community) are perforce limited, particularly within the parameters of New York City politics. The role they generally play is to gently prod the immigrant community in the direction of increased participation in American politics. They organize registration and naturalization tables on street corners and fairs, endorse local politicians and work for their campaigns, hold fund-raisers for candidates they like, and try to get out the vote. In short, they act much as other emerging interest groups do in American politics, the one difference being that they still identify themselves as members of the Latin American immigrant community. Along with the maturing second generation, they represent the rise of a new ethnic politics in the city.

Unable to appeal for aid from the larger first-generation immigrant community, and confronted by steep costs of entering New York City's hegemonic electoral political arena, Latina activists and their allies must look elsewhere for support. Coalition partners within the city are either unlikely or problematic; immigrant activists in Queens have little to offer in return for inclusion in any coalition except their presence. Unlike their male counterparts in the mainstream organizations, they have little leverage. However, by stepping outside the local political realm, and turning to an outside arbiter, these activists were able in the crucial instance of electoral redistricting in the city and state in 1991 to play a role in shaping a Latino district in Queens. By appealing for federal government intervention they were able to trump resistance at the city level.

Redistricting in 1991 was different from any political reapportionment in New York City within recent memory because it actually was a redistricting – the drawing of entirely new districts – and not simply incremental fiddling with district boundaries, as is usually the case. The actual task of redistricting was assigned to a commission appointed by the mayor to expand the number of City Council districts from 35 to 51. The possibilities for significantly expanding minority representation on the council prompted Latino organizations – in particular the Puerto Rican/Latino Coalition for Voting Rights, the Puerto Rican Legal Defense and Educational Fund, and the Puerto Rican/Latino Voting Rights Network – to enter the process and present a number of proposals to the commission for an increase in the number of Latino districts.

The expansion of minority districts was actively resisted in Queens. On the one hand, incumbent politicians were nervous because redistricting meant changes that were not fully under their control; on the other, the possibility of minority districts in Queens was perceived by white ethnics as a threat to their sense of community as neighborhood. They argued that ethnic redistricting was a break with tradition. Neighborhood boundaries were being violated by newly drawn district lines, with neighborhoods partitioned among various political districts in order to group minorities together. The two arguments paralleled and reinforced one another.

At public hearings held by the state's task force for reapportionment, the refrain of native-born blacks and whites (there were few immigrants present) was: "They're dividing the community." Jackson Heights and Corona, with their mixed populations of new immigrants and long-term residents, were at the center of the controversy in Queens. White ethnic residents in Jackson Heights mobilized so that "traditional" borders wouldn't be broken and the "community" split. They used the language of the civil rights movement to defend themselves against implicit charges of discrimination and to undercut the arguments for creating separate minority districts, charging that minority districts would lead, in fact, to resegregation along racial lines. For example, Rudy Greco, president of the Jackson Heights Beautification Group, testified: "Jackson Heights shouldn't be penalized for having desegregated [in the 1960s and 1970s]. We worked hard to integrate and now Jackson Heights is being resegregated. . . . The people of the community are up in arms. . . . You have created a black district, a white, and an Hispanic district where blacks, whites, and Hispanics used to live together."

[. . .]

In their public rhetoric, white ethnics defended Jackson Heights as a place, protesting the elevation of ethnicity above locality as the deciding principle of political community.

The city commission, in drawing the lines for new council districts, did not entirely agree with the "neighborhood as community" argument. Many of their lines crossed "traditional" boundary lines. But they did use census figures to make the argument that Latinos in certain areas of the city – primarily in Queens, where no one of Latin American origin had ever been elected to political office – were not sufficiently concentrated residentially to be able to draw majority-Hispanic districts. The commission presented its final maps on June 3, 1991. The redistricting plans still had to be approved by the U.S. Justice Department, however, since three out of five of New York's boroughs are covered by the 1965 Voting Rights Act and its amendments – Manhattan, Brooklyn, and the Bronx (but not Queens). It was at this juncture that the mostly female activists from the Queens Hispanic Coalition and the Queens Hispanic Political Action Committee joined other Latino interest groups to circumvent the city's political establishment by appealing directly to the Justice Department.

The Justice Department's Civil Rights Division was sympathetic to the appeal. Despite its lackluster record under a Republican administration, John Dunne, the assistant attorney general in charge of the division, indicated that he was concerned about the maps drawn up by the city, particularly in parts of Brooklyn and the Bronx, and, unofficially, in Queens. On July 19, 1991, after hearing the protests and testimony of various Latino groups involved, the Justice Department turned down the city's proposed redistricting plan, saying it "consistently disfavored" Hispanic voters in violation of the law. It took another two weeks of negotiations before the Justice Department approved a final version of the districts on July 26th. The new maps included alterations favoring Latino representation in the city, including the formation of a majority-Latino district in the Jackson Heights Corona area of Queens. In the new maps Latinos made up 55 percent of the population of the 21st council district in Queens (from 49.6 percent previously). Ironically the creation of these new boundaries actually reduced the number of registered minority voters: the district changed from having 44 percent black registered voters and 33.5 percent Hispanic registered voters, to 22.3 percent black voters, and 24.5 percent Hispanic voters.

In the end, when the electoral districts were drawn and the plan was approved by the city, white ethnic voices were drowned out. Although they had the ear of the local political machinery, they couldn't convince the Justice Department that their vision of neighborhoods as the basis of political community was adequate, that it included everyone equally, without discrimination. They couldn't argue against the fact that while Latinos are 20 percent of the borough's population, no Latino had yet held elected office. Though the vast majority of Latin American immigrants had little interest in the apportionment dispute, it was relatively easy for a small group of Latino activists (most of them women involved in U.S. electoral politics in the Jackson Heights-Corona area) to go to Washington and lobby the lawyers at – the Department of Justice. Once there, they persuaded the Justice Department counsels that the city's redistricting lines were discriminatory and that the relevant definition of "community" should not be the municipally defined neighborhood but the ethnic origin of the population. Only in this way, they argued, could a Latino be elected to political office in Queens.

The activists' task was successful, first of all, because the federal government's definition of who counts as Hispanic corresponded with and reinforced some facets of Latino immigrant activists' self-identification. Of course, it also helped that the Justice Department's political motivations – the desire for a Republican administration to show up the failings of Democratic machine politics in New York City, and build political support among the growing Latino population – momentarily coincided with the activists' own. But the 1991 City Council redistricting indicated both the strengths and weaknesses of the activists' approach. A relatively small group of men and women were able to appeal for intervention from the federal government for a "Hispanic" district in Queens, but were unable to run a candidate of their own for the district. Helen Marshall, an African American of Caribbean descent, and a former New York State assemblywoman, ran with backing from the Democratic Party in Queens, and won the seat in November 1991.

THE TURN TO ELECTORAL POLITICS

It's clear the situation for Latin American immigrants in Queens and in New York City is changing rapidly, particularly following the reapportionment of electoral districts. Immigrants and immigrant groups – particularly those already on the margins of the immigrant communities – are beginning to redirect their efforts toward their communities in the United States, rather than anticipating return to their countries of origin. The change in women's long-term strategies, especially, is reflected in their higher naturalization and participation rates in American politics. It is reflected as well in the willingness of Latina activists to do more than simply mirror the ambivalence of mainstream immigrant organizations.

It is uncertain, though, how this initiative will translate in electoral politics. Until now Latin American women have been likely to play the role of ethnic intermediaries, serving as facilitators or negotiators between immigrants and the American political and bureaucratic systems, and Latin-American men have taken charge of the main-stream immigrant organizations. It's still an open question as to which leadership cluster will be better-positioned to enter the electoral sphere. On the one hand, immigrant women have more experi-ence in dealing with American institutions, and so might seem to be the natural choices to step into electoral politics, as elected positions are to some degree an extension of that mediating role. On the other hand, activist women have been able to succeed in constructing the mediating role because this role has been undervalued by the rest of the immigrant population; men are still seen by many immigrant men and women as providing the more appropriate leadership for the community. Until now male leadership has focused its energies in first-generation organizations, where men can maintain their social status even if their economic status declines. However, over time, as participa-tion in the American sphere becomes more valued, men may begin to supplant women as the most visible intermediaries between immigrants and the broader society.

In one scenario, then, men will run for electoral office, crowding women to the background. Some women activists are convinced that men will want those leadership positions regardless of whether they are qualified, and are concerned that women might cede them these positions, if that is the price for men's participation: "Women participate more because they have a more responsible vision of what is going on. They are more ready to par-ticipate. Men only participate if they have a position of leadership. You have to give men a position [with a title]. Women are readier to just be members." One woman activist noted cynically that if a seat opened up, men would want to run for it, and more qualified women would be shunted aside. Even if there are women who would be more qualified candidates, the expectations are that men will likely step forward to take any open positions. Four Latino men did run in City Council district 25 after the new redistricting in 1991. None had any political experience, and all lost. No women ran. . . .

But events may turn out otherwise. Nydia Velasquez and Elizabeth Colon, the two leading candidates against Stephen Solarz in the 1992 primary elections for congressional district 12 (which includes parts of Elmhurst, Jackson Heights and Corona in Queens, and also stretches into Brooklyn and Manhattan) were both Puerto Rican women who had come into politics through com-munity organizations and mediating institutions between government and Puerto Rican migrants. In an upset victory, Nydia Velasquez won the primary, and went on to be elected the first Puerto Rican congresswoman. Though both women are Puerto Rican and not immigrants per se, they indicate the potential women may have of using the intermediary positions they occupy to launch electoral political careers, and the support they may be able to garner from Latino voters will-ing to accommodate themselves to the idea of female leadership. So things may be changing. A Colombian respondent captured the mood: "It's time," she stated, "that we stop letting ourselves be used. If [men] aren't going to give us what they promised, then we aren't going to help them or support them either."

Regardless of whether future Latino electoral can-didates in Queens are women or men, their efforts will have been greatly facilitated by the ground-work laid by a group of first-generation immigrant activists, for the most part women. These activists have worked to lower the costs of participation for new Hispanic voters by encouraging voter regis-tration and participation, building ties with elected officials within the Democratic and Republican Parties, laying the foundations for independent Latino political organizations, and working for majority-Hispanic districts in the area.

"Black, White, and Blurred"

from *Governing* (2001)

Rob Gurwitt

Editors' Introduction

Browning, Marshall, and Tabb's analysis, based on case studies from the 1970s, assumes that race is the key characteristic shaping urban voting behavior in U.S. cities. Is this still the case? Can we deduce the outcome of a city's mayoral election simply by noting the race of the candidates and the race of the majority of voters? Or has race become less salient as voters choose candidates, and as candidates shape their appeal to voters?

Rob Gurwitt does not make the claim that race has ceased to matter in big city elections. In most cities, most people still vote for candidates of their race most of the time. But he argues that it is less prominent than it once was. We find cities in which a largely white electorate votes in a black or Hispanic candidate (for example, Los Angeles's election of Tom Bradley, and Seattle's choice of mayor Norm Rice). Chicago's (white) Mayor Richard Daley has won re-election with support from many African-American leaders. And in majority African-American cities like Detroit, Atlanta and Newark, where black candidates compete among each other for black votes, it is not enough for candidates to mount an appeal to racial solidarity. More frequently, then, big city election campaigns are about something other than race.

Other cases, however, suggest that race is still an important lens through which urban voters view local elections. In Milwaukee's 2004 mayoral race, in which a white congressman defeated the African-American acting mayor, voting was substantially along racial lines. In some cities (New York and Philadelphia are two examples) black candidates did get significant numbers of white votes, but in these strongly Democratic cities, their white Republican contenders did unusually well, suggesting that many white voters had crossed party lines to vote for the white candidate. Even in Newark's 2002 mayoral election, between two African-American candidates, race became an issue, as the incumbent claimed that the challenger, who had reached out to the city's minority white and Latino populations, was not sufficiently "black."

Perhaps the only lesson that may be drawn is that race often matters, but to different degrees under different circumstances. Karen Kaufman's insightful analysis of mayoral politics in New York and Los Angeles *The Urban Voter* (Ann Arbor: University of Michigan Press, 2004) shows that racial perceptions have a complex relationship with voting choices, gaining or diminishing in importance depending on a host of local issues. Apart from larger demographic and economic factors, there is always room for contingency in any election campaign, as candidates can show more or less skill reaching out to new voters. As Browning, Marshall, and Tabb point out, the key to winning and exercising power in most cities is the formation of cross-racial coalitions, and these depend as much on leadership ability as they do on simple demographics.

Gurwitt focuses on Cleveland, a city whose white population had, as of the 2000 Census, become the minority (at about 40 percent). There, in 1967, Carl Stokes became one of the first African-Americans to be elected mayor of a large U.S. city, winning an election marked by high levels of racial polarization. Gurwitt finds it striking that, in 2001, as four-term (African American) mayor Michael White stepped down, race did not seem to be the key prism through which voters and political leaders would choose from among the ten candidates vying to succeed him. Faced with continued job loss and deteriorating living conditions (Cleveland was found to be the poorest large city in the U.S. after the 2000 Census, surpassing Detroit and Newark), the city's voters were most interested in electing a mayor who could bring about economic improvements. (Jane Campbell ultimately won the 2001 Cleveland mayoral race.)

Rob Gurwitt has written for *Governing* magazine since it began in 1987. He is now a freelance writer focusing on how change affects American communities. He lives in Norwich, Vermont.

It's normal urban politics, you might say, except that it's not normal at all in Cleveland. More than 30 years ago, this was the first major city in America to elect a black mayor, and it has been fixated on its racial divisions ever since. For decades, major municipal elections in Cleveland have been about race and very little else. The last time City Hall was genuinely up for grabs, in 1989, the city split straight down the color line, even though the two major contenders were both African American. A blurred racial landscape like the current one represents dramatic political change for Cleveland. But it is not a change unique to Cleveland. This relaxation of racial politics has been taking place for some time around the country. Cities everywhere seem eager to confound old expectations. There are now nearly 500 black mayors in the United States, a substantial increase over the 314 a decade ago. And of the eleven black mayors in cities with more than 400,000 residents, most were not chosen by black majorities.

Houston, Dallas, San Francisco, Denver and Minneapolis – none with a black population greater than 30 percent – all have [had] black mayors. Baltimore, about two-thirds black, voted in 1999 for Martin O'Malley, a white city councilman, to replace the African-American Kurt Schmoke. St. Louis, which is 51 percent black, this year elected a white mayor, Francis Slay, after a campaign that was far less racially charged than the 1997 contest between two black candidates, one of whom had the overwhelming support of white voters.

In Philadelphia [in 1999], John Street – who first made his political name in the 1970s as an activist within the African-American community – became mayor after a contest against a white opponent in which race hardly figured as a subject at all. And [in 2001], Los Angeles elected James Hahn, a white candidate who won in large part because of his solid support in the black community. It isn't

difficult to make the case for the headline on a story in the Christian Science Monitor a while back, "Racial politics subside in cities."

There is a difference between subsiding and disappearing. In both St. Louis and Philadelphia, the campaigns may have been free of racial opportunism, but the voters themselves split mostly along racial lines; the winner was the one who got more crossovers. And as the Los Angeles mayor's race suggests, it doesn't take much to inflame racial sensitivities: The white candidate won in Los Angeles because of solid support among blacks and moderate and more conservative white voters, which he consolidated by tying his Latino opponent to the cocaine trade, in one of the more blatant racial stereotyping ads in a recent campaign.

So it's not so much that racial politics have abated in America's big cities, it's that they've changed, taken on a new format and new language. Cleveland is as good an example as any. During the tenure of the outgoing mayor, Michael R. White, city government made great progress in discarding its old racial prisms. Still, poverty, unemployment, low-performing schools, an aging population, housing segregation and other urban ills continue to hamper the city, and these are matters that community leaders and city residents do not wish to have forgotten down at City Hall. What they haven't been able to decide yet is whether the race of the person occupying the mayor's office matters in getting the problems solved.

It took [the 2001] campaign in Cleveland an inordinately long time to get started. That is because the entire city had to wait upon the decision of Mayor White, who became the city's second African-American mayor in 1989 and now, after three terms, is the longest-serving mayor in its history. It was initially assumed that there was a black heir apparent – Stephanie Tubbs Jones, who used to be the county prosecutor and now represents much of Cleveland in the U.S. House.

But Tubbs Jones decided not to run, and her decision served to accentuate the racial blurring. There are two announced black candidates, state Representative John Barnes Jr. and Raymond Pierce, a former mid-level official in the Clinton administration, but neither has much of a political base in the city. Barnes won legislative office largely on the strength of voter loyalty to his father, a former city councilman; his only attention-getting role has been as chairman of a roving statewide Commission on African-American Males. Pierce, though he has won respect as a good speaker and hard campaigner, remains essentially unknown to much of the city.

That has left three white candidates free to scramble for black votes, largely on the basis of their history of concern for inner-city social welfare issues. Cuyahoga County Commissioner Jane Campbell, a liberal Democrat, won broad popularity for her oversight of the county's transition to welfare reform and, as a former state legislator, has long-standing ties to the city's East Side, where most African Americans live. Tim McCormack, another county commissioner, also has made a name for himself as a child welfare advocate. Bill Denihan, a public administrator who had been director of public safety under White and director of natural resources at the statewide level, became head of the county's child welfare department in 1999. He won broad respect in that position for setting the politically troubled department on a more even keel.

In the last days before the October 2 [2001] primary, no one is quite sure what role race is likely to play in determining the outcome. But the suggestion that racial issues may be just a sideshow is a striking development in Cleveland's political history. To understand how it came to this juncture, it is important to know something about Mike White's twelve years in office.

When White first ran for mayor, no promise he made carried more weight than his insistence that he would treat the city's east and west sides equally – or, as he put it, that he would say the same things on the East Side that he said on the West. For generations, Cleveland was, in essence, two cities separated by the Cuyahoga River. The West Side was white and ethnic, the East African American,

except for some neighborhoods along its fringes. Not surprisingly, after Carl Stokes was elected as the first black mayor in 1967 and city government began opening up – before Stokes, City Hall had pretty much been off limits to the African-American community – struggles over allocation of the city's resources took on the convenient shorthand of east versus west.

When White first ran in 1989, his opponent in the general election was the African-American president of the city council, George Forbes. Forbes – who now heads the NAACP in the city – was an autocratic leader, and both on the council and in his regular radio show he displayed little interest in treating the concerns of West Siders diplomatically. So when it came time to vote that year, white voters were drawn to White on more than the strength of his argument that there was no place for racial division in running City Hall; it was also, simply put, the fact that he was not Forbes. As Norman Krumholz, a professor of urban affairs at Cleveland State University and the city's former planner, puts it, "People on the West Side were waiting in the weeds to vote against George Forbes."

A starkly race-based election such as that could not happen these days in Cleveland. And this is due in no small part to White's years in office. He demanded solid performance from his employees, and if they didn't deliver, he came down hard on them, regardless of their race. When issues arose that had the potential to push racial hot buttons, White did the unexpected. He lobbied to end the city's long and debilitating experiment with court-ordered school busing, and even allowed the Ku Klux Klan to hold a rally in downtown Cleveland.

As a self-described moderate-to-conservative, White was part of a generation of mayors – along with counterparts such as Steve Goldsmith of Indianapolis, John Norquist of Milwaukee and Dennis Archer of Detroit – who shared a belief in running city government efficiently and with a broad public in mind, rather than using it to broker the demands of competing identity groups. This was a collection of politicians who understood, as Swarthmore College political scientist Keith Reeves puts it, that "when you look at folks in these cities, African Americans and Latinos and the whites who are coming back into the cities, at the end of the day they want to make sure their cities are safe, their kids are taken care of, they're going to get a quality education, and the quality of services are such that they believe their tax dollars are well spent." From the beginning, White spoke of Cleveland as a single city and focused his attention on issues with broad appeal: developing its downtown, redeveloping its neighborhoods and fixing its schools.

The result is a place that looks quite different from the one he took over. There is, of course, the famous downtown makeover: new sports stadiums, the Rock and Roll Hall of Fame, a revitalized warehouse district. There is also progress on the school front. In 1998 White gained power to appoint the school board – fighting off charges that he was undermining African-American voters by doing so – and three years ago brought in a new administrator, Barbara Byrd-Bennett, from New York; Byrd-Bennett is now widely considered one of the most effective and popular public officials in the city.

Just as important, there are signs that some of the city's neighborhoods are regaining their health. Since 1992, says Community Development Director Linda Hudecek, private lenders have put $4.3 billion into Cleveland's neighborhoods, thanks in part to pressure from City Hall, and the results are evident on both sides of the river: New housing developments dot the city, some of them scatter-site, some of them much like suburban tracts, and they in turn are beginning to spark commercial revitalization.

Not surprisingly, all of this has begun to transform the city's politics. When a city council coup in 1999 ousted President Jay Westbrook and replaced him with Mike Polensek, a council veteran from a once solidly white ethnic ward on the far East Side, the central issue was whether the council had been too compliant in going along with White, whose strong personality and my-way-or-the-highway attitude had alienated other politicians. Race was essentially irrelevant: Each side was racially mixed.

Slowly but noticeably, Cleveland voters have begun dismantling their habit of looking at everything through the filter of race. The progress is subtle: When White ran for his last term in 1997, he did well citywide but lost most of the white precincts on the West irrelevant. "Our highest

unemployment rates," says Ed Rybka, a white city councilman, "are in the African-American community – we need to develop job training, day care facilities, the ability to transport people to where the jobs are – but as glaring as the needs are there, the people who need to have this addressed are black, white and Hispanic."

The Reverend Marvin McMickle, one of the city's most politically active black ministers, says that mayoral candidates sometimes bring up racial issues with him, and he tells them to talk about something else. "That's not the issue," says McMickle, whose Antioch Baptist Church sits in the midst of a mixed poor and lower-middle-income black neighborhood not far from Hough. "There's no other race to relate to if you live in this neighborhood. These neighborhood residents want to talk about jobs, medical care, functional public education, things like that. When was the last time the curbs were replaced here? When was the last legitimate time a small business went in along the main street through Hough? What will it cost to keep neighborhood business owners who are trying to stabilize neighborhoods one block at a time? These issues are citywide."

Yet it's also true that McMickle worked hard to persuade Stephanie Tubbs Jones to run for mayor, and was one of about 100 African-American community leaders who met to plot strategy just after she'd announced she didn't want to leave Washington. To understand this, it helps to see the results of a Gallup Poll Social Audit released in July. That survey of 2,000 people across the country found whites to be considerably more optimistic about race relations than blacks, and blacks growing more pessimistic about progress in such areas as equal housing and equal educational opportunity. In Cleveland itself, there remains an intense awareness among black politicians that there is a limit to how high they can aspire – there are ten city council positions, a few seats in the state legislature, Tubbs Jones' seat in Congress, and the mayoralty of Cleveland. Winning countywide is difficult, and statewide, they believe, impossible.

It should be no surprise, then, that some African-American leaders still have an abiding conviction that a black mayor will, quite simply, be more attuned to black sensitivities. "What's at stake is not just the prestige of a black mayor but also the power of that mayor to appoint," McMickle says.

"It's not just the one position of mayor that's at stake but the many places where the hand of the mayor reaches."

Even these, however, are ambiguous currents. When Tubbs Jones withdrew as White's heir apparent, some black leaders were troubled that there would be no black candidate of citywide stature. On the other hand, they had trouble reaching a consensus on how to proceed, other than by listing a set of issues they wanted addressed – improving emergency services, restoring an inner-city trauma center, creating youth programs, developing a biomedical technology program. A group of white, West Side ministers could have come up with pretty much the same list.

It is a fact of political life in Cleveland that whoever does replace Mike White will have to do it on his or her own. There is no white-run political machine in town, and the one attempt to build a political organization in the black community – begun by former Congressman Louis Stokes – petered out after he retired. White himself is a political loner, and never groomed a replacement – as McMickle points out, "I think Mike could say, 'I won this office the hard way, and whoever does it next will have to do the same.'"

As the October 2 primary approaches, there's no question that all the candidates are looking forward to a stronger showing among voters of their own race than across racial lines. But they aren't looking forward to it with anything like the certainty they would have felt just a few years ago. As Charles See puts it, "Of course race matters. But folks rise to a level of sophistication at a time like this: They want to know what is the likelihood of a candidate delivering on what he or she promised. Everyone knows what to SAY these days. We want to know where you've been."

"Ethnic Stability and Urban Reconstruction: Policy Dilemmas in Polarized Cities"

from *Comparative Political Studies* (1998)

Scott A. Bollens

Editors' Introduction

Students of political science might assume that when they choose to study local-level politics they are choosing to study conflicts over local issues, such as land use, public services, or economic development. Those interested in nationalism or geopolitical power struggles may – too quickly, in our view – assume they will need to pursue their interests in courses on international relations or world politics.

But politics of all kinds are, ultimately, spatially grounded – in other words, all political activity has to take place *somewhere*, and most often that *somewhere* is an urban landscape. The Cold War, for instance, represented a global clash of competing ideologies and economic systems, but it also played out in distinct locations, most notably Berlin, where urban planners learned to live first with, and then without, the Wall (see Strom, Part 3). The conflicting national aspirations of Israelis and Palestinians become fodder for global conflict, but most directly they are felt as disputes over sacred sites, over housing and over resources in a city like Jerusalem. You can probably think of other cities where "local" politics and policy have been shaped by broader national and international conflicts: Bollens mentions Sarajevo, Beirut, and Hong Kong, among others.

Scott Bollens is interested in how planners and local governing institutions function in cities that are at the epicenter of such deeply entrenched national and ethnic conflicts. In his article, "Ethnic Stability and Urban Reconstruction," he notes that students of the city have long been familiar with the problems of racial strife, segregation, and inequality, but in the urban settings that are typically studied, such conflicts are ultimately resolvable with a better distribution of resources. In the cities he studies, however, urban conflicts are reflective of profound national disputes that are barely manageable, and certainly not reconcilable, on a local level. This particular study focuses on two cities: Belfast, in Northern Ireland, where conflicts pit majority Protestants, who seek to maintain the region under British rule, and Catholics, who seek to ally with the independent Republic of Ireland; and Johannesburg, the largest city in South Africa, in the years after the end of apartheid. Planners may adopt different strategies as they make urban policies in the face of these fault lines. They may attempt to remain neutral, relying on technical competence and professional best prac- tices and trying to prevent racial and ethnic tensions from influencing their work. Or, they can play more of an advocacy role, working actively to challenge (or maintain) the systemic inequalities that have produced the current conflicts.

Like most scholars, Bollens uses an array of data sources, including secondary analysis and census figures, but his work is distinguished by its reliance on open-ended, in-depth interviews with decision makers – current and former government officials, and members of nongovernmental organizations – in his case study cities (Bollens identifies these officials by name in his original article). He is thus able to shed light on the thinking of professional planners in these cities.

Scott Bollens is Professor of Planning and Design in the University of California-Irvine's School of Social Ecology. Interested in many facets of urban governance and planning, he has written extensively on the role of urban policy and urban planners in the most conflicted societies. His books include *On Narrow Ground: Urban Policy and Conflict in Jerusalem and Belfast* (New York: SUNY Press, 2000) and *Urban Peace-building in Divided Societies: Belfast and Johannesburg* (Boulder, CO: Westview Press, 1999).

This article investigates the role and influence of urban planning in ameliorating or intensifying deeply ingrained ethnic conflict. It assesses, based on more than 70 interviews with urban planners and opposition groups, the relationship between urban policy and inter-communal strife in the ethnically polarized cities of Belfast (Northern Ireland) and Johannesburg (South Africa). It explores how urban planning and policy is affected by, and itself affects, extra–local ethnic conflict. More specifically, by examining two cities embedded within larger peacemaking processes, it investigates the lessons – both positive and negative – that city management of ethnic conflict provides for inter-group stability and reconstruction at national and cross-national scales. The nexus between city planning and ethnic conflict is increasingly salient because a disquieting number of cities in the world are prone to intense inter-communal conflict and violence reflecting ethnic differences. Cities such as Jerusalem, Nicosia, Algiers, Sarajevo, New Delhi, Beirut, Montreal, Hong Kong, Brussels, Belfast, and Johannesburg are urban arenas penetrable by deep inter-group conflict. Common to many of these cities are strong ethnic identities and nationalistic claims of sovereignty, which combine to create pressures for group rights, autonomy, or territorial separation. In the most intense cases, these cities are battle-grounds between homeland ethnic groups, each proclaiming the city as their own. The machinery of government is often controlled by one ethnic group and used to discriminate against competing and threatening groups. In other cases, a third-party mediator may be brought in to govern the urban setting. In either case, the legitimacy of a city's political structures is commonly challenged by ethnic groups who either seek an equal or proportionate share of power or demand group-based autonomy or independence.

DIVIDED AND POLARIZED CITIES

Cities are frequently divided geographically by ethnicity, race, income, and age. Yet, urban conflicts in these so-called divided cities are addressed within an accepted political framework. Questions of what constitutes the public good are debated but largely within a sanctioned framework. Both in the 1960s and 1990s, for instance, African American grievances have focused more on the distribution of urban services and economic benefits than on territoriality and sovereignty. In divided cities and societies, there is a belief maintained by all groups that the existing system of governance is capable of producing fair outcomes, assuming political representation of minority interests. Coalition building remains possible across ethnic groups and cross-cutting cleavages defuse and moderate inter-group conflict.

A deeper, more intractable type of urban conflict . . . occurs in cases in which ethnic and nationalist claims combine and impinge significantly . . . on distributional questions at the municipal level. In such a circumstance, a strong minority of the urban population may reject urban and societal institutions, making consensus regarding political power-sharing impossible. Governance is perceived by at least one ethnic community as either illegitimate or structurally incapable of producing fair societal outcomes to subordinated ethnic groups. Allocational and housekeeping policies often become politically conflictual, viewed by the subordinate ethnic group as an intrusive imposition of one culture or political claim onto another. The urban arena, and public actions within it, become saturated with ideological, ethnic, and nationalistic meaning.

URBAN PLANNING AMID POLARIZATION

The role of urban policy in ethnically polarized cities is problematic because city officials must contend with both broad ideological imperatives and the particular exigencies of daily urban life. . . . Yet, cities are not . . . simple reflectors of larger societal tensions and dynamics; rather, they are capable . . . [of exerting] independent effects on ethnic tension, conflict, and violence. There are qualities of the urban system – proximate ethnic neighborhoods, territoriality, economic interdependency, symbolism, and centrality – that may bend or distort the relationship between ideological disputes and the manifestations of ethnic conflict. As such, urban policy and planning approaches, and their effects, are not essentially predetermined by ideological goals and parameters. Much as a prism deviates light from a straight line projection, the physical, social, and political conditions of a city may modify the relationship between the more broad causes of ethnic strife – political disempowerment and cultural deprivation and urban inter-group relations. Not the primary cause of strife, cities and their policy makers may nonetheless be capable of activating or moderating extant inter-group tension.

Urban policy has substantial potential effects on material and psychological conditions related to inter-group ethnic stability or volatility. The relative deprivation theory of ethnic conflict posits that disparities in such conditions, and unmet human needs, are primary motivational forces underlying oppositional action. . . . Urban policy most directly affects interethnic relations through its significant influence on control of land and territoriality. Common techniques of territorial control amidst ethnic tension aim to alter the spatial distribution of ethnic groups or territorial boundaries. Urban policy also shapes the distribution of economic benefits and costs across antagonistic ethnic groups and can have substantial effects on the distribution of local political power and access to policy making. Finally, maintenance of group identity is critical to the nature of interethnic relations in a polarized city and can be affected by urban government actions. Collective ethnic rights, such as education, language, press, cultural institutions, and religious beliefs and customs contain potent ideological content; their exercise is viewed as a critical barometer by an out-group of an urban government's treatment of their rights.

In this article, I examine four urban planning strategies that urban regimes can adopt under conditions of political and ethnic polarization. A neutral urban strategy employs technical criteria and distances itself from issues of ethnic identity, power inequalities, and political exclusion. Planning acts as an ethnically neutral, or color-blind, mode of state intervention that is responsive to individual-level needs and differences. This is the traditional style of urban management and planning, which is rooted in an Anglo-Saxon tradition and is commonly applied in liberal democratic settings. It seeks to de-politicize territorial issues by framing urban problems as value-free technical issues that are solvable through planning procedures and professional norms. Disagreements between ethnic groups would likely be channeled by government toward day-to-day service delivery issues and away from larger sovereignty considerations. A partisan urban strategy, in contrast, chooses sides. It furthers an empowered ethnic group's values and authority and rejects claims of the disenfranchised group. Strategies seek to entrench and expand territorial claims or to enforce exclusionary control of access. Monopoly or preferential access to the urban policy-making machinery is provided for members of the dominant group. City residents are identified through their ethnic group affiliation, the main lens through which urban policy is directed.

An equity strategy gives primacy to ethnic affiliation to decrease intergroup inequalities. Criteria such as an ethnic group's relative size or need are used to allocate urban services and spending. Equity-based criteria will often be significantly different from the functional and technical criteria used by the ethnically neutral professional planner. An equity planner is much more aware than a neutral planner of group-based inequalities and political imbalances in the city (both historic and contemporary) and recognizes the needs for remediation and affirmative action policies based on group identity. [This model] assumes that the conditions of ethnic conflict and tension can reside, at least partially, in the objective economic disparities of the urban landscape. The final model – a resolver strategy – seeks to transcend urban-based symptoms by emphasizing solutions to the root causes of urban polarization (i.e., power imbalances, competitive ethnic group identities, and disempowerment). It seeks to re-conceptualize the planning of cities and urban communities to facilitate mutual empowerment and tolerable urban coexistence. Urban professionals would play a significant role in documenting and validating each side's territorial and resource needs, which are required for coexistent community vitality. The development of alternative physical plans that refute and challenge government intent, lobbying, and legal challenges at national or international levels are forms of resistance that can benefit most significantly from urban professional input. Planners-as-resolvers are essentially confrontational of the status quo in attempts to link scientific and technical knowledge to processes of system transformation.

STUDYING URBAN POLICY AND CONFLICT

Belfast is the most populated and dominant city in Northern Ireland, whereas the Johannesburg urban region is the most populated and economically powerful metropolis in South Africa. Both cities have encapsulated deep-rooted cleavages based on competing nationalisms and arguments about sovereignty or state legitimacy. Each provides multi-decade accounts of urban planning and management in contested bi-communal environments. Each urban system functions as a physically unpartitioned whole, although the perceived environment is one of polarization. In addition . . . each was engrossed during my research in a transition process that was tied to progress on a [broader] political front. The Belfast interviews occurred in early 1995, about five months after ceasefires were announced first by the Provisional Irish Republican Army (IRA) and then . . . by loyalist paramilitary groups. . . . The Johannesburg interviews occurred mid-1995, about fifteen months after the historic democratic elections at national and provincial levels and amidst preparations for local government elections, which were successfully held in November 1995.

[. . .]

BELFAST. NEUTRAL PLANNING AND ETHNIC STABILITY

Sectarian issues don't intrude into our considerations. We don't particularly plan for one color or the other-orange or green. We do land-use planning, that's it. What difference would they make in land-use planning terms in any event. Catholics need all the housing, schools, churches, shops, and facilities, just like Protestants do.

(G. Worthington, head of the Belfast Planning Service)

Belfast is an urban setting pervaded by an overlapping nationalist (Irish/British) and religious (Catholic/Protestant) conflict. . . . The urban arena is hyper-segregated and of strict sectarian territoriality, with antagonistic groups both proximate and separate. Intercommunity hostilities have required the building of fifteen "peacelines" – ranging from corrugated iron fences and steel palisade structures to permanent brick or steel walls to environmental barriers or buffers. The city of Belfast, similar to Northern Ireland as a whole, has a majority Protestant population. The 1991 city population of 279,000 was about 57 percent Protestant and 43 percent Catholic. The Catholic percentage has been increasing over the past few decades due to higher birthrates and Protestant out-migration to adjoining towns. . . .

Religious identities coincide strongly with political and national loyalties. The allegiances of Protestant unionists and loyalists are with Britain, which since 1972 has exercised direct rule over Northern Ireland. In contrast, Catholic nationalists and republicans consider themselves Irish and commit their personal and political loyalties to the country of Ireland to its south. Since the imposition of British direct rule in the midst of sectarian conflict in 1972, legislative power for the province has been held by the British House of Commons, of which only seventeen members come from Northern Ireland. . . . Ministers in charge of Northern Ireland governance take their political cues from Westminster and tend toward an inherently conservative, non-risk-taking approach to the province's controversial issues.

A significant alteration of Northern Ireland governing institutions and constitutional status is specified in the 1998 political agreement. Much of the legislative and administrative authority in the province is to be transferred from Britain to a new, directly elected Northern Ireland Assembly, in which Protestants and Catholics will have shared power. The accord states that Northern Ireland will remain within the United Kingdom as long as a majority in the province wants to remain there. In addition, the new assembly and the Irish Parliament are to form a North–South Council to coordinate and encourage cross-border cooperation. Despite these potential momentous changes, the challenges of governance and policy making that have been faced by policy makers under third-party direct rule are likely to remain highly salient under any future alternative governance arrangement.

The dominant urban policy maker in Belfast . . . has been the DOENI [Department of the Environment in Northern Ireland]. Within or connected to the DOENI are three major entities involved in Belfast urban policy. The Town and Country Planning Service is responsible for creating the policy framework within which growth takes place and for regulating development. Belfast Urban Area plans have the force of law and establish a broad policy framework within which more detailed development proposals can be determined. Almost all planning and project applications, both private and public, are reviewed by the Planning Service for consistency with the area plan. The Belfast Development Office (BDO) promotes and coordinates physical regeneration and implements revitalization grant programs. In addition, the Northern Ireland Housing Executive is responsible for the construction of public housing, the rehabilitation and maintenance of existing units, and the allocation of public housing units to needy households and individuals.

BELFAST URBAN POLICY SINCE 1972

The operative principles for Belfast's urban policy makers and administrators have been to (a) assure ethnic stability through government policy that manages ethnic space in a way that . . . does not exacerbate sectarian tensions and (b) maintain [government] neutrality . . . so that it is not biased toward either orange (Protestant) or green (Catholic). Since 1972 . . . efforts have been made to base policy decisions on rational, objective, and dispassionate measures. However, the imperatives of containing urban violence dictate that policy makers condone the strict territoriality of the city,

one which imposes tight constraints on the growing Catholic population while protecting the underused land of the declining Protestant majority. These political realities have significant distorting effects on urban policy making. For example, although objective need dictates it, housing planners "simply cannot say there is to be a Catholic housing estate in an area that is traditionally Protestant" (interview).

Planning efforts since the 1960s for the Belfast urban area have commonly emphasized physical and spatial concerns, separating them from issues of localized ethnic conflict. The Belfast Regional Survey and Plan 1962 made no mention of the ethnically divided nature of Belfast. . . . A 1977 regional plan . . . recognized the segregative forces and inter-communal territorial competition within Belfast but . . . supported a government role accommodating of ethnic demarcations. . . . The 1987 Belfast Urban Area (BUA) Plan 2001 neglects issues of sectarianism by defining them outside of the scope of planning. DOENI, in its [1989] plan adoption statement "notes the views expressed on wider political, ecological, social or economic matters" (p. 2). However, it states that "it is not the purpose of a strategic land use plan to deal with the social, economic, and other aspects involved" (p. 2). . . . Even the bread and butter of land-use planning work – the forecasting of total and sub-group populations – is excluded from the plan, due likely to its ethnic and political sensitivity.

In contrast to town planning policy, the development-oriented NIHE and BDO address sectarian realities more directly. The [Northern Ireland Housing Executive] acknowledges interfaces and peacelines as "locations where conflict can quite frequently occur and where the Housing Executive is seeking to manage and maintain homes on an impartial basis". In building housing near these areas, the NIHE uses pragmatic tactics on a case-by-case basis within the limits set by sectarian geographies. At times, the NIHE has built walls or other physical barriers as part of a housing development if they are deemed necessary by national security agencies to stabilize inter-communal conflict. The BDO also, by necessity, confronts sectarian issues more directly than the planning service. Two main physical tactics have been used: the creation of neutral land uses between antagonistic sides and the justification of physical alterations in interface areas based on the forecasted economic benefits of BDO-sponsored projects.... Whereas the first method seeks to distance opposing sides through neutral infrastructure, the second method seeks economic gains for both sides and could facilitate nontrivial alterations to sectarian territoriality.

Planning's neutral, hands-off approach to ethnicity has sacrificed the development of a strategic plan that could guide housing and development decisions. In accordance, public actions by government units such as NIHE and BDO have primarily been ad hoc tactics rather than strategic acts, have been project based rather than area based, have been reactive rather than proactive, and have emphasized vertical rather than lateral government approaches. Planning in the strategic and comprehensive sense has been marginalized. Although town planning could play a central role in providing a reality check and frame of reference that development agencies could use when interacting with contested territoriality, there has been "no coherent and strategic planning response to the troubles" [according to an interview respondent].

The Psychology of Belfast Planning

[...]

Government officials operating amidst ethnic polarization do not want to be seen as social engineers; they see such a role as producing more harm than good. Thus, benign engineering by government to artificially bring people together is viewed as stimulative of intercommunity tensions. [One respondent] states that in Belfast's sectarian complexity "government should not impose a top-down macro view of how the city should work; rather, it should be responsive and sensitive to the needs and abilities of local communities." [Another] claims that "we must recognize the realities of the situation. If we shifted color, the end result would clearly not work. We're not about making social engineering decisions, or ones that would be perceived as such." Government officials stick as close as possible to objective standards and must watch the meanings behind their language in public documents because "words can cause a lot of trouble here".…

Town planners in Belfast defend their neutral ideology of land-use-based technical competence. [A] town planner ... views the stance as beneficial: "Planning works quite well behind the scenes"; more deterministic actions by government are best left to others. In contested public discussions, [according to another respondent] "it can be useful for planners to adopt the technical and professional role because it allows them the ability to avoid confrontation".... Town planners, then, become arbitrators rather than players in the sectarian realities of Belfast.

In the end, government's approach to urban policy in Belfast is characterized by a set of self-limiting features. There is both substantive separation of the town planning function from ethnic management responsibilities and the fragmentation of policy along division and department lines. Combined, these factors decrease government's ability to mount an ethnically sensitive strategy that would be multidimensional (physical, social-psychological, economic, and human development) and interdivisional (integrating planning, housing, and development units).…

JOHANNESBURG: EQUITY PLANNING AND URBAN RECONSTRUCTION

Planners have grown up providing services for a well-understood and familiar client – White and affluent.
(Tim Hart, SRK Engineers)

Johannesburg anchors a[n] ... urban region of enormous economic and social contrasts. The region presents dual faces: one is healthy, functional, and White; the other is stressed, dysfunctional, and Black. The most luxurious suburbs on the African continent and downtown skyscrapers of iridescent modernity coexist with townships and shantytowns of intentionally degraded living environments.... With the establishment of the multiparty Government of National Unity, and the national democratic elections in April 1994, hope ... for urban change exist[s] alongside an awareness by policy makers of the difficulties of bettering the stark conditions of many Black Africans.

The Johannesburg (central Witwatersrand) metropolitan region contains at least 2 million people. The population is approximately 60 percent

Black and 31 percent White. Racially segregated townships, cities, and informal settlements/shanty-towns characterize the urban landscape. Income distribution is grossly skewed in Johannesburg's province of Gauteng. . . . [B]asic needs pertaining to housing, land tenure, and water and sanitation facilities [remain] unmet. . . . [T]here is a[n estimated] 500,000 formal housing unit shortfall in the province alone. . . . The two primary [concentration of Black Africans] are Alexandra and Soweto townships, the latter an amalgamation of 29 townships more than 10 miles southwest of, and spatially disconnected from, Johannesburg. Formal bricks-and-mortar housing was intentionally under-built because urban Blacks were considered temporary and unwanted. . . . [T]ownship [housing is] . . . characterized by near-inhuman conditions of living, lack of secure tenure, inadequate standards of shelter and sanitation, and lack of social facilities and services. . . .

Urban apartheid policy, anchored by the 1950 Group Areas Act, divided towns and cities into group areas for exclusive occupation by single racial groups. Buffer strips of open land, ridges, industrial areas, or railroads separated races to minimize inter-group contact. City centers, environmentally stable areas, and otherwise prestigious areas were zoned White; peripheral areas were zoned non-White and restricted in scope. . . . Town planning's traditional emphases on efficiency, order, and control were not at odds with notions of ethnic segregation and ordering. . . . Thus, the town planning profession in Johannesburg went down "the long road of coercion and domination" (interview). "Apartheid planning was terribly effective in achieving its goals," an interviewee said. Yet, the success of this partisan planning erected functionally and economically unsustainable urban conditions that contributed over time to the downfall of the apartheid system. . . .

Reconstructing Urban Policy since 1991

[. . .]

From 1991 to 1995, urban leaders engaged as resolvers of core political issues in helping to transform local and metropolitan governance. Officials of the old regime, nongovernmental groups, and representatives of formerly excluded Black com-

munities collaborated in a . . . process that changed the basic parameters of representation, decision making, participation, and organizational structure. City-building issues dealing with day-to-day consumption concerns and the Black boycotting of payments for housing rent and urban services were successfully connected by nongovernmental and opposition groups to root issues of political empowerment and local government reorganization. Discussions transcended sole emphases on the urban symptoms of racial polarization and targeted the need to radically transform apartheid-based urban governance. After difficult negotiations, local and metropolitan government in Johannesburg was restructured to politically consolidate formerly White local authorities with adjacent Black townships. In late 1995, absolute Black majorities were voted in for all four local governments and the Johannesburg metropolitan council.

Concurrent with this political restructuring of local governance, there was the formulation of alternative, equity-based urban policies for a democratic Johannesburg. The [Central Witwatersrand Metropolitan Chamber] was established in 1991, in part, to develop an overall vision for future development in the Johannesburg region. . . . Equity-based, post-apartheid city-building principles aspire to stitch together apartheid's urban discontinuities and integrate the torn parts and peoples of Johannesburg. Key facets of this city building include: (a) densification and infill of the existing urban system and (b) upgrading and renewal of those parts of the urban system under stress. The densification approach seeks to encourage growth inward around already developed areas that are close to employment opportunities, have access to high levels of services and facilities, or fill in urban discontinuities intentionally created by apartheid planners. This "compact city" approach would be a primary means to increase opportunities for Blacks to enter the residential and economic fabric of the White city . . . [reversing] the decades-long practice of apartheid policy makers of . . . isolating non-White[s] far from urban opportunities. The second policy approach focuses on the upgrading and renewal of those areas . . . under stress due to inadequate housing, poor or nonexistent water and sanitation services, public health hazards, and inadequate social facilities. Whereas the first policy approach seeks to create sustainable metropolitan land use patterns, an upgrading approach aimed at alleviating the many crisis-related needs out on the urban remote fringes may over time solidify apartheid geography. That there may be tension between these two types of policy responses is brought out by one interviewee who asserts that "there may be a case for subsidies to Soweto on reconstruction grounds, but there is not a case for it in terms of economic growth". Another vexing problem in efforts to reconstruct Johannesburg is that although the old centralized apartheid state is gone, land market, economic, and class-based interests now shape urban geography in ways that may produce similar spatial outcomes. In particular, high inner-city land costs obstruct efforts to incorporate the majority into a compact city of urban opportunity, profitability criteria used by private developers likely exclude the marginalized poor from the benefits of urban reconstruction, and upper-class communities resist the integration of low-income communities into their urban network.

The Dual Faces of Post Apartheid Planning

. . . There is . . . debate among urban policy makers about how best to engage in Johannesburg reconstruction. . . . The traditional model of town planning in South Africa, derived from British and European foundations, has been focused on regulatory control and spatial allocation, administered in a centralized and hierarchical fashion and lacking in community consultation. Today, not only is this blueprint . . . discredited due to its alignment with apartheid but there appears a disconnection between the socioeconomic and empowerment needs of Black areas and this model of development control. Where Black needs seek transformation of basic conditions of livelihood, the traditional planning model offers reform-minded, yet ultimately conservative, prescriptions. In response, a new paradigm of development planning has emerged. It represents a fundamental re-conceptualization of, and challenge to, traditional town planning. Development planning is defined by three main characteristics: (a) it integrates traditional spatial planning with social and economic planning; (b) it attempts to restructure the general budget to meet development policy objectives that cut across governments, sectors, and departments; and (c) it includes a participatory process aimed at empowering the poor and marginalized.

Development planners have distinctly different personal histories than traditional town planners. Many are Black Africans who are not trained in the technical, legal, and regulatory foundations of physical development control. Rather, they have experience in NGOs in which they developed a set of skills related to community development, social mobilization, and negotiation that were directed at both antiapartheid resistance and the improvement of basic living conditions for marginalized communities and people. Development planning, in South Africa's use, connotes . . . the empowerment of the deprived majority. Traditionally trained town planners fall short here. The lack of community consultation in the town planning model meant that such planners worked in closed rooms in developing spatial frameworks. "You did 'what was best for society' and society had to accept whatever you did," [one respondent] recalled. In contrast, development planners emphasize their role as mediators in the development process between community needs and government resources. They see their key characteristic as an ability to speak two languages – that of the community in helping them identify their needs and that of the bureaucrat having knowledge of governmental processes and realities. Inside knowledge of governmental processes was gained initially as the communities for which they worked used various tools to engage the state and its transformation; subsequently, it has been developed during their tenure as public administrators. Development planning, however, remains embryonic in South Africa, and its methods appear only broadly articulated. "Nobody has been trained in doing the work that we do," [said a development planner].

[. . .]

The Psychology of Johannesburg Planning

Town planning and development planning in many ways represent uneasy bedfellows in their common pursuit of a more humane Johannesburg. Town planning must contend with its image as "old guard," its past links to apartheid implementation, and its lack of connection to community. At the same time, it provides a methodology and technical capacity that is essential to city building. Development planning, meanwhile, is ascendant from community-based struggle and newly knighted as the way forward for urban South Africa. Yet, it is a young practice whose techniques are not clearly developed and that is burdened by demands for it to be all things to all people. When the two faces of post-apartheid planning come in contact, one can detect a clash[:] . . . town planners [are] rooted to existing systems, rules, and regulations, whereas development planners are more proactive and sympathetic to experimentation. In the face of a clearly ascendant development planning function, traditional town planners are reacting in ways ranging from defensive rigidity, to counter-attack, to uncertainty, to productive acceptance of the need to change. . . .

[One planner] points to a thick statute book specifying the contents of town planning and zoning schemes and states that "many planners cannot cross the river of change because of this little bible that they have." Professional biases toward spatial control are also impediments to change: "It is very difficult for many planners to get out of the groove of doing up nice maps and pictures on the wall. It is part of the education system they carry with them." Other town planners, however, defend traditional planning's value and criticize any wholesale move toward development planning. [One] asserts that criticism of traditional planning too simplistically positions planners as technicians worthy of marginalization in the face of emergent community activists. . . .

Traditional planning's defense of its unique contribution to city building is brought out in other observations. [One planner] said that "community specialists and social workers are needed for communication purposes, but at the end of the day someone else must come in to deal with technical issues such as water provision and engineering capacity." . . . [Another notes] an additional and not insignificant contribution of traditional planning – an ability to maintain property values and municipal tax bases and to assure protection of property rights and investment. Some town planners who were surveyed expressed professional uncertainty amidst institutional transformation. Other traditionally trained planners in government are

rising to the challenge. For them, it is an invigorating time to develop new techniques of community consultation or to question assumptions and theories of the past.

[. . .]

The new paradigm of development planning represents an historic attempt to create a system of social guidance that uses the legacy and lessons of social mobilization. Town planning practice has a vital role to play in supporting this movement from mobilization to management because development planning, alone, does not yet have the methodologies or systematic knowledge bases to fully engage in city building. If these two faces of post-apartheid planning were effectively combined, the result would likely be an altered and Africanized practice of community-based urban planning, encompassing both social mobilization and rational governance. Only this re-conceptualized form of African planning can meaningfully contribute to a remaking of metropolitan Johannesburg's apartheid landscape.

URBAN PLANNING AND ETHNIC STRIFE

[. . .]

Three of the four urban planning strategies – neutral, resolver, and equity – are represented in the contemporary policy making of the case study cities. (The fourth strategy discussed earlier – partisan planning – was represented by Johannesburg's planning during the apartheid years.) In Belfast, the British government's stabilization strategy is to deal pragmatically on a neutral basis with the day-to-day symptoms of sovereignty conflict. Inter-group equity issues are excluded from metropolitan plans; public housing allotment formulae use color-blind procedures; and town planning separates its spatial concerns from the more broad social issues of housing, social services, and ethnic relations. Johannesburg illustrates two roles that urban policy makers have assumed during the demise of apartheid. First, as resolvers of basic political issues, local leaders linked urban symptoms to political causes, recognizing that restructuring of urban governance was a necessary prerequisite to effective equity policy making. Since local elections of late 1995, the urban policy strategy in Johannesburg has been more focused on equity objectives, seeking to address the horrendous urban symptoms of past racial conflict by lessening the gross disparities in urban opportunities and outcomes across races. The effectiveness of such an equity approach in Johannesburg, however, faces multiple constraints – the sheer magnitude of urban Black needs, market-based obstacles to spatially restructuring urban housing and economic opportunities, restricted ability to pay for basic services on the part of Black customers, a culture of government illegitimacy, and budgetary limitations.

[. . .]

The institutional structure of city planning in both cities is, or has been, significantly shaped by national imperatives. The racial cleansing of Johannesburg required a centralized planning apparatus capable of bypassing possible municipal-level concerns about sustainability and functionality. Ironically, such centralization also appears needed when trying to remove urban policy neutrally above the partisan fray, as in Belfast. In apartheid Johannesburg, local initiative regarding land-use regulation was subsumed within strong central state-led spatial planning. Today, in designing intergovernmental relations in the new Johannesburg, a centralization of urban policy needed to reconstruct and equalize a torn society seems at odds with pent-up needs for local empowerment. Belfast represents an extreme intergovernmental case, with direct rule having removed decision-making authority from local policy-making bodies. Because it emphasizes vertical and tactical policy rather than comprehensive, or lateral, planning, urban policy becomes the net outcome of a set of uncoordinated, single-function, centralized interventions.

THE CHALLENGE OF URBAN COEXISTENCE

. . . [U]rban operationalization of political imperatives is not a straightforward exercise. In both cities, efforts to stabilize or reconstruct strife-torn polarized environments present difficult policy dilemmas that can endanger state urban goals themselves. More specifically, stabilization policy in Belfast may be creating spatial impediments to urban peace, whereas reconstruction policy in Johannesburg may solidify, rather than transcend, apartheid geographies. In Belfast, the main means

toward conflict containment – the condoning and formalization of ethnic separation through housing, planning, community development, and peaceline tactics – appears to provide short-term stability at the expense of long-term opportunities for intergroup negotiation and reconciliation. The hardening of Protestant–Catholic territorial identities allowed by government as an indispensable ingredient to short-term stability might paralyze the city long after the occurrence of any future political restructuring in Northern Ireland. . . . The reactive protection of the status quo by a neutral policy aimed at stability and maintenance defends a dysfunctional and sterile territoriality, reinforcing the physical and psychological correlates of urban civil war.

In the case of South Africa . . . [r]esolution of the root political causes of city-based conflict coexists with distressing conditions of unmet basic needs. . . . [E]ffective urban implementation of reconstruction and democratization goals will need to provide tangible benefits to the long-neglected majority. Here, though, two policy dilemmas confront reconstruction planners. First, policy makers must decide between responding to crisis-level needs of the Black poor in a way that reinforces apartheid geography or [broader, longer-term] efforts that would more equitably reshape the urban region but that would [not] be . . . as responsive to immediate needs. Second, policy makers must decide whether the reconstruction of the urban area will be more dependent on private market mechanisms or state intervention. . . . Private sector economic and development processes will likely apply major structural brakes to Johannesburg's movement toward equity; yet, assertive state urban policies to counter these effects may not be fiscally possible. Spatially, market-based normalization of Johannesburg – for instance, the decentralization of employment to high-income suburbs – will likely produce intra-regional disparities that eclipse urban equity and spatial compaction efforts. Economically, market corrections to apartheid will produce a society deeply cleaved by class. To the extent that non-ethnic, market-based factors replace ethnic engineering as allocators of goods and resources in the new South Africa, the vitality and sustainability of post-apartheid democracy will likely be threatened by mass poverty and new forms of inequality.

National and local policy makers may play critical roles in whether these somber trajectories are fulfilled or whether alternative futures of sustainable mutual coexistence are possible. In Belfast, urban policy should be formulated so that it is a constructive part of peace building. . . . It would need to be sensitive to the unique needs of each community while keeping in mind the overall good of the city. This would necessitate an engagement in equity policy that would target greater material resources to the disproportionately more deprived Catholic population while, at the same time, responding to the social-psychological needs of the Protestant majority [for] neighborhood vitality, personal security, and ethnic identity. Decline in the Protestant population would be more strategically managed to produce a vital but geographically consolidated Protestant population. The goal would not be the maintenance of Protestant territories but the viability of Protestant communities. In Johannesburg, urban policy making has a significant role in the post-resolution, re-construction phase. Resolution of ethnic conflict must be nurtured by reparative on-the-ground policies that deepen and sustain political change. The chances will increase that democracy in Johannesburg (and South Africa) will be sustainable if a collective sense of social justice underlies public policy, not non-racialized market dictates that keep the poverty-stricken poor and on the urban periphery. This may require a more expansive state role in the creation of inner-city housing to overcome high land value impediments to affordable housing, close state oversight over large projects in key locations to assure that equity objectives are not displaced by profitability criteria, and state mediation of class-based neighborhood opposition to assure that apartheid-constructed concentrations of poverty are attenuated. Cities such as Johannesburg not only constitute spatial and economic keys to reconciliation but also are important arenas in which a participatory and functional post-apartheid relationship between government and community can be nurtured and sustained.

Cities, regions, and nations

Plate 6 Viennese public housing, circa 1927.

INTRODUCTION TO PART SIX

Although the contemporary study of urban politics began by seeing the city as a separate, distinct, and independent entity, something resembling the Greek *polis*, it has been clear, at least since the rise of the modern nation state, that we can only understand urban politics by taking account of larger contexts. In the contemporary period, these contexts must include metropolitan regions, nations, and indeed the international and global system. Some contextual dynamics limit what cities can do – for example, their legal subordination to state governments, the small amount of funding that U.S. cities receive from national governments, or their competition for revenue-generating investments – but others may enhance their scope of action, at least if they pursue shrewd local strategies. Saskia Sassen, Neil Brenner, and others have argued, for example, that globalization has weakened the capacities of nation states while potentially enhancing those of metropolitan regions.

The readings in Part 6 set forth many of the main themes being discussed by scholars of the changing settings of urban politics. Providing a book-end to Margaret Weir's opening essay on how the city has fared in the evolution of the American welfare state, policy historian Alice O'Connor begins with a careful review of federal policy toward big cities. She shows how weak the national political consensus has been for such policies, in contrast to the broader coalition behind national policies that send benefits to people, as opposed to places. Indeed, as Richard Sauerzopf and Todd Swanstrom's contribution points out, the national political influence of the large old central cities has diminished tremendously, while that of the newer, more conservative cities and metropolitan regions of the South and West have grown. Geopolitical trends within the U.S. suggest that a central city-based national political strategy simply cannot be successful on its own. Buzz Bissinger, a journalist, closely followed Ed Rendell, then mayor of Philadelphia and now governor of Pennsylvania, through his trials and tribulations as he sought to influence the actions of higher levels of government that might hurt his city. He provides concrete examples of just how difficult it can be to navigate this terrain, but also shows the importance of having a mayor with strong political skills.

If city advocates will have a hard time achieving the kinds of national policies that will help them on their own, then they will have to forge new coalitions and develop new policy ideas to broaden their political base. One obvious place to begin is by drawing on the resources within their own metropolitan regions as a building block for these coalitions and policies. Political scientists Peter Dreier, John Mollenkopf, and Todd Swanstrom evaluate the past practices and current debates about political cooperation and regional governance within metropolitan areas. They find that past efforts may offer important clues about future directions, but that bolder and more far-reaching institutional changes are needed. Pietro Nivola, a leading student of comparative urban government, draws on policy experiences outside of the U.S. to show just how important national frameworks in support of regional cooperation can be in dampening some of the negative effects of the fiscal competition so evident in the U.S.

The destruction of the World Trade Center in New York City on September 11, 2001, the bombings of the Madrid commuter railway system on March 11, 2004, and the attack on the London Underground on July 7, 2005, show a new way in which cities are vulnerable to developments in global politics. While the fate of entire cities has been at risk since the deployment of carpet bombing and nuclear weapons in World War II, at least this risk was subject to some control by the system of nation states. Today, cities have become a target for non-state terrorists in their campaign to win influence in global politics. This fundamental change in the urban political environment is given thoughtful consideration by political scientist Peter Eisinger.

The Reader concludes with a look forward. Bruce Katz is director of the Center on Urban and Metropolitan Policy and Vice President of the Brookings Institution in Washington, D.C. During the Clinton administration, he was chief-of-staff for Henry Cisneros, Secretary of the U.S. Department of Housing and Urban Development, and previously served as staff director for the U.S. Senate Committee on Housing and Urban Affairs. While the Clinton administration had a hard time enacting many of its ideas, under Secretary Cisneros, it did some highly creative thinking about the future of federal urban policy. As such, Katz combines a sophisticated understanding of the technical details of virtually all aspects of urban policy with a keen appreciation of the political realities and possibilities facing urban advocates. He reviews the economic, social, and political trends of recent decades, critically analyzes the state of policy under the current administration, and proposes a more cohesive program for metropolitan transportation and housing that could command broad support.

"Swimming Against the Tide: A Brief History of Federal Policy in Poor Communities"

from Ronald F. Ferguson and William T. Dickens (eds), *Urban Problems and Community Development* (1999)

Alice O'Connor

Editors' Introduction

Although there is a long U.S. tradition of localism and self-help, federal policies explicitly aimed at encouraging local, community-based initiatives have been inconsistent. This reflects the short history of federal involvement at this level: before the twentieth century, the federal government had a negligible role in financing or mandating state and local programs. Even as the federal government became more involved in the Great Society period, a long-standing debate has taken place over whether federal aid should be directed at *people* (regardless of where they live), or at *places*. Alice O'Connor is interested in tracing the history of place-based federal aid and she examines programs that have ostensibly been designed to improve poor communities.

Perhaps more than most policy areas, community development policy has experienced dramatic shifts in scope and purpose. Whereas programs with widely dispersed benefits, such as the mortgage interest tax deduction or Social Security, have thrived for decades, place-based community development efforts have always been contentious, and have been continually restructured as new partisan alignments and political philosophies have come to dominate. So place-based policies have at times been robust and at other times weak; under more liberal administrations they have sought to mobilize new voters and challenge the political and economic status quo, while in more conservative times they have sought to encourage the market.

Nonetheless, O'Connor has identified persistent patterns in the now seventy-year history of federal community development policy. For instance, she notes that place-based strategies have usually been marginal, and have lacked a solid political constituency. Perhaps for that reason, community development policies, usually operating at the micro-level, inevitably run counter to the thrust of other, more compelling policy trends. For example, targeted urban renewal efforts, aimed at preventing businesses and middle-class residents from fleeing the city, were unlikely to succeed when federal subsidies for highways and single-family homes facilitated the flight to the suburbs. Presumably her title, "Swimming Against the Tide," refers to the unstable footing such programs have against the forces of decentralization. Place-based policies are unlikely to disappear, however, as long as we have a Congress based on geographically defined districts (in which each re-election-seeking member needs to be able to "bring home the bacon" in some form), and an established interest network of place-based organizations advocating for the continuation of such policies in some form.

O'Connor's chapter is primarily an effort to synthesize some seventy years of policy history, but O'Connor also writes as an advocate, seeking to identify ways in which academic researchers in general, and historians in particular, can contribute to a revitalization of community development policies. Thus far, she argues, community development policies have failed to exhibit any real policy "learning curve"; new policies "replicate, rather than learn from" the (often failed) policies of the past. She concludes by exhorting policy analysts and advocates to make a case for federal support of place-based antipoverty programs. Although critics of federal urban programs have been quick to point out their many failures, O'Connor notes that the federal government has been quite successful in meeting its goals when providing benefits for the middle class (the success of Social Security at eliminating poverty among the elderly, and of the federal income tax deduction for mortgage interest as a means of promoting home ownership are two examples). With effective grass-roots mobilization, consistent support from philanthropies, and an academic research agenda that does more than bemoan the inevitability of urban decline, it is possible that a new generation of federal urban policies may not have to "swim against the tide."

Those interested in learning more about twentieth-century federal urban policies can choose from many sources. A few of the more comprehensive reviews include: Robert Halpern, *Rebuilding the Inner City: A History of Neighborhood Initiatives to Address Poverty in the United States* (New York: Columbia University Press, 1995); R. Allen Hays, *The Federal Government and Urban Housing: Ideology and Change in Public Policy* (Albany, NY: SUNY Press, 1985); Kenneth T. Jackson, *Crabgrass Frontier: The Suburbanization of the United States* (New York: Oxford University Press, 1985); John H. Mollenkopf, *The Contested City* (Princeton: Princeton University Press, 1983); and Michael J. Rich, *Federal Policymaking and the Poor: National Goals, Local Choices, and Distributional Outcomes* (Princeton: Princeton University Press, 1993).

Alice O'Connor is an Associate Professor in the History Department of the University of California, Santa Barbara. Her research focuses on U.S. public policies, most notably in the areas of urban history, poverty, and welfare. She is the author of *Poverty Knowledge: Social Science, Social Policy and the Poor in Twentieth Century U.S. History* (Princeton: Princeton University Press, 2001) and co-editor of *Urban Inequality: Evidence from Four Cities* (with Chris Tilly and Lawrence D. Bobo) (New York: Russell Sage Foundation, 2001).

Community development is a time-honored tradition in America's response to poverty, but its meaning remains, as other observers have pointed out, notoriously hard to pin down. Defined variously as social and cultural uplift, integrated social service provision, local economic development, physical renovation, and political empowerment, the term has come to encompass a large number of different place-targeted interventions that have never quite added up to a coherent, comprehensive strategy. Nor have efforts to establish a federal community development policy been of much help. Instead, the historical evolution of policy has been disjointed and episodic, starting from ideas that first emerged in private, local reform efforts during the Progressive Era, moving through an extended period of federal experimentation from the New Deal to the Great Society, and devolving to an emphasis on local, public–private initiative beginning in the 1980s. The result has been a sizable collection of short-lived programs, many administered by agencies since disappeared, that seem continually to replicate, rather than learn from, what has been tried in the past. Although certainly not alone in lacking coherence or institutional memory, federal community development policy is notorious for reinventing old strategies while failing to address the structural conditions underlying community decline.

And yet, the push for place-based policy continues, as it has for the better part of the past sixty years. No doubt this has something to do with the geographic basis of political representation: naturally, members of Congress will support programs to stem decline and depopulation back home. In the wake of ghetto uprisings since the 1960s, federal aid for community development has also become a political quick fix, a palliative for communities on the verge of revolt. But the continued appeal of community development is not simply a matter of political expediency or social control. Equally important in keeping the idea alive has been a loosely organized grouping of grassroots activists, neighborhood groups, community-based providers, national "intermediary" institutions, and philanthropic foundations, a kind of community development movement that has made a business of improving poor places as a way of helping the poor. Geographically dispersed and internally conflicted though it may be, over the years this movement has been largely responsible for keeping the idea of community development alive. It has had a significant effect on the shape of federal initiatives in poor communities and, despite recent decades of worsening local conditions and government retrenchment, it shows little sign of going away. What does the historical record have to say to this movement and its continuing effort to carve out a federal role? Does the past offer any guidance for the future of community development policy? This chapter is an attempt to find out.

HISTORICAL PATTERNS IN FEDERAL POLICY: CONTINUITY AMIDST CHANGE

At first glance it may seem there is little to learn from a history of policies with origins in the New Deal political order. After all, policymakers are operating in a much circumscribed environment, now that the era of big government is over. And poor communities are struggling against much steeper odds in a globalized economy that values mobility and flexibility more than place. But the plight of poor communities does have instructive historical continuities. Like the abandoned farm communities and industrial slums of an earlier era, the depressed rural manufacturing towns and jobless inner-city ghettos on the postindustrial landscape represent the products of economic restructuring and industrial relocation, of racial and class segregation, and of policy decisions that have encouraged these trends. The historical record also points to recurrent patterns within community development policy, which help explain its limitations in combating the underlying causes of decline.

First, government works at cross-purposes in its treatment of poor places. Small-scale interventions are intended to revive depressed communities while large-scale public policies undermine their very ability to survive. Nowhere are these policy contradictions more clear-cut and familiar than in the case of central cities, which were targeted for limited amounts of assistance and renewal beginning in the late 1940s even as more substantial federal subsidies for home mortgages, commercial development, and highway building were drawing industry, middle-class residents, and much needed

tax revenues out to the suburban fringe. Rural farm communities faced a similar plight during the Depression and post-World War II years, when federal aid for local readjustment paled in comparison with support for the large-scale mechanization, commercialization, and industrialization that transformed the agricultural economy.

More recent community-based interventions have also been undercut by economic policy, which has favored flexible, deregulated labor markets and left communities with little recourse against wage deterioration and industrial flight. Public policy was similarly instrumental in the intensification of racial segregation in residential life by encouraging redlining practices in mortgage lending agencies, maintaining segregationist norms in public housing projects and by uneven commitment to the enforcement of federal antidiscrimination laws. Thus, having encouraged the trends that impoverish communities in the first place, the federal government steps in with modest and inadequate interventions to deal with the consequences – job loss, poverty, crumbling infrastructure, neighborhood institutional decline, racial and economic polarization – and then wonders why community development so often "fails." In its attempts to reverse the effects of community economic and political decline, federal policy has been working against itself.

A second pattern is that while the historical record is replete with examples of place-based strategies, they have always occupied a marginal position in the nation's antipoverty arsenal. In part this is because investing in declining communities runs counter to the dominant conventions of social policy analysis, which since at least the 1960s have been based on economic concepts and norms. Place-based policies are inefficient, even quixotic, according to conventional economic wisdom, in comparison with policies emphasizing macroeconomic growth, human capital, and individual mobility. Community investment also goes against the individualized model of human behavior underlying policy analysis, which presumes that people are principally motivated by rational self-interest in making life decisions. For those stuck in places with little hope of revival, the more rational choice is out-migration, according to economic calculation. Thus policy should promote "people to jobs," not "jobs to people" strategies. The analytic framework further denigrates community development for its inability to define and achieve clear-cut quantifiable goals and outcomes. After all, "building local capacity," "mending the social fabric," "cultivating indigenous leaders," and, most of all, "encouraging community empowerment" are amorphous objectives and difficult to measure. Nor does community development come out well in traditional cost-benefit analysis. Among other things, it takes time and experimentation, and its benefits are largely indirect.

Opposition to place-based programs is not simply analytic however; it is grounded in politics and ideology as well. Community development meets continual resistance from those reluctant to interfere with the "natural" course of economic growth. It has also generated animosity among local politicians when it threatens to upset the local power base. And the debate over investing in place versus people has become artificially polarized in the politics of fiscal austerity since the 1970s. In a system structured principally to meet the needs of families and individuals, place-based programs have routinely lost out.

A third pattern in the movement advocating federal community development policy has been its reliance on unlikely or tenuous political alliances for support. In 1949, advocates of public housing reluctantly lined up with downtown real estate developers to help pass urban renewal legislation, an alliance that proved disastrous for poor and minority neighborhood residents. Several years later, policy analysts in the Budget Bureau joined forces with an assorted group of activists, philanthropists, and social scientists ("kooks and sociologists," President Lyndon B. Johnson is rumored to have called them) to make "community action" the centerpiece of the War on Poverty, only to discover that they had widely varying definitions of action and, especially, of "maximum feasible participation" in mind. Community development corporations took the idea from anticolonialist, anticapitalist ghetto activists and remolded it into a form of "corrective capitalism" with government and foundation support. When forged at the local level, these types of alliances have been praised as expressions of community-based consensus, the idea, as Industrial Areas Foundation founder Saul Alinsky said, that the common pursuit of neighborhood interests can inspire "unusual sympathy and understanding between organizations which had previously been

in opposition and conflict." At the national level, however, they reflect a basic political reality: the most likely constituency for community development policy – the resident base – is mobile, unorganized and, especially as the two major parties compete to capture the suburban vote, diminishing in political power at the national level. Building national coalitions for change, then, has been a continual process of compromise with interests outside the community, often at the expense of the residents that community development seeks to assist.

A fourth pattern is that precisely because they cut across so many different policy domains, community development policies have suffered more than most from administrative fragmentation and bureaucratic rivalry. Even when administered by a designated community development agency, federal initiatives have drawn most of their funding from scattered sources, ranging from the Department of Housing and Urban Development to the Department of Defense, each with its own bureaucratic culture and priorities, and each eager to protect its turf.

This administrative fragmentation, to some extent a characteristic of the federal welfare state, also mirrors divisions within the community development movement. Integrated services, planning and economic development, infrastructure rehabilitation, and political organizing might in theory complement one another, but in reform circles they have historically been promoted as alternative if not competing strategies. Urban and rural development networks have also operated along separate intellectual and bureaucratic tracks, a division that has been heightened by the increasingly urban bias in antipoverty thinking throughout the postwar years. Periodic federal efforts to overcome fragmentation by designating lead agencies have yet to produce an effective coordinating mechanism. More often the real burden of coordination is left to the localities, where, particularly in the wake of reduced budgets since the 1980s, the competition between programs to tap into available funds has grown even more fierce.

[. . .]

A [fifth] pattern is that in its treatment of poor communities federal policy has operated within the two-tiered system of provision that marks U.S. social policy. In this system poor communities, like poor individuals, are assisted through an elaborate concatenation of means-tested programs, while their wealthier counterparts are subsidized through essentially invisible, federalized, non-means-tested subsidies such as highway funds, state universities, home mortgage assistance, and tax preferences. Poor communities are targeted as places for public assistance – public housing, public works, public income provision – while the middle class is serviced by nominally private but heavily subsidized means. Thus the retreat from the public in all walks of life has been doubly dangerous for poor communities. It has brought not only a loss in funds but the stigma of having been designated as "public" spaces in a society that equates "private" with quality and class.

Finally, despite its race-neutral stance, community development policy has continually been confounded by the problem of race. Minorities were routinely excluded from the local planning committees established in early federal redevelopment legislation, and their neighborhoods were the first to be bulldozed as a result. The programs of the 1960s were subsequently caught up in the politics of racial backlash. Race is deeply embedded in the structural transformations that beset urban and rural communities as well. Poverty and unemployment are more concentrated in minority than in white neighborhoods, and poor minorities are more likely to live in high-poverty areas than are poor whites. Yet race is rarely explicitly acknowledged in community development policy, and then only when it can no longer be avoided: within the confines of racial uprising and violence in the late 1960s and again in 1992.

One lesson from historical experience, then, is that community development policy has been undermined by recurring patterns in the structure of policy. Internal contradictions, marginalization, weak political coalitions, fragmentation, associationalism, second-tier status, and institutionalized racial inequality have kept community development policy swimming against the tide. As a closer look at the historical record will show, these patterns are not the product of immutable ideological or structural forces but of the political processes through which policy choices have been negotiated and made. Many can be traced to the very beginnings of the community development movement in the decades before place-based policy had become a part of the federal welfare state.

[. . .]

FOUNDATIONS OF FEDERAL POLICY: THE NEW DEAL AND BEYOND

During the 1930s the Roosevelt administration's New Deal made a massive investment in shoring up distressed communities with direct job creation, public works, and infrastructure building, while also recognizing the plight of displaced rural communities with land distribution, planned resettlement, and even the construction of model greenbelt communities. At the same time, the New Deal also laid the foundations for an indirect form of community development in two of its most far-reaching measures: the mortgage insurance system that would later help underwrite the postwar suburban housing boom and the investment in regional economic modernization that would transform the political economy of the South. By the end of the New Deal these hidden forms of federal community investment were on the verge of major expansion, while most of the direct job creation, public works, and resettlement policies had either fallen to opposition or been allowed to die. In their stead was the combination of public housing assistance, cash grants and services, and localized planning that would constitute the foundation for federal aid to poor or declining communities for the next four decades.

Perhaps the most significant New Deal measure in terms of future community policy was not specifically place-oriented at all. The Social Security Act of 1935 established the basic approach to social welfare provision that would regulate the federal approach to communities as well: individualized and income-oriented. This strategy implicitly rejected the environmentalist efforts of the community reform tradition. Despite a network of social work professionals in New Deal agencies, services were relegated to a relatively minor position in the Social Security Act. That services were something of an afterthought is also suggested by the administrative fragmentation built in to the bill, which reserved oversight responsibility for social insurance, Aid to Dependent Children, and other cash programs for the Social Security Board, while parceling out child and maternal health, welfare, and other services among different agencies. From the start, then, the federal welfare state created a fragmented administrative structure for providing cash and services and set up hurdles that future reformers would perpetually try to overcome.

In its reluctance to interfere with private markets, the Social Security Act also set the pattern for federal aid to communities. Although recognizing the risks and vicissitudes generated by exclusive reliance on the market, the Roosevelt administration was eager to work within and undergird the private enterprise system and, above all, to get the federal government out of the business of job creation and direct relief. . . . Perhaps most important, the Social Security Act set the pattern for the two-tiered structure of federal social provision: on the top tier, a federalized, contributory, non-means-tested social insurance program for protection against income loss in old age and unemployment; on the bottom a localized, means-tested system of public assistance for poor women and children. Poverty, whether addressed at the individual or community level, would hereafter be treated separately from the problems of old age and unemployment.

A second New Deal measure, the Housing Act of 1937, created the basis for public housing, a mainstay of federal assistance to poor communities for decades to come. It also established a complicated political infrastructure for housing programs, based on an uneasy mixture of private profit and public purpose, that reflected the administration's hope of achieving several not always compatible goals at once. One, shared by most New Deal programs, was to put the unemployed to work. More controversially, federal housing programs were also used for slum clearance, which made them appealing to urban developers but generated criticism from advocates for the poor. Federal construction projects administered by the Public Works Administration managed to serve both goals directly, creating thousands of government jobs on construction sites located in cleared-out slum areas.

With the Housing Act of 1937 the administration moved from direct government provision toward a more decentralized system of market subsidy and local control. It also incorporated another major goal: stimulating the private construction industry. Under the terms of the legislation, local housing authorities were created to issue bonds, purchase land designated for slum clearance, and contract with private builders to construct public housing. Thus they provided the public with affordable housing, the unions with jobs, and the construction market with a subsidy from the federal government.

Although fiercely opposed to the Housing Act of 1937, local real estate developers soon found that they, too, could get in on the benefits of public housing. They recognized that federal funds for slum clearance offered a rich public subsidy for potentially valuable downtown real estate that could be developed for more profitable purposes. Thus, by the end of the 1930s public housing was tied into a broad-based constituency that included labor, urban interests, and reform groups as well as private builders and developers. Meanwhile, by tying public housing almost exclusively to the goal of slum clearance and leaving locational decisions up to local initiative, the act essentially guaranteed that public housing would remain concentrated in central cities.

The overarching goal of New Deal housing policy, however, was to promote home ownership among working- and middle-class Americans, a goal it achieved largely at the expense of poor and minority city dwellers and the neighborhoods they inhabited. In 1933 the Roosevelt administration created the Home Ownership Loan Corporation (HOLC) to protect home owners from the threats of foreclosure and high interest rates. In 1934 home ownership got a bigger federal boost when President Roosevelt signed legislation creating the Federal Housing Administration (FHA). Neither program provided benefits directly. Instead, by insuring long-term loans made by private lenders, they stabilized the home mortgage insurance market, made mortgages and home improvement loans more accessible to the middle and working classes, and provided a permanent stimulus for the private housing market. The benefits of these policies did not extend to slum dwellers, however, or to families with incomes too low to meet even subsidized mortgage requirements. Blacks and other minorities were also systematically excluded through officially sanctioned redlining, neighborhood covenants, and other forms of discrimination.

[. . .]

The New Deal established the foundations for federal aid to declining communities, but its legacy was decidedly mixed. For the next several decades politicians concerned about community deterioration could look to federal housing and planning programs for local rebuilding and development. New Deal policy also forged the political alliances that would help keep those programs alive. Perhaps most important the New Deal linked its efforts at local economic revival to the creation of stable jobs at decent wages. At the same time, New Deal policies laid the basis for a growing political, economic, and racial divide between middle-class and low income communities. The insurance policies created by the Social Security Act provided economic security for millions. Mortgage subsidies put home ownership within popular reach. Their benefits were substantial but largely hidden, and they enjoyed a legitimacy that publicly subsidized welfare programs could never hope to achieve: social security because its benefits were partly financed by individual contributions; mortgage assistance because its benefits were mediated through the private market.

These benefits were simply unavailable to millions of marginally employed workers, tenant farmers, and minorities, who instead relied on visible, public, and regularly contested sources of federal support.

FROM SLUMLESS CITIES TO AREA REDEVELOPMENT: AID TO COMMUNITIES IN POSTWAR PROSPERITY

During the postwar decades the federal government made two massive investments in community development. Both relied on expansion of the hidden forms of federal subsidy initiated during the New Deal. One was the growth of suburbs, with the help of highway funds, business tax incentives, and home ownership subsidies now extended to returning war veterans as well as other groups. The other was the continued investment in defense and related industry that transformed once underdeveloped regional economies, particularly in the South. By the late 1950s the American suburb was the symbol of prosperity, while budding high-technology centers promised the triumph of American know-how during the cold war.

There were serious problems beneath the veneer of prosperity, however. The economy suffered a series of recessions in the late 1940s and 1950s, and in 1960 unemployment rates hovered between six and seven percent. Joblessness brought about by economic change and automation led to growing concern about a structural unemployment problem

that would not respond to macroeconomic growth alone. And the benefits of home ownership, while remarkably widespread, were far from evenly distributed. Beginning in the 1950s, analysts raised fears that the distressed areas in America's older cities and rural communities were becoming permanent "pockets of poverty." Working within the New Deal policy framework, the federal response to these communities revolved around housing, local redevelopment, and subsidies for private industry, without significantly redirecting market forces. This response was reflected in two programs: urban renewal and area redevelopment, whose limitations contributed to the upsurge in community-based activism and reform in the 1960s.

Urban renewal came about in response to what journalists, academic urbanologists, and planners were beginning to refer to as a "crisis of metropolitanization" in the 1940s and 1950s. The combination of industrial decentralization, property blight, middle-class out-migration, and minority-group in-migration was changing the face of postwar cities, they warned, while newly incorporated suburbs were reaping the benefits of metropolitan growth. . . . One answer . . . was to expand federal assistance for slum clearance, housing construction, and redevelopment in blighted inner cities. Urban Renewal, as the policy established by the Housing Act of 1949 came to be known, promised to clear out the slums and revive the downtown economy by attracting new businesses and middle-class residents back to the urban core. For much the same reason, urban rebuilders also aggressively sought out federal subsidies for highway building, thinking to make the city friendlier to the age of the automobile along the way. The objective, the "slumless city" in the words of New Haven Mayor Richard Lee, had little to do with meeting the needs of the urban poor.

The strategy behind urban renewal emerged out of negotiations among public housing advocates, private builders, big-city mayors, and real estate developers who had been active in debates over the 1937 Federal Housing Act. Crucial to its operation was eminent domain, the power to amass land tracts for slum clearance, which the courts had determined was reserved for localities. Since 1937, eminent domain had been exercised by local housing authorities, which would buy or reclaim land and then contract with private developers to construct public housing. Following the Housing Act of 1949 it was exercised by local redevelopment authorities for purposes that went well beyond housing. In the debates leading up to passage of the act, developers lobbied for and won generous federal subsidies (two-thirds of the costs) of local land acquisition, and also demanded the flexibility to use reclaimed land for nonresidential purposes – all in the name of reviving the ailing downtown economy for the greater good of the community. Although skeptical of the motivation of developers, the public housing advocates were willing to go along with the arrangement as the price they had to pay for getting a public housing bill passed. They came to regret this decision, or at least their own failure to get enough in return. The 1949 legislation specified that a designated proportion of cleared land be used for residential purposes and that the bill include provisions for relocating displaced residents to "decent, safe and sanitary housing." As private developers were quick to discover, however, "residential purposes" did not necessarily mean housing for the poor. In subsequent amendments the balance between housing construction and redevelopment was steadily shifted to the latter as Congress loosened the requirement that cleared land be used for housing construction. Evaluation studies also confirmed that requirements to help the displaced relocate were barely enforced. Provisions for consulting residents and cooperative planning, another component of the bill, were also meaningless in practice.

By the late 1950s, public housing advocates had come to see the program as little more than a generous public buyout of land for private real estate interests. Among black urban residents, it became widely known as "negro removal." Highway building projects brought similar results, consistently displacing or breaking up low-income neighborhoods and encouraging rather than stemming the middle-class migration to the suburbs. After a decade of what conservative critic Martin Anderson called the "federal bulldozer," one conclusion was hardly contested: urban renewal was a boon for private developers and for the mayors who brought in the federal funds, and an unmitigated disaster for the poor.

[. . .]

[The omitted section discusses the Area Redevelopment Act]

COMMUNITY ACTION, MODEL CITIES, AND THE SPECIAL IMPACT PROGRAM

Federal aid to communities entered a new phase in the mid-1960s, turning from the bricks-and-mortar focus of urban renewal to the "human face" presented by the problems of urban economic decline and from upholding the segregated norms of local residential patterns to a more forthright integrationist agenda. Supporting these policy shifts was an upsurge in liberal activism at the national level, which reached a height in the declaration of the War on Poverty by President Lyndon Johnson in 1964. Organized citizen activism was also on the rise, much of it inspired by the gains and innovative strategies of the civil rights movement throughout the 1950s and 1960s. Later in the decade, liberal policies also became caught up in the social turmoil of antiwar protest and racial unrest, symbolized nowhere more powerfully than in the use of federal troops to quell violence in the nation's dark ghettos. The popular imagery of poor places had taken on a new, more urban and minority face by the late 1960s.

It was thus in a context of federal reform, citizen action, social protest, and heightening racial tension that the Johnson administration launched a rapid succession of federal programs and demonstration projects with the goal of comprehensive community renewal. These programs . . . attempted to push federal community policy beyond the New Deal framework by using federal power to alter existing political, economic, and racial arrangements in poor communities.

A centerpiece of the War on Poverty, the Community Action Program (CAP) [a program that provided federal funds directly to community-based service organizations, and stressed citizen participation] was created during an intensive period of planning leading up to the Economic Opportunity Act of 1964, but the thought and action that gave it shape had been emerging at the local level for several years.

[. . .]

CAP was initiated in a burst of activity and enthusiasm that was almost as quickly halted by the political controversy it caused. Suddenly denied direct access to the federal funding pipeline, urban mayors, whose loyalty was crucial to the Democratic party, threatened to revolt, earning CAP the enmity of Lyndon Johnson. Infighting among local organizations for control of antipoverty funds hurt the cause even further, and the meaning of "maximum feasible participation" remained subject to debate. CAP then suffered devastating blows in the summer of 1965 when the Conference of Mayors threatened to pass a resolution against it and congressional opponents claimed that the program was responsible for the racial uprising in the Los Angeles neighborhood known as Watts. Dissatisfaction also welled up from communities. Despite its innovations in services and service delivery, CAP could not deliver one badly needed ingredient for development: jobs for the residents of the low-income neighborhoods it served. The Johnson White House continually rejected proposals for a targeted job creation program for ghettos on the grounds that it was unnecessary and, as spending for the Vietnam War escalated, too expensive. Instead, seeking to stem its political losses and prevent further "long hot summers" like that in 1965, federal policymakers responded with two additional programs: Model Cities and the Special Impact Program, which were aimed principally at communities with concentrations of poor minorities.

On one level Model Cities was an attempt to make up for the failures of federal antipoverty initiatives: it combined services with bricks-and-mortar programs while giving control of local planning to city officials, thus avoiding the political liabilities of CAP. But Model Cities was also part of a longstanding movement involving urban legislators, liberal philanthropists, social scientists, and labor officials to establish a national urban policy. Despite several legislative setbacks, this movement had been gaining ground since early in the Kennedy administration and had achieved a major breakthrough with the creation of the Department of Housing and Urban Development (HUD) in 1965.

Emerging from the administrative task force appointed to create a blueprint for the new agency, Model Cities brought together many of the ideas that had been operating in the foundation experiments of the 1950s and early 1960s but drew immediate inspiration from a memo written to President Johnson by labor leader Walter Reuther in the spring of 1965. Reuther called for making a massive investment in six demonstration sites for a comprehensive plan of rebuilding, economic revitalization, and service provision as a national expression of "our ability to create architecturally beautiful and socially meaningful communities in large urban centers" and "to stop erosion of life in urban centers among the lower and middle class population." The plan called for massive slum clearance to make way for the most up-to-date design and technology in construction. It also envisioned a more integrated healthy environment in the inner city, with a full array of public and private services for a mixed-income base of residents. Invoking the coalition of interests forged in New Deal housing programs, Reuther described his plan as a kind of urban Tennessee Valley Authority that would stimulate jobs for labor while also tapping into the great reservoir of expertise built up over decades of liberal experimentation and reform. Johnson appreciated its Rooseveltian scale. And in the wake of the Watts riots, he needed a plan to prevent more uprisings among inner-city blacks. In October the White House formed a task force to develop the idea further.

After more than a year of task force planning and congressional deliberation, the Demonstration Cities and Metropolitan Development Act was passed in 1966 with a more circumscribed mission than the one Reuther had initially proposed. More narrowly targeted on poor inner-city neighborhoods, it relied on the familiar mechanisms of local planning and federal agency coordination for a comprehensive attack on physical, social, and economic problems. The legislation called for the creation of local demonstration agencies under direct supervision of the mayor's office and made them eligible for existing federal human service, job training, housing, and infrastructure-building programs on a priority basis. The demonstration cities were also eligible for grants and technical assistance to generate redevelopment plans in poor neighborhoods. Participation by the poor was strongly encouraged but not directly supervised by federal authorities. Nor was there any designated agency with authority to enforce cooperation and coordination among agencies at the top. In a repeat of previous experience, even this limited plan was watered down in the legislative process. By the time it was passed, Model Cities was to serve twice the number of areas recommended by the task force with less than half the recommended budget in one-third the recommended time. Limited from the start, Model Cities also indicated how the meaning of urban policy itself had changed since the 1930s, signifying not so much an . . . effort to restore cities to a central role in the national economy as a series of programs to help the urban poor.

[. . .]

[The omitted section discusses the Special Impact Program (SIP)]

By 1967 the Johnson administration had amassed an array of policies aimed at poor communities: more, better, and integrated services; physical and human renewal; local economic development; community organizing; and empowerment. These policies in turn provided support for local activism and institution building, creating jobs and political opportunities for thousands of neighborhood residents and leaving community health centers, neighborhood service organizations, law centers, community development corporations, Head Start centers, and local action agencies in their wake. The initiatives also gave rise to a new network of nonprofit providers and intermediary organizations committed to community-based antipoverty intervention that would sustain the community development movement in decades to come. Expanding the scope of President Kennedy's antidiscrimination executive order, the Fair Housing Act of 1968 added another significant dimension to the federal capacity to combat place-based poverty.

For all their promise and ambition, however, the Great Society programs remained just that – programs, not a coherent community policy, that remained separated within different federal agencies. They were too limited in scope and funding to alter the political inequities or combat the structural economic shifts that continued to segregate poor places as the "other America." Nor did policymakers overcome a basic ambivalence over whether their aim was to build up communities or help people leave them. The conflict between those two strategies would only become more sharply defined as local conditions deteriorated in the 1970s and 1980s.

THE ROOTS OF RETREAT: COMMUNITY POLICY IN THE 1970s

The 1970s brought dramatic changes in the economic and political context for community development policy. Unemployment and inflation rose

sharply, while growth, productivity, and real wages stagnated. . . . The prospects for national policy were further diminished by the politics of racial backlash, working-class resentment, and sentiment against big government, which moved the political center steadily to the right and undermined the New Deal urban–labor–civil rights coalition that had supported community development in the past. Equally important, changes introduced under the banner of Richard Nixon's New Federalism profoundly altered the infrastructure of policy, in effect abrogating the special ties between the federal government and poor communities that had been forged in earlier eras. The result of these changes was a renewed emphasis on localism, fiscal austerity, and neighborhood ethnic solidarity in community development policy. This emphasis was meant to broaden community policy's appeal to the white working class, but it also marked the beginning of a steady decline of federal government involvement.

The Nixon administration was instrumental in bringing about this changed policy environment, as can be seen most directly in its attempts to rein in and restructure the liberal welfare state. Underlying these efforts was a distinctive philosophy of social provision, known as the New Federalism, that sought to give states greater power and responsibility and to lighten federal restrictions in determining how public funding would be spent. It also envisioned a more efficient federal bureaucracy, reorganized to eliminate government waste. But Nixon's reforms were also based on a more clearly partisan agenda through which he aimed to forge a new electoral majority based on white working-class resentment of the black welfare poor and free the federal bureaucracy of its New Deal influences by bringing it under more direct presidential control. Unveiling his New Federalist agenda in the summer of 1969, Nixon promised to get rid of "entrenched programs" from the past and replace them with a system based on "fairness" for the "forgotten poor" and working classes.

During the next two years the administration introduced measures to achieve the New Federalist agenda, with far-reaching consequences for community development policy. . . . They introduced a new, less redistributive and centrally regulated way of providing federal aid to localities. Revenue sharing, enacted in 1972, provided funds to states and localities automatically rather than through categorical grants. In this way Nixon sought to reduce the federal role in determining how funds would be allocated and to end the New Deal tradition of establishing direct links to poor communities to offset their political weakness in state and federal legislative bodies. The new federalism also "obliterated," in the words of one observer, "an entire art of grantsmanship" that had galvanized local political organizing in pursuit of federal funds. The administration's adoption of block grants, which came to fruition in the creation of Community Development Block Grants (CDBG) in 1974, gave localities still broader discretion in allocating funds and brought the flagship programs of the War on Poverty to an end. By the mid-1970s, Model Cities, CAP, and SIP were slated to be replaced by block grants. . . .

Following the changes introduced during the Nixon and Ford administrations, actual spending levels for community development, while remaining less than 1 percent of federal expenditures, rose fairly steadily for the rest of the decade. These expenditures were spread over a much larger number of communities and used for a broader range of purposes, however. Revenue sharing and block grants also brought a significant change in the overall distribution of funds, both within and between different kinds of communities, increasing funding in the suburbs and away from central cities and rural areas, providing more services and benefits to middle-class recipients, and moving a greater proportion of funds away from traditional Democratic strongholds in the Northeast and Midwest and toward the South and West. Meanwhile, the political relationship between federal government and poor communities deteriorated rapidly, symbolized nowhere more clearly than in the looming fiscal collapse of several major cities at mid-decade, while Washington stood by.

[. . .]

For all the setbacks and reversals in national policy, the legacy of the 1970s was not necessarily one of defeat for the community development movement. The increased emphasis on local initiative pushed community-based organizations to strengthen institutional capacity, while the vacuum created by federal withdrawal from housing construction opened up a market niche for community development corporations (CDCs). Community

activists used the momentum of the 1960s to launch a new phase of organizer training and national network building that could be applied to a diverse range of community-based consumer, environmental, and antipoverty concerns. Taking advantage of the emergence of attention to public interest issues among legislators and in the courts, these groups realized a major victory with the passage of the Home Mortgage Disclosure Act of 1975 and the Community Reinvestment Act of 1977. This legislation provided for public scrutiny of lending records and recognized the obligation of banks to lend in communities where they do business. Promoted as a weapon against discrimination and redlining, it also gave community groups a powerful tool in their own negotiations with local lending institutions. Although the successes of local organizations did not necessarily make up for the losses in federal support, they proved increasingly important in the decade ahead.

THE END OF THE NEW DEAL ERA?

In the 1980s the Reagan and Bush administrations greatly reduced the already diminished federal presence in poor communities. Playing on antigovernment sentiment and fiscal fear, Republicans eliminated revenue sharing, UDAGs, and most other remaining development programs, cut Community Development Block Grants in half, and left a much diminished welfare and services sector as the only source of direct federal assistance to poor communities. The resources and mandate for enforcing housing discrimination law all but disappeared. The Reagan revolution also introduced a much more radical framework of decentralization and privatization than the president's predecessors had envisioned – in fact, it threatened to dismantle the federal policy infrastructure for community building altogether. Judging from the reductions in place-targeted federal funding, the revolution was a success. But the expansion in the number and size of high-poverty neighborhoods during the decade tells a different story.

Two initiatives emerged from federal retrenchment, both premised on the belief that the absence of government was the key to community revitalization. The first, enterprise zones, promised to introduce free market principles and restore entrepreneurial activity to low-income communities through a combination of government deregulation and generous tax breaks for businesses. Modeled on programs promoted by British Prime Minister Margaret Thatcher, Reagan's proposals were also consistent with the supply-side philosophy embraced in his economic policies: allowing entrepreneurs to keep more of their profits, the reasoning went, would stimulate new investment and eventually trickle down to community residents. In keeping with their anti-interventionist premises, the proposals also rejected the components of local planning and supplemental government assistance that had characterized programs such as area redevelopment. Despite repeated legislative attempts, however, the enterprise zone idea was never adopted as national policy. But it was adopted in a number of states during the 1980s, where the designated zones were assisted by substantial government investment and planning and came to resemble earlier development policies more closely.

The other major initiative to emerge from the free market framework was a proposal to privatize and promote residential ownership in public housing, this time in the name of individual empowerment in low-income communities. Like the proposal for enterprise zones, this proposal never got off the ground, due partly to the fallout from political scandals in the Reagan administration's Department of Housing and Urban Development.

The administration was unable to eliminate or privatize all the social welfare programs it targeted for attack. But the Reagan revolution did succeed where it mattered most – redirecting federal fiscal and economic policies – and the impact on low-income communities was devastating. In addition to the withdrawal of federal aid, the communities suffered from the increased income inequality, capital flight, labor setbacks, and crippling budgetary deficits that resulted from Reagan-era policies. Hit hard by recessions at either end of the 1980s, poor communities were politically marginalized as well. And the very idea of community development policy, premised as it was on collective well-being and supportive government policies, was challenged by a harsh, individualistic ideology positing that no intervention would work. Ironically, for the first time since the 1930s, federal policy in poor communities was actually in harmony with the direction of social and economic policy writ large.

Reagan era changes did not devastate the community development movement, however, and in at least one sense they could be turned into a source of strength. Pushed to do more with less, CDCs moved aggressively to become more efficient operators and to tap into local and private sources of development support. Foundations created new intermediaries to provide support for existing and emerging community-based organizations, particularly in housing and economic development. The movement for comprehensive, integrated service delivery gathered momentum as a new generation of multiservice and systems reform initiatives got under way. And community organizers, galvanized by growing inequality and federal cutbacks, created training intermediaries and focused on strengthening national networks. Impressive as these achievements were, local initiatives were heavily absorbed in making up for lost ground and could only imagine what could have been achieved in a more supportive policy environment.

REVISING THE PAST: CLINTON'S COMMUNITY POLICY

Promising "a new way of doing business for the federal government," in 1993 the Clinton administration launched an initiative to revive declining communities. Known as the Empowerment Zone/ Enterprise Community (EZ/EC) initiative, this effort was originally considered part of Clinton's "new covenant" with the cities and was eventually extended to rural communities as well. In December 1994 the administration designated eleven empowerment zones, each eligible for grants and tax breaks of up to $100 million, and 95 enterprise communities eligible for smaller grants and business incentives. In most places the initiatives were just getting under way as of the late 1990s.

The EZ/EC is different from past efforts, according to administration officials, in its rejection of old ways of thinking about the problems in urban communities. The program proposes to move beyond a focus on countercyclical grant-in-aid programs to an emphasis on enabling cities to compete in the global economy, Clinton's national urban policy report noted. It also seeks to invest in people and places, recognizing the old dichotomy as false. EZ/ EC marks another innovation in its metropolitan framework for economic development. And unlike past efforts, it is designed to foster locally initiated, bottom-up strategies that connect the public, business, and neighborhood sectors in community-building partnerships for change.

In fact, as both its title and the rhetoric accompanying it suggest, the Empowerment Zone/Enterprise Community initiative contains much that is familiar to veterans of the community development movement. Indeed, one of its most striking features is the extent to which it combines, in best postmodern fashion, some disparate programs from the past. Like enterprise zones, it relies heavily on tax incentives to promote private sector investment – only this time the tax breaks are tied to hiring residents of the zone rather than realizing capital gains. Like Model Cities it draws on existing housing, education, job training, and service programs for most of the funds that will be given to the designated areas. . . . It designates two tiers of recipient communities, presumably as a way of sharing the wealth. Community planning boards also figure prominently in the EZ/EC legislation, which combines the experience of CAP and Model Cities to require evidence of participation from all sectors in the community, including government and the poor. . . . And operating within a Nixonian New Federalist framework, it offers waivers from categorical program requirements and channels all federal grants through the states. It even borrows a note from organizer Saul Alinsky in its rhetorical appeal to the consensus ideal. "Across the country, from the Mississippi Delta to Detroit," the program literature reads, "the existence of a common goal has brought together diverse groups that rarely, if ever, cooperated in the past." Most striking from historical perspective is EZ/EC's endorsement of "four fundamental principles" that restate the essential themes that have defined community development from the start: economic opportunity in private sector jobs and training; sustainable community development characterized by a comprehensive coordinated approach; community-based partnerships that engage representatives from all parts of the community; and "strategic vision for change based on cooperative planning and community consultation."

Clearly, the EZ/EC plan rests on the hope that this time the federal government will be able to overcome interagency conflict, weak investment incentives, competition among local political interests, and racial inequity that have plagued community development policy in the past. In this hope it is banking on the expertise of the people who have been working in low-income communities for decades and on the willingness of industries to locate and hire in areas they have traditionally stayed away from. Unfortunately, EZ/EC also repeats other patterns that have left many wondering whether it, like its predecessors, is promising much more than it can possibly deliver. Even supporters agree that the funding is inadequate given the size of the task. Its associationalist tenor leaves critics skeptical about how much investment or job creation can be expected from the private sector and the extent to which community residents will be able to expect corporate responsibility. Like past federal demonstrations, it begs the question of what happens to the thousands of communities not chosen for support, and what happens to the EZ/EC sites once initial funding runs out. It also smacks of symbolic politics at a time when poor urban and rural communities command little more than rhetorical attention on the national agenda. Most of all the plan represents a very modest investment in community revitalization, especially in the face of an overarching policy agenda that encourages footloose capital, low labor costs, reduced social spending, and persistent wage inequality, and that brings about "the end of welfare as we know it" with little thought for the policy's effect on communities.

CREATING A NEW POLICY ENVIRONMENT

The historical record offers important insights about the intellectual origins, political frustrations, and recurring patterns of federal policy, but the challenges it poses to the community development movement are even more immediate and direct.

The first is to make a case for investing in communities as part of an antipoverty policy that focuses on income inequality, job opportunities, and racial exclusion as well. Such a policy would strengthen the position of residents with better wages and training while taking steps to stem the geographic dispersal of industry and jobs. It would enforce antidiscrimination regulations to stimulate lending in poor neighborhoods and ensure access to housing and jobs. And it would challenge the myth that mobility and community development are either/or choices. Most of all it would begin with the recognition that targeted community development – no matter how comprehensive, well planned, or inclusive – cannot reduce poverty all by itself. This is not to suggest that community development is futile without these larger changes, but with them it stands a much greater chance of success.

The second challenge for community development is to reassert the importance of the federal government's participation. This is no easy task in light of historical experience or the current political climate. It begins from an understanding that past failures do not prove that revitalization is impossible; few programs enjoyed the funding, time, or sustained political commitment necessary to make community development work. Indeed, the federal commitment to middle-class and affluent communities has been much more substantial and comprehensive, including housing, infrastructure, and tax incentives among its forms of support. It is also unrealistic to expect a revival in poor communities without both federal resources and direct public provision. Two decades of federal withdrawal sent neighborhood poverty soaring. And past efforts to stimulate private market development have not trickled down.

The third challenge is to reconstitute and strengthen the political coalition behind community development policy. This will take collaboration with labor, civil rights, and other traditional allies, but it can begin by addressing the barriers to mobilization within the community development movement itself. Particularly important is to examine how funding practices affect political mobilization by tightening the tensions between outside providers and communities and discouraging the kinds of activities that can help community-based organizations become more effective politically. Foundations are rarely willing to provide the long-term undesignated funding that organizations need to build capacity and institutional stability. Nor do they generally fund local organizing, advocacy, or coalition building among community organizations. . . . As an initial step toward more effective political mobilization, then, foundations need to be willing to examine and alter these practices

and organize themselves into a more coherent and persistent voice for changes in policy.

The fourth challenge is to acknowledge not only how race has contributed to the problems in poor communities, but to explore how it may be part of the solution. A race-conscious strategy would identify how race continues to shape the policy decisions affecting political representation, housing location, transportation, social services, and access to jobs. It would move beyond the simplistic black–white dichotomy to investigate how racial barriers operate across ethnic, class, and gender lines. And it would make an explicit commitment to ending institutionalized as well as individual acts of racial exclusion.

Perhaps the most important and overarching challenge from history is to reverse the policy contradictions that keep community development swimming against the tide. Meeting this challenge requires focusing not only on community interventions but creating the economic and political conditions within which community development can actually work. One starting point is to think of communities as part of metropolitan and regional political economies and influence the regional practices that determine the geographic distribution of work and residential opportunities, services, and political representation. Meeting the challenge also requires reorienting research away from the problems and pathologies that plague low-income communities and toward a better understanding of how these conditions are maintained or made worse by contemporary economic and social welfare policies. We need to know more, for example, about how federal and state retrenchment, devolution, and new requirements in the welfare system affect communities as well as individuals, and how policies designed to open up global markets affect local job opportunities and labor markets. The larger point of such research is to reintroduce a sense of human and political agency into discussions of these conditions rather than accepting them as structural inevitabilities not subject to change. Research should also look toward developing more proactive policies by asking how government investments in economic restructuring, labor market regulation, metropolitan development, and social infrastructure can work for rather than against the community development effort.

"The Urban Electorate in Presidential Elections, 1920–1996"

from *Urban Affairs Review* (1999)

Richard Sauerzopf and Todd Swanstrom

Editors' Introduction

Alice O'Connor notes the rise and fall of targeted, place-based community development policies. From the 1930s through the 1960s, policies aimed at ameliorating the difficulties faced by cities were high on the national agenda. Since then, however, cities have become the forgotten stepchild of U.S. national policy making.

How do we explain these ups and downs? As authors like O'Connor note, the reasons are complex. But one of the major factors explaining America's vanishing urban agenda is simple demographics: there are fewer people living in, and voting from, cities, so urban voters comprise a smaller percentage of the electorate than in the past.

Why does this matter? There is an old expression, "You dance with them that brung you" (variations on this have appeared as book titles and in song lyrics). It is often cited in politics to remind elected officials that their policy proposals should reflect the interests of the constituents who put them into office. For U.S. presidents like Franklin Roosevelt in 1932, and Lyndon Johnson in 1964, there was little question who had "brung" them: the large plurality of urban Democratic voters. Similarly, many members of the Congress would have represented urban constituencies. Together, they shared an interest in producing policies that would reward the urban voters who would, they hoped, return them to office.

However, as an earlier selection by Wyly, Glickman, and Lahr made clear, the percentage of Americans who live in cities has been shrinking since the 1930s, and most dramatically since the 1960s. This shift in population becomes reflected in a shifting electoral geography. Today, the percentage of the population defined as "urban" is below 30 percent, and, just as importantly, urban populations no longer dominate the legislatures of such states as Illinois, New York, or Pennsylvania. This changes the strategic calculus of presidential candidates, who are less concerned with wooing the urban vote. It also changes the make-up of Congress, as each decennial redistricting results in a loss of urban seats in favor of suburbs and exurbs. For example, in 1910 the city of Philadelphia boasted a Congressional delegation of eight members, and as recently as 1980 the city elected five members to Congress, but after the most recent redistricting, the city was down to two full representatives (two other representatives include pieces of the city in their suburban districts).

Sauerzopf and Swanstrom offer evidence of the scope and magnitude of these changes. They argue that the shift in the urban vote is more complex than simple population change. For one thing, they find that the urban vote remains crucially important to Democratic candidates. Even if city voters have declined as a percentage of the electorate, they have become more loyally Democratic, and no Democratic candidate can win without this vote. They also note that the notion of the "suburban voter" remains undefined. In some metropolitan areas, older suburbs may resemble cities in racial composition and socio-economic status; voting habits among suburbanites can be very diverse indeed. They call for further research to tease out the factors that link place and voting. Are voting habits linked to individual characteristics (e.g., issues such race, income, and education will shape your voting habits regardless of where you go)? Or does "place" exert an independent influence, so that a person might have Democratic leanings while living in the city, but shift toward Republican voting after relocating to a more conservative suburb?

Historical perspectives on the link between urban voting and urban policy may be found in some of the following sources (additional references may be found at the end of the selection): C.N. Degler, "American Political Parties and the Rise of the City: An Interpretation," (*Journal of Urban History* (1964) 51); K. Anderson, *The Creation of a Democratic Majority, 1928–1936* (Chicago: University of Chicago Press, 1979) and T. Edsall and M.D. Edsell (*Chain Reaction: The Impact of Race, Rights and Taxes on American Politics* New York: Norton, 1991).

Richard Sauerzopf is on the Urban Planning Faculty at Wayne State University, where he continues to research the role of urban voters in national politics. Todd Swanstrom is Professor of Political Science at St. Louis University. He is the author, co-author or co-editor of numerous books, among them the award-winning *The Crisis of Growth Politics: Cleveland, Kucinich, and the Challenge of Urban Populism* (Philadelphia: Temple University Press, 1985), and *Place Matters: Metropolitics for the 21st Century*, co-authored with Peter Dreier and John Mollenkopf (Lawrence: University Press of Kansas, 2001). His scholarly articles span a range of interests, including regionalism, electoral politics, and housing policy.

This study of voting in presidential elections in twelve central cities from 1920 to 1996 shows that cities played a crucial role in the New Deal realignment that dominated presidential elections from 1932 to the 1960s. Since then, cities have declined as a share of the total electorate, but they still provide crucial votes for successful Democratic presidential candidates. As cities have increasingly deviated from national voting trends, however, their turnout rates have increasingly fallen behind the national rates. A call is issued for researchers to break down the suburban vote and to examine contextual effects on voting behavior.

Many scholars have documented the rise and fall of the urban electorate in national elections. The rise of the political fortunes of cities followed the migration from farms to cities that, by the middle of the twentieth century, made cities the population centers of the country. The Democratic Party relied heavily on urban votes to build the New Deal coalition that dominated national elections from the 1930s to the 1960s. Similarly, the migration outward from cities to suburbs, which accelerated in the 1950s, fueled the rise to power of the Republican Party in national elections beginning in 1968 and signaled the progressive marginalization of urban electorates in national politics.

Accounts of the urban electorate in presidential elections rely heavily on the theory of electoral realignment. According to realignment theory, approximately once in a generation (1860s, 1890s, 1930s), the existing party system reaches a crisis when "emergent tensions in society . . . not adequately controlled by the organization or outputs of party politics as usual, escalate to a flash point" (Burnham 1970, 10). Political entrepreneurs take advantage of the crisis with new appeals, ideological differences widen, the coalitions behind the parties are shuffled, and a new party system is formed with a new dominant coalition. The realigned party system incorporates political demands that had been ignored by the old party system.

Franklin Roosevelt's victory in 1932 is viewed as a quintessential critical election. Roosevelt was able to swing urban ethnics, blacks, and Jews into the Democratic Party, forming the nucleus, along with the "solid South," of the Democratic coalition that dominated presidential elections for the next generation. The flight to the suburbs, which accelerated after World War II, weakened the urban base of the Democratic Party. With suburban voters perceiving that liberal Democratic policies siphoned off their tax dollars to expensive and wasteful programs targeted on central cities and blacks,

Republicans successfully used "wedge issues" to split off elements of the New Deal coalition. By the 1992 election, Schneider (1991, 1992) estimated that suburbanites cast an absolute majority of the votes, and he proclaimed "the dawn of the suburban era in American politics."

The above account of the rise and decline of urban electorates provides the backdrop for Bill Clinton's vaunted "suburban strategy" that supposedly won him the 1992 and 1996 elections. According to Clinton pollster Stanley Greenberg, the "forgotten middle class" of white suburbanites, as exemplified in Greenberg's (1995) research on Macomb County, Michigan, had abandoned the Democratic Party in the face of what they saw as its preoccupation with urban blacks. The key to Democratic presidential victories was to challenge the Republican Party for suburban votes, Greenberg argued, enticing back into the fold the so-called Reagan Democrats (Greenberg 1995).

To attract suburban votes, Clinton avoided identification with policies targeted to cities or to minorities, emphasizing instead policies that benefit people wherever they live. Clinton's well-publicized attack on Sister Souljah in the 1992 campaign distanced him from Jesse Jackson and the liberal wing of the Democratic Party centered in cities. The highly successful Clinton/Gore bus tours avoided the inner cities and were resplendent with small-town, even rural, imagery. Other than enterprise zones, Clinton proposed few policies targeted on cities. Instead, he proposed broad-based policies, such as national health insurance and investments in education that would benefit cities and suburbs, blacks and whites. Clinton's two victories were widely credited to his ability to contest Republicans for the swing votes in the suburbs. Even though the conventional account, described earlier, of the rise and decline of the urban electorate is widely accepted, there is relatively little published analysis of city voting trends overtime to back it up. The main reason is that voting statistics in the United States are reported by county, not by city. Unless the city boundaries correspond to the county boundaries, city election returns are not widely available. Before 1950, county voting returns usually corresponded closely to city returns, but with massive suburbanization over time, county returns have increasingly deviated from city returns.

One exception to the dearth of urban voting studies is a 1949 article by Samuel Eldersveld, who painstakingly collected data on presidential elections for twelve cities from 1920 to 1948. In this article, we update Eldersveld's data for the presidential elections since 1948. Eldersveld chose all cities with populations more than 500,000 according to the 1940 census, eliminating Washington, D.C., and Buffalo. The twelve cities are the following: New York, Chicago, Philadelphia, Pittsburgh, Detroit, Cleveland, Baltimore, St. Louis, Boston, Milwaukee, San Francisco, and Los Angeles. This sample of cities is by no means representative of the nation as a whole. It is biased toward older northern cities, although it does include border cities such as Baltimore and St. Louis, as well as the West Coast cities of Los Angeles and San Francisco. Our main justification for choosing these cities is that it enables us to piggyback on Eldersveld's work and thus track city voting over a long time period. Moreover, the ten states represented by these twelve cities still control 215 electoral votes and thus are intrinsically important. Admittedly, our work is only exploratory, and further research is needed on urban voting in key sunbelt states such as Florida and Texas.

We use this aggregate voting data to examine the voting behavior of central-city and non-central-city electorates over time and to suggest some revisions in the conventional accounts, both scholarly and in the mass media, of the rise and decline of urban electorates. Our data are limited, however, because they only allow us to show how the urban electorate voted, not why it voted the way it did. The striking voting patterns that we uncover could be the result of the characteristics of individual voters who live in cities (e.g., their race or class), or voting behavior could be due to a "contextual effect" of living within central-city municipal boundaries. Our data do not allow us to decide between individual and contextual effects. However, municipalities are important units within our federal system, with billions of dollars of federal money targeted to them. Moreover, municipalities may be important units for shaping political identities. Therefore, in the conclusion, we call for further research on our hypothesis that contextual effects arising from central-city residence significantly affect voting behavior.

RISE OF THE URBAN ELECTORATE: NEW DEAL REALIGNMENT

Eldersveld's (1949) main argument in his original article was that the traditional way of understanding American presidential elections in terms of sectional divisions (e.g., the South vs. the North) was no longer adequate. Rather, this view needed to be supplemented by an appreciation of a powerful urban–rural division in the national electorate. That is, rather than looking at presidential elections as being won by a candidate's appeal to base and key swing regions of the nation, Eldersveld argued that we should look at elections as hanging on a candidate's ability to appeal to cities. This was the case because the most powerful predictor of how individuals voted had clearly become whether they lived in urban areas. By 1949, to a great extent regardless of region, small-town and rural area voters chose Republican candidates, and big cities voted Democratic. Eldersveld showed that twelve of the nation's largest cities in ten states consistently turned out victories for Roosevelt and Truman.

. . . The shift of the urban electorate to the Democratic Party that drove the New Deal realignment actually began in 1928 with the nomination of Al Smith, the first non-WASP ever nominated by a major American party. As Lubell (1965, 49) put it, "Before the Roosevelt Revolution there was an Al Smith Revolution." A Catholic of immigrant background whose career was based on the New York Democratic machine, Tammany Hall, Smith's campaign was replete with urban imagery. Especially important was Smith's opposition to the Klu Klux Klan and prohibition, stands that were popular with urban ethnics. Key (1955) noted that the 1928 election had all the earmarks of a critical realigning election, with voter turnout rising 26.5 percent from the previous election and the coalitions behind the parties undergoing major change.

If Smith began the transition of urban voters to the Democratic Party, clearly Roosevelt completed it in the 1930s. Having grown up in the country, Roosevelt was not partial to cities. Furthermore, the Democratic Party prior to 1932 was more closely identified with Jeffersonian localism than with an activist federal government. Roosevelt ran for office in 1932 promising to balance the budget. But once in office, Roosevelt recognized the need for federal action and enacted a series of urban-targeted programs, "spending nearly $20 billion in new federal funds to put nearly one-fifth of the unemployed to work at transforming the urban infrastructure" (Mollenkopf 1983).

[. . .]

THE DECLINE OF THE URBAN ELECTORATE: SUBURBANIZATION

As shown in Figure 1, the relative size of the city presidential vote reached its highest point in 1948. For our twelve cities, the city proportion of the total national vote reached an impressive 21.3 percent in 1948, subsequently falling to only 5.9 percent by 1996. In 1948, for example, New York City cast more than 50 percent of the statewide vote; in 1996, New York City represented only 32.1 percent of the state vote. Similarly, the city of Chicago fell from 46.5 percent of the Illinois vote in 1948 to only 20.9 percent in 1996 (see Table 1).

Notwithstanding the relative decline of the urban electorate . . . one can see that city voters have continued to play important roles in presidential elections. With the exception of Al Smith, probably no presidential candidate of a major political party was more identified with cities than John Kennedy. The Democratic Party platform in 1960 attacked the Republicans for having turned their backs on cities and contained specific planks to meet the problems of cities as governmental units, calling for the creation of a cabinet-level department to coordinate urban programs. According to Eldersveld's (1949) criterion, our twelve cities were responsible for swinging 146 electoral votes to Kennedy in a very close election. Kennedy's massive plurality in Chicago was just enough to overcome the Republican advantage downstate (see Table 2). Republicans complained that voting fraud in Chicago swung Illinois to the Democrats.

Kennedy's victory precipitated an agonizing debate within the Republican Party over what to do to close the "big-city gap." Senator Barry Goldwater publicly asserted that the Republican Party should stop trying to attract bloc support from Negro and other minority groups concentrated in cities. With Goldwater's nomination in 1964, the Republicans pretty much wrote off the urban vote. Indeed, since 1956, when Eisenhower won pluralities in six out of the twelve cities in our sample,

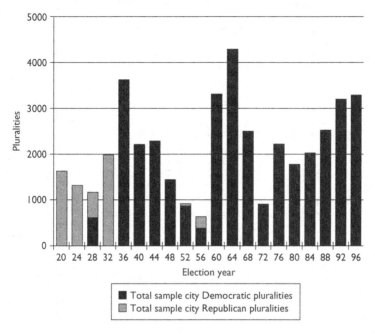

Figure 1 Sum of sample city pluralities, 1920 to 1996.

Table 1 Sample city proportion of actual state electorates

Election year	1920	1932	1940	1948	1956	1964	1972	1980	1988	1996
New York	44.1	46.8	51.2	50.6	44.7	41.7	34.6	31.0	31.2	32.1
Chicago	37.5	40.9	44.9	46.5	37.7	34.2	28.0	23.5	23.4	20.9
Philadelphia	22.6	21.2	21.8	23.5	19.5	18.9	17.0	15.8	14.9	11.7
Pittsburgh	5.6	5.6	7.5	7.2	6.0	5.4	4.3	3.8	3.5	2.9
Detroit	26.8	25.7	26.7	31.8	25.5	21.3	14.8	10.4	8.4	8.0
Cleveland	8.3	9	11.7	10.6	9.1	7.6	5.4	4.3	3.6	2.9
Baltimore	51.4	48.3	47.9	42.3	34.2	28.4	19.5	17.2	13.6	9.6
St. Louis	21.2	22.2	21.9	21.8	18.1	14.7	10.4	8.1	7.2	5.5
Boston	16.7	16.3	16.9	16.2	13.2	10.9	8.6	7.1	7.1	6.8
Milwaukee	17.8	15.7	19.0	20.3	18.9	17.7	14.1	12.6	12.2	9.5
San Francisco	15.6	9.8	9.5	8.4	6.2	4.6	3.6	3.0	2.8	1.1
Los Angeles	27.3	42.8	43.3	42.9	18.5	15.3	12.5	10.7	10.4	7.5
Sample city votes as % of national electorate	15.8	18.3	20.7	21.3	15.7	13.2	10.2	7.6	7.1	5.9

none of our sample cities has voted Republican in a presidential election.

Notwithstanding the importance of urban votes to the Democrats, Table 2 demonstrates that the power of cities to influence presidential elections has declined since the Roosevelt–Truman victories.

Our sample cities were important for the victories of Carter in 1976 and Clinton in 1992 and 1996, but the number of electoral college votes "won" by cities has generally declined (see Figure 2). In every election from 1936 onward, with the exception of six cities in 1956, positive figures in Table 2

Table 2 Sample city percentage of state party pluralities, 1920 to 1996

	1920	1924	1928	1932	1936	1940	1944	1948	1952	1956	1960	1964	1968	1972	1976	1980	1984	1988	1992	1996
New York	40.5	15.8	0	145.9	163.3	320.6	244.9	0	0.29	0	206.2	51.8	187	0	248.2	0	0	253.9	88.7	64.4
Chicago	41.5	41.2	4.7	55.5	77.8	326	298.6	926.5	0	R5	5,151	75.8	0	0	0	0	0	0	89.7	74.1
Philadelphia	30.5	29.5	14.6	R44.9	31.6	63.3	142.9	0	0	R20.5	285	29.6	160.4	0	207.7	0	0	0	67.8	77.3
Pittsburgh	7.6	2.3	0.7	0	15.4	24.9	54.3	0	0	0	78.6	8.3	41.5	0	33	0	0	0	16.6	14.2
Detroit	31.9	30.3	15.5	65.2	57.7	0	908.7	0	0	0	466.4	38	95.4	0	0	0	0	0	98.5	51.4
Cleveland	12.7	0	0	56.7	26.3	104.8	0	1,228	0	0	0	19.3	0	0	901.8	0	0	0	101.1	28.5
Baltimore	70.9	71.4	11.7	62.3	71.1	75.5	231.8	0	0	R20.1	115.5	47.7	483.9	0	111.4	294.2	0	0	53.7	38.8
St. Louis	37.5	58.9	0	22.4	32	74.7	152.2	38.2	0	1,811	1,011	29	0	0	85	0	0	0	31.4	50.3
Boston	7.9	4.3	582.4	161.9	65.5	65.7	71.9	58.2	0	0	28.3	15	18.5	0	12.1	465.2	79.1	29.1	14.1	10.8
Milwaukee	7	M20.2	0	27	32.7	280.8	0	96.4	0	R4.3	0	28.9	0	0	201.5	0	0	120.2	77.3	38.7
San Francisco	15.9	1.6	0	15.5	14.1	12	15.8	55.6	3	R2	0	10.7	0	0	0	0	0	0	11.9	11.5
Los Angeles	30.8	58.8	55.5	37.8	43	47.7	46.2	44.4	5.6	R2.7	0	24.2	0	0	0	0	0	0	28	25.9
Mean city %	27.8	31.4	17.2	64.1	52.5	126.9	216.7	347.7	3	266.6	917.9	31.5	164.6	0	225.1	379.7	79.1	134.4	56.6	40.5
Number of cities	12	10	6	10	12	11	7	7	3	7	8	12	6	0	8	2	1	3	12	12

Note. This table shows sample city presidential party pluralities as percentages of sample state pluralities. Zeros indicate instances when state pluralities and city pluralities were for different parties. For 1920 to 1928, all figures indicate sample city support for Republican pluralities, except where indicated by an M for LaFollette Progressive and D for Democratic pluralities. For 1932 to 1996, all nonzeros indicate sample city support for statewide Democratic pluralities, except where indicated by an R for Republican pluralities. All zeros for this period indicate instances when states produced Republican pluralities despite sample city Democratic pluralities.

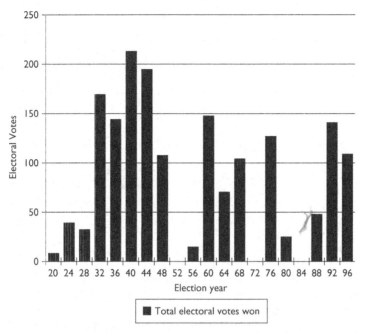

Figure 2 State electoral college votes won by sample city pluralities, 1920 to 1996.

represent instances when cities made a contribution to the margin of victory of a Democratic candidate within the state. All of the zeroes that appear in Table 2 since 1928, except in 1956, indicate instances when cities produced Democratic presidential pluralities while their respective states went Republican. In 1972, every one of Eldersveld's (1949) sample cities voted Democratic, and all of their respective states went Republican. In 1984, Ronald Reagan nearly duplicated Nixon's feat, losing every city but winning every state in our sample, except Massachusetts.

REVISING THE CONVENTIONAL WISDOM

When we update Eldersveld's (1949) data and analyze it, the results at first seem to confirm the conventional explanations of the decline of the political significance of cities. The conventional account emphasizes that the main reason for the declining electoral influence of central cities is the flight to the suburbs. In 1950, according to the U.S. Bureau of the Census, the population of all of Eldersveld's sample cities accounted for 57.8 percent of the total population of their metropolitan

areas; by 1990, these central cities accounted for only 27.7 percent of that figure. It is well understood that suburbs tend to favor Republican presidential candidates, but the opposite is true for cities. Increasingly, since 1948, large Democratic pluralities in central cities have not been able to make up for the rising tide of suburban and rural Republican votes, although it seems that this has not been for lack of effort. As central-city electorates have shrunk, they have become proportionally more Democratic. As Schneider (1992) noted, however, suburbs are simply growing faster than cities are growing more Democratic. The vaunted "suburban strategy" that Clinton followed in his recent victories is rooted in basic demographics.

A closer examination of the data, however, suggests that the conventional wisdom is too simplistic. Although the population of our twelve cities has declined, the total Democratic presidential votes coming out of them since 1968 has remained remarkably stable (see Figure 3). There are two main reasons for this. First, four of our cities located on the two coasts (New York, Boston, Los Angeles, and San Francisco) reversed the postwar downward trend and gained population in the 1980s, largely because of immigration. In 1990, for

Figure 3 Sample city electorate as a percentage of the national electorate, 1920 to 1996.

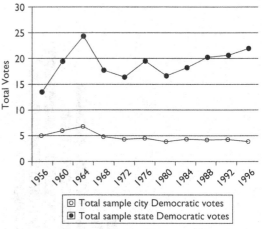

☐ Total sample city Democratic votes
■ Total sample state Democratic votes

Figure 4 Total democratic presidential votes for sample cities and states, 1956 to 1996.
Note: Vote tallies are in millions of votes cast for president.

example, 28 percent of New York City's population was foreign born. The rate at which immigrants naturalize, of course, varies, and studies show that immigrants who become citizens tend to vote at low rates. In the long run, however, just as immigration fueled the political rise of cities at the beginning of the twentieth century, it is doing the same thing at the end of the century (New York City's proportion of state actual electorate actually increased between 1980 and 1996).

A second reason why Democratic votes have remained stable in cities is because the urban electorate has voted progressively more Democratic even as the rest of the nation has tended to vote more Republican in presidential elections. A striking trend that emerges from our data is the increasing divergence between the voting behavior of cities and the rest of the nation. We can see this clearly here when we examine the index of difference (see Figure 4). The index of difference measures the magnitude of the difference between the partisan division of the vote in cities and in the rest of the state. The higher the final score in a given year, the greater the difference in partisan voting behavior between the city and non-city vote. For instance, if there were no difference between the partisan presidential voting patterns of city and non-city voters, the index would be zero. If, on the other hand, all city voters voted Democratic and voters in the remainder of the state voted Republican, then the index of difference would be 200.

Figure 4 shows that since 1952, the index of difference has increased markedly. In comparatively good years for Republicans, the index of difference is quite high. This illustrates dramatically that Republican presidential victories have occurred,

since 1968, in utter opposition to the preferences of the urban electorate.

One would expect the index of difference to fall in years when the Democrats do comparatively well, winning more votes outside the cities. However, even in elections when the Democrats win most or all of the states covered by our data, the index generally remains high, reaching its highest point in the 1996 election. Urban voters are increasingly "eccentric," responding differently to political cues than voters in the rest of the nation. "Solid cities" have largely replaced the "solid South" as the bedrock of the Democratic coalition. Contrary to conventional wisdom, which emphasizes that elections are won or lost in the suburbs, central-city votes are still crucial to Democratic victories in presidential elections.

According to our analysis, our sample of twelve cities was controlling (contributing more than 50 percent of the statewide plurality) in states representing 138 electoral votes in 1992 and 107 votes in 1996. In 1992, for example, New York City provided 89 percent of Clinton's nearly one million vote statewide margin. Indeed, Clinton lost (narrowly) to Bush in suburban Long Island (Nassau and Suffolk counties). Obviously, if we went beyond our limited sample to look at, say, all central cities with populations more than 100,000, cities would have been decisive in delivering many more electoral votes into the Democratic column. Although it would be exaggerating our point, it would be possible to turn the conventional wisdom on its head (at least for the 1992 and 1996 elections): How could the Republicans have expected to win the presidency when they were incapable of seriously contesting for votes in the big cities?

Given the fact that central-city voters increasingly vote differently from the rest of the nation and increasingly find themselves on the losing side, perhaps it is not surprising that the propensity for people in cities to vote has declined relative to votes in the rest of the state. The declining electoral influence of cities is not just due to suburbanization but to demobilization of the urban electorate. The effect is significant. Our data show that in 1996, although the total population of voting-age persons in our sample cities represented a healthy 17.4 percent of the sample state figure, our city voters accounted for only 13.5 percent of the sample state total.

City pluralities contributed significantly to both of President Clinton's victories, city turnout rates declined sharply in 1992 and 1996 relative to their respective states and the nation as a whole. These trends, taken together, suggest diminishing returns for an electoral strategy that does not engage central-city voters.

In a more catastrophic scenario for the Democrats, the alienation of the urban electorate could be mobilized by a charismatic third-party candidate. According to a New York Times/CBS voter survey, in 1992 Clinton actually lost the white vote 39 percent to 41 percent. Nationwide, Clinton won 82 percent of the black, largely urban, vote. If Jesse Jackson had run as an independent in 1992, George Bush probably would have been elected to a second term.

In short, our data suggest that cities are both more important to the Democrats and more volatile than the conventional wisdom suggests. There may be costs to the Democrats' suburban strategy: declining turnout among potential Democratic voters in the cities.

A CALL TO EXAMINE THE SUBURBAN VOTE AND CONTEXTUAL EFFECTS

Although central cities no longer have the power to determine most presidential elections, they are still essential to Democratic presidential victories. If anything, our research on voting in central cities underestimates their importance. Our dichotomy between the central-city electorate and the rest of the state, however, fails when examining the most rapidly growing segment of the statewide vote: the metropolitan voters outside central cities. At the same time that cities have declined in relative importance, metropolitan areas have grown. Nardulli, Dalager, and Greco (1996) found that in 1990, the nation's 32 largest metropolitan areas constituted more than 40 percent of the national electorate. Therefore, investigating ways in which metropolitan areas structure elections is critical to understanding national electoral politics.

The first issue that should be examined is where to draw the line between the "urban" and "suburban" voters. The conventional wisdom naturalizes this division, assuming that all suburban voters share something essential in common and are fundamentally distinct in political orientation from urban voters. Reed and Bond (1991, 735) referred to this as the "mystified notion of the suburbs as a new, coherent political constituency." In fact, "cities" and "suburbs" are problematic social constructions. The current dividing line between cities and suburbs is arbitrary – being based, as it is, on the jurisdictional boundaries between the original central city and everything else in the metropolitan area. Although some cities are elastic, expanding their boundaries as their metropolitan area grows, others, including most of the cities in our sample, are inelastic, hemmed in by independent suburban municipalities. In these cases especially, as metropolitan areas continue to become more urbanized, characteristics associated with central cities spill over into neighboring suburbs.

Although cities increasingly resemble the conventional view, with minority populations, social problems, and physical decline concentrated within their borders, suburbs are becoming more varied. When estimating the role that metropolitan place plays in electoral politics, one too often fails to consider the fact that many suburbs do not resemble the conventional image of leafy, white, middle-class, residential refuges. As the dichotomous assumptions of the conventional view are challenged, so are the consequences of this perspective for cities, metropolitan areas, and the Democratic Party.

S I X

Weiher (1991) explored the fantastic prolifera-tion of incorporated municipalities surrounding most of the nation's large central cities. As he showed, these suburbs over time have developed increasingly distinct economic and social char-acteristics. These characteristics structure the cognitive map, Weiher argued, that prospective residents use to inform their moving decisions. Individual residential choices accumulate to further cement the identities of specific suburbs. The result is a sociopolitical fracturing of metro-politan areas. This fragmentation, as opposed to stratification, is characterized by substantial varia-tion among suburbs. However, as variation among suburbs is increasing, many suburban municipali-ties are becoming more homogeneous internally. The result of this fragmentation may have serious political consequences as identification with municipalities supersedes understandings of metropolitan areas as interdependent polities.

The suburbs that represent the most critical problem for the conventional view are those that more closely resemble central cities, typically inner-ring suburbs. Many inner-ring suburbs – such as Harvey, Illinois; Lackawanna, New York; Compton, California; and East Cleveland, Ohio – have more in common with their central cities than they do with the stereotypical middle-class suburbs. Orfield (1997) has hypothesized that the "swing" votes in recent presidential elections have not been in the suburbs generally but in the inner-ring, distressed (largely white, working-class) suburbs.

Our analysis provides support for those scholars who argue that the current period is not charac-terized so much by a Republican realignment as by de-alignment, the inability of either party to assemble a permanent winning coalition. Under de-alignment, the ability of parties to mobilize voters weakens, and voters withdraw from politics as such. Our results, which show declining voter turnout both by urban and suburban electorates, support those who argue that the current period is characterized more by de-alignment than realign-ment – at least in metropolitan areas outside the South. Future voting research needs to disaggregate the suburban vote. Inner-ring voters may be cross-pressured between economic concerns that pull them toward the Democratic Party and social con-cerns, especially about race and crime, that draw them toward the Republicans. By breaking out the inner-ring suburban vote, researchers could test the hypothesis that this is where voter volatil-ity (switching parties) and withdrawal (declining turnout) are centered differentiating them from central cities and outer-ring suburbs.

The second major thrust of electoral research on urban areas should be to determine whether there is an independent "contextual effect" of place on voting behavior. The aggregate data used here only allow us to describe voting patterns by place but not to determine why people vote or do not vote the way they do. The results we have reported could be an artifact of the individual-level characteristics of central-city and suburban electorates (e.g., their class or racial background). On the other hand, the results could also be a result of the social networks, sources of information, and reference groups rooted in places. Scholars have hypothesized, for example, that the move from rural areas to cities caused people to become more conscious of the interdependencies of society and the need for government to protect people from the destabilizing and damaging effects of industrial capitalism. Similarly, others have argued that the suburban context has a conservativizing effect, pulling erstwhile Democratic voters toward the Republicans: As white ethnics migrated to the suburbs, they became home owners and lived more privatized lifestyles, making them more suscept-ible to Republican appeals that favor the market over government.

More research needs to be done on the con-textual effects of different metropolitan residential environments. Aggregate data can be used to test for contextual effects by controlling for income, education, race, and other individual-level variables and seeing if place still has independent effects on voting behavior. The best data, however, for examining contextual effects are individual-level data (e.g., from sample surveys in which each case has been geo-coded so that researchers can move back and forth between individual and contextual levels of analysis).

In one of the only studies of contextual effects of central cities, Wolman and Marckini (1997) found that after controlling for a wide range of other characteristics, central-city congressional repre-sentatives voted differently than non-central-city representatives. Apparently, the central-city context makes representatives vote more liberal. In one of the few studies of the effect of suburban contexts on voting behavior, Oliver (1994, 1996) has used geo-coded individual-level data to show that political participation declines, controlling for individual-level variables, in homogeneous suburban contexts. Oliver's research suggests, beyond selection and rational interest structuring, that residence within a specific kind of place has a transformative effect. For Oliver, not only do some middle-class residents select suburbs as a form of escape from dynamic urban political environments (as Schneider 1991, 1992 argued) but residents are also de-politicized by the socioeconomic homogeneity and lack of significant political conflicts associated with these places and their reform-style municipal governments.

If one takes seriously the potential of metropolitan places to exert politically transformative effects on their residents and extend it to other metropolitan municipal types, then understanding how metropolitan areas structure political attitudes and behaviors becomes a critical project, central to understanding national political divisions.

REFERENCES FROM THE READING

Burnham, W.D. 1970. *Critical elections and the mainstreams of American politics*. New York: Norton.

Eldersveld, S.J. 1949. The influence of metropolitan party pluralities in presidential elections since 1921: A study of twelve key cities. *American Political Science Review* 43 (6): 1189–1206.

Greenberg, S.B. 1995. *Middle class dreams: The politics and power of the New American majority*. New York: Times Books.

Key, V.O., Jr. 1955. A theory of critical elections. *Journal of Politics* 17: 3–18.

Lubell, S. 1965. *The future of American Politics*. New York: Harper & Row.

Mollenkopf, John H. 1983. *The contested city*. Princeton: Princeton University Press.

Nardulli, P.F., J.K. Dalager, and D.E. Greco. 1996. Voter turnout in U.S. presidential elections: An historical view and some speculation. *PS: Political Science and Politics* 29: 480–490.

Oliver, J.E. 1994. City size and political participation in the fragmented American metropolis. Paper presented at the annual meeting of the American Political Science Association, New York, September 1996.

Oliver, J.E. 1996. The influence of social context on patterns of political mobilization. Paper presented at the annual meeting of the American Political Science Association, San Francisco, September.

Orfield, M. 1997. *Metropolitics: A regional agenda for community and stability*. Washington, DC: Brookings Institution.

Reed, A., Jr. and J. Bond. 1991. Equality: Why we can't wait. *Nation*, December 9, 733–737.

Schneider, W. 1991. Rule suburbia. *National Journal* 28.

—— 1992. The suburban century begins. *Atlantic Monthly* July, 33–44.

Weiher, G. 1991. *The fractured metropolis: Political fragmentation and metropolitan segregation*. Albany, NY: SUNY.

Wolman, H., and L. Marckini. 1997. Changes in central city representation and influence in Congress. Paper presented at the annual meeting of the Urban Affairs Association, Toronto, Canada, April.

SIX

"Urban Sacrifice"

from *A Prayer for the City* (1997)

Buzz Bissinger

Editors' Introduction

When Ed Rendell was elected mayor of Philadelphia in 1991, journalist Buzz Bissinger sought and received permission to shadow the mayor throughout his four-year term. During heated meetings with union leaders and private conversations with political advisors alike, Bissinger was the "fly on the wall," observing the inner workings of one city's political leadership. The resulting book, *A Prayer for the City*, sheds light on the way in which the mayor of a large, diverse, embattled city balances the demands of constituents, negotiates among political leaders, and tries to realize his vision for an improved city center.

Like many cities that had been heavily dependent on their industrial bases, Philadelphia's decline began somewhere mid-twentieth century, and accelerated in the wake of suburbanization and economic restructuring. Between 1970 and 1990, the city had lost nearly 20 percent of its population and 10 percent of its employment. There had been a dramatic shift in the sorts of jobs available: whereas in 1970, 192,000 residents had worked in manufacturing, in 1990, only 88,000 were so employed. Twenty percent of the city's population lived below the official poverty line. A dwindling tax base led to growing fiscal strains, so that when Rendell assumed office in 1991 he faced a potential $1.5 billion gap between income and expenditure.

Rendell was an energetic, politically savvy mayor (he has since been elected governor of the State of Pennsylvania) whose major accomplishments over his two terms in office include addressing the city's fiscal crisis (largely by cutting or privatizing services and challenging labor unions) and sparking new development in the city's downtown (while, critics argue, residential neighborhoods continued to deteriorate). Bissinger describes him as a passionate advocate willing to do anything to promote the city – at one point he was photographed hugging Mickey Mouse (in a failed attempt to lure a Disney development) and at another, scrubbing a City Hall bathroom floor. Despite his popularity and political skill, however, he failed to staunch the city's continued demographic and economic destabilization.

Most academic literature emphasizes the impersonal nature of forces that shape our cities. Bissinger's close-up of Rendell's mayoralty offers a different perspective. We see that individuals and their daily actions do make a difference, as Rendell uses charm and political skill to negotiate a budget deal, persuade donors to back a performing arts center, and perhaps, as demonstrated in the following reading, convince a president to keep military jobs in the city. This study, then, allows us to reflect on the nature of leadership in a way that less closely observed studies do not. But individual agency has its limits: the structural forces undermining industrial cities like Philadelphia prove resistant, and while Rendell is succeeding in keeping 8,000 military base jobs in the city, thousands more beyond his control were lost.

The chapter excerpted here shows the many dimensions of the relationship between the U.S. federal government and its cities. Bissinger summarizes a few of the ways in which earlier federal policies, perhaps inadvertently, served to disadvantage central cities in their competition with the suburbs. He then illustrates the ongoing and often frustrating efforts of big cities to command attention and resources from Washington. As Sauerzopf and Swanstrom note in a previous reading, the declining size of the urban electorate, and the partisan differences frequently found between urban voters and the Washington political leadership, make it increasingly difficult for mayors to make their case.

Bissinger relies on several sources to put Philadelphia's experience into historical perspective. These include: Kenneth Jackson, *Crabgrass Frontier* (New York: Oxford University Press, 1987), and William Schneider, "The Suburban Century Begins" (*The Atlantic Monthly*, July 1992). Those who are interested in learning more about Philadelphia can consult Sam Bass Warner's history, *The Private City* (Philadelphia: University of Pennsylvania Press, 1987), or *Philadelphia: Neighborhoods, Division, and Conflict in a Postindustrial City*, by Carolyn Adams *et al.* (Philadelphia: Temple University Press, 1993).

Buzz Bissinger has enjoyed a varied career, reporting for the Philadelphia *Inquirer* (where he won a Pulitzer Prize) and the Chicago *Tribune*. He is probably best known for his 1990 bestseller, *Friday Night Lights*, since made into a movie, which chronicles the role of high school football in small-town Texas. His most recent book is *Three Nights in August*, about major league baseball, published in 1995 by Houghton-Mifflin.

Mayor Rendell was going to the airport to meet [President Clinton]. It was a matter of protocol, but it was also more important than that. As the motorcade made its way to City Hall, where the president would give a speech, Rendell would have twenty minutes of precious private time in which to bring up the issues that he believed were of the utmost importance to the city and the ways in which the federal government could help realize that agenda. The list was potentially endless, and everybody had an idea of how best to maximize the time. . . . But in the modern era of politics, where photo ops and sound bites fed to the masses set the standard, twenty minutes of private time with the president was a millennium, and the mayor had to be judicious.

The weeks preceding the president's visit had been particularly grim and unremitting, even for the city. The Senate had killed a jobs bill President Clinton had proposed, meaning a loss to the city of as much as $70 million. The failure of the bill in the Senate came in the wake of news that the city had lost 17,700 jobs in the previous year. On top of that, there was a report by the National Association of Realtors that housing prices had dropped more precipitously in the metropolitan area than in any other region of the country in the past year – by 9 percent. And on top of that, there was a report that seven million square feet of office space in Center City was vacant. Ten entire buildings had been mothballed, and the vacancy

rate in the downtown area was at its highest in nearly fifteen years.

As if in response to the grim economic news . . . [a]t the beginning of May, on a quiet Sunday, a man named Santiago Pineda, tired of the drug dealers in his neighborhood, tried to chase them from the front of his home on North Eighth and Pike. As a result of this act of protecting his family and his neighborhood, Pineda was ambushed, shot twice, and beaten.

Six days later four juveniles robbed a man over on Torresdale. There was nothing unusual in that, but then, after fighting over who was going to share in the meager spoils, they went back to their victim several minutes later, realized he was intoxicated, and sensed an opportunity: they poured some clear liquid over him, set his hair on fire, and then ran off laughing as the flames quickly spread to his head and shoulders. The coup de grace – the lighting of the first match – said the police, was administered by a fourteen-year-old.

A day after that, Mother's Day, Rendell spent the bulk of his time in Newark testifying before the federal government's Base Closure and Realignment Commission, the same commission that had made the decision to close the [Philadelphia] navy yard [several years earlier]. Rendell was there to fend off yet another potential round of military-installation closures in the city that had been recommended by the Pentagon and would mean the loss of an additional ten thousand jobs. "I understand the need to reduce the [federal] deficit," he told the commission. "In sixteen months, we have eliminated a four hundred-and-fifty-million-dollar deficit. But the city of Philadelphia will lose more jobs than forty-seven states. This would cut us off at the knees at a time when we're just getting on our feet."

There would be no word on the fate of those jobs until the end of June, but twelve days after Rendell's Mother's Day testimony, the city was told that construction of a proposed $200-million (federal) General Services Administration building had been canceled. The news made Rendell livid. Not only would the construction project have been an enormous boon for the moribund building and trades industries, but the cancellation also raised the question of whether the city, beyond the obstacles it already faced, was being willfully pushed closer and closer to a state of impotence. It wasn't a matter of complete extinction because there would always be a downtown. Given the building boom of the 1980s, there would be enough skyscrapers to last until well into the twenty-first century. There would be quaint shops along Walnut and a renaissance of restaurants and the newly constructed Pennsylvania Convention Center and jaunty shuttle buses painted purple to take conventioneers to the city's favored attractions and men and women in uniform cleaning the streets. Mayors everywhere, sensing the changing tide of the economics of cities and seizing on tourism and entertainment dollars from conventioneers and suburbanites as the new solution, had restocked their downtowns so they never looked better. But it was often deceptive, a brocade curtain hiding a crumbling stage set.

[. . .]

In anticipation of the official opening of the Pennsylvania Convention Center the following month, the city was planning an eleven-day extravaganza called Welcome America! Leading up to Independence Day, it would be replete with laser shows, fireworks displays, giant hoagies, and visits by both President Clinton and Vice President Gore. "I feel like the Roman emperor," Rendell said during a planning session. "I can't give decent city services. I want to close [city] health centers, and I want to cut back on [city] library hours, and here I am giving bread and circuses to the people."

[. . .]

The mayor arrived at the airport tarmac to await the president, and he was well aware of the machinations that had taken place in preparation for the visit, the constant changes in locations and itinerary like something out of *Mission: Impossible*. "What a fiasco!" he whispered, but he was in a buoyant mood. Only an hour earlier, shining his shoes at his desk, he had seemed quite nervous, not about what he was going to say to the president – he seemed quite secure about that – but about what he was going to say at a wedding he had agreed to officiate at right before meeting the president. Fortunately the bride had cried in joy, and the mayor of course had started getting weepy too, just like a damn bridesmaid, and the whole thing had been beautiful, and as he stood on the hot asphalt waiting for personal seance with the most powerful leader in the world, he seemed positively aglow.

Some of his aides had thought he should push the issue of the navy yard, pressing the president on what the federal government could do to ease its transition from a massive defense installation to an industrial and economic center, particularly since word was out that Mercedes was shopping around for a site for its first U.S. plant. But rather than fog the president's head with a myriad of issues, Rendell thought it was better to focus on one issue, the base closures recommended by the Pentagon, which would cost the city nearly ten thousand jobs.

For the ride from the airport back to the city, he slid into the limo, when he and the president, as he later put it, "bullshitted" for three or four minutes. Then he spent the remaining seventeen minutes on the base closures. He pointed out that with the Pentagon's proposal, the city would lose more jobs than just about any other place in the country. He showed Clinton a copy of an executive order issued by President Carter, noting that urban areas should be favored by the federal government in cases of consolidation or relocation. He knew these arguments carried some weight, but not the weight that mattered the most, so he gave Clinton two sets of numbers:

Two point nine, four, and thirty-one.

Nine, twenty-five, and seventy-four.

The numbers seemed cryptic, one of those silly logic games the object of which is to figure out what each set had in common. The first set, Rendell explained, was for Cumberland County,

Pennsylvania, where the jobs would go under the proposed relocation and consolidation plan: 2.9 percent was the unemployment rate; 4 percent was the poverty rate; and 31 percent was Clinton's share of the votes in the presidential election. The second set was for Philadelphia: 9 percent was the unemployment rate; 25 percent was the poverty rate; and 74 percent was Clinton's share of the votes in the presidential election.

Clinton laughed.

[. . .]

What Ed Rendell didn't say was that in working so desperately to keep his city afloat, he wasn't simply fighting the effects of a stingy Congress or the effects of the Reagan–Bush years, when the incidence of poverty in America's cities had multiplied. He was also fighting one hundred years of history and federal policy that had willed his city and other cities like it into such a condition that the best he could do, the best any mayor could do, was beg from a speeding motorcade and depend on the strength of his charisma.

It was a map, and it lay buried and folded in a musty box of documents in the National Archives in Washington, untouched at this point except by a smattering of scholars driven enough to find it. It was color coded in shades of green and blue and yellow and red, and it was marked with the letters A, B, C, and D to correspond to the colors. It had held up well, and the colors looked as good now as they had in 1937, when a team of appraisers had spread over the city, quietly doing their work for the federal government. The map had divided the city into about twenty-five different sections, each section delineated by boundary marks, each section given a particular color and grade, depending on the appraisers' determinations. The map was easy to read, even down to the street names. The appraisers had done their work well. It was hard to believe that in this map, in the careful and deliberate choice of colors and grades for each section of the city, like the inverse of a secret treasure, lay startling evidence of the seeds of the city's destruction.

But it was true, for the map, along with the other documents in that box, provided shocking insights into the reason the American city had seemed destined to fail as far back as the 1930s and the way the federal government, in terms that were blatant, racist, and unsanitized, had aided and abetted that destruction while opening up the floodgates of the suburbs.

Location well-protected against encroachment . . .
Lower part of section is threatened with Italian expansion . . .
Colored forcing way in some spots . . .
Influx of Jewish has discounted values . . .
High grade Americans. Professional and Executives. . . .

These terse descriptions of Philadelphia had been typed succinctly across the neighborhood survey form, NS Form 8, without a single note of shock or regret. Headings such as "Negro," "Infiltration of," and "Foreign Born" were all part of the standard criteria to determine the level of risk in granting a mortgage. The more "Negro" there was, the more "Infiltration of" there was, the more "Foreign Born" there was, the less chance someone living in a particular neighborhood would get a mortgage.

These survey forms and assessments, which in turn formed the basis for the color-coordinated map of the city, had been prepared by a federal agency called the Home Owners' Loan Corporation. In the plethora of programs started by President Roosevelt in the New Deal era, it was hardly the most noteworthy. But as pointed out by Kenneth T. Jackson in his remarkable book *Crabgrass Frontier*, HOLC ended up affecting every facet of American life, from the future shape of cities to the future shape of suburbs to the blockade that kept minorities from the American dream of home ownership.

Signed into law by Roosevelt in June 1933, HOLC refinanced tens of thousands of mortgages that were in danger of default because of the Depression. But the real value of HOLC, Jackson wrote in his book, a history of the suburbanization of the United States, was the way it "introduced, perfected and proved in practice the feasibility" of the modern mortgage. In the 1920s, a mortgage had an average life of five to ten years and then was subject to renewal. With the advent of HOLC, mortgages were extended to twenty years, and the payments were fully amortized. But as a precaution, the federal agency established exhaustive appraisal procedures to determine which areas of a city or suburb were more suitable for lending than others. Metropolitan areas all over the country were analyzed, including Philadelphia. The resulting maps may have looked clean and almost lovely with their different colors, but they were often brutal in their findings when it came to the future of the cities.

"First Grade" areas, which were marked by the color green and the letter *A*, signified areas of a city that were well planned, virtually free of blacks and what HOLC appraisers referred to as "foreign-born white," and therefore ripe for mortgage funds in the maximum amount available.

"Second Grade" areas, which were marked by the color blue and the letter *B*, signified areas of a city that were still good but were beginning to fade around the edges a little bit. Mortgage lenders in these areas were advised to make loans 10 to 15 percent below the maximum limit.

"Third Grade" areas, which were marked by the color yellow and the letter *C*, were characterized, according to HOLC literature, by "age, obsolescence, and change of style; expiring restrictions or lack of them; infiltration of a lower grade population. . . ." Mortgage lenders in these areas were advised to be very careful in making any loans.

"Fourth Grade" areas, which were marked by the color red and the letter *D*, were characterized by "detrimental influences in a pronounced degree, undesirable population or an infiltration of it." Mortgage lenders in these areas might well refuse to make loans altogether.

For a city like Philadelphia, not only were the HOLC surveys devastating, but they also told an enormous amount about what the federal government privately thought of big cities (the surveys themselves, because of their inflammatory nature, were confidential). Philadelphia had older housing stock to begin with, and HOLC surveyors saw considerable risk in investing in much of it, particularly given what they saw as the combustible elements of blacks and foreign-born whites. Of the thirteen neighborhoods in the Philadelphia metropolitan area that received first-grade, or A, designations, none had a black presence. . . . The bulk of the city received fourth-grade designations and was therefore awash in red on the survey map. . . .

. . . . Section D6, in North Philadelphia, was dismissed because of what surveyors listed under "detrimental influences": "Heavy concentration of Negro-properties in poor condition." Section D13, also in North Philadelphia, was painted red not simply because of "nominal Negro population" but also because of "concentration of Jewish in Northern part." . . . If the HOLC appraisers damned the city, they delighted in the suburbs that ringed it, not simply because of the amount of undeveloped land there, but also because of the ethnic and racial purity. Virtually every suburban area that was appraised received a first- or second-grade designation. . . . The area comprising Brookline, Oakmont, South Ardmore, and Lanerach received high praise as well because its inhabitants were "all Americans." Under the heading "Favorable Features," the surveyors noted both "good transportation" to the city and "no colored."

[. . .]

The irrevocable legacy of HOLC, Jackson and other historians have noted, lies in its appraisal system and the way in which it was copied and modeled by the greatest tool of housing ever developed, the Federal Housing Administration. By insuring long-term mortgages from private lending, the FHA made home ownership more accessible than ever. Buoyed by the guarantee of the federal government, lenders were willing to shave more interest points off mortgages than ever before, amortize payments over twenty-five or even thirty years, and require a down payment of only 10 percent. The FHA, founded in 1934, was intended to help revive the nation's dormant housing industry during the New Deal. But the ultimate influence of the FHA and its housing cousin, the Veterans Administration, went far beyond that, making the dream of home ownership available to millions of middle-class Americans, just as long as it was a dream that largely confined itself to the suburbs and not to the older cities.

[. . .]

From 1934 to 1972, the percentage of families owning homes in the United States increased from 44 percent to 63 percent. The majority of these families lived in the suburbs and never would have realized the dream of home ownership without the policies of the federal government. The majority of these home owners weren't born in the suburbs. As census data clearly show, they moved by the millions from cities that had been all but written off by the federal government in terms of the lasting viability of residential investment.

There was something ironic about the seeds of the cities' destruction being planted in the era of the New Deal, since Roosevelt was the first U.S. president to pay any deliberate attention to cities whatsoever. But much of what Roosevelt did was enact entitlement programs and dispense federal aid to keep cities from the brink of collapse and insolvency. He had little direct interest in the social viability of cities and even less love for them, and he did what he did because he needed the urban vote and the influence of urban legislators if he was going to push the New Deal through Congress. Like the vast majority of Americans, he found cities dirty and grimy and far too big for their own good. As he said in a speech in 1937 dedicating the Bonneville Dam in the Pacific Northwest: "Today many people are beginning to realize that there is inherent weakness in cities which become too large for their times and inherent strength in a wider geographical distribution of population."

In the 1950s, despite considerable downtown renewal, ten of the country's fifteen largest municipalities lost population, and the percentage of people living in central cities in metropolitan areas dropped from 59 percent to 51 percent. Since the beginning of the Republic, the 1950s marked the first decade in which cities lost a significant share of their population, and this trend would only accelerate. In the age of the automobile, the only physical impediment to the suburbs' domination was an effective way of getting to them from the city, and President Eisenhower took care of that in 1956, with the passage of the Highway Act, which created a forty-one-thousand-mile highway system that eviscerated city after city while making exodus to the suburbs easier than ever. City planners actually welcomed the expressway system because they thought it would hasten the return of shoppers downtown, but they were wrong, underestimating the degree to which Americans would embrace the flat and predictable flow of the suburban shopping mall and the antipathy that Americans have always had for the city, not to mention the degree to which cities were already beginning to buckle.

[...]

If the long-range goal of federal policy has always been to help cities grow and adapt to the ever changing dynamics of their populations, then federal policy has failed despite the goodness of its intentions. But if the long-range goal of federal policy has been the very opposite – to slowly and deliberately defrock cities, diminish their influence, and promote instead a distinct and separate suburban culture based on race and socio-economics and privacy – then federal policy has been enormously successful. "The lasting damage done by the national government was that it put its seal of approval on ethnic and racial discrimination and developed policies which had the result of the practical abandonment of large sections of older, industrial cities," wrote Kenneth Jackson. "The financial community saw blighted neighborhoods as physical evidence of the melting-pot mistake. To them, cities were risky because of their heterogeneity, because of their attempt to bring various people together harmoniously. Such mixing, they believed, had but two consequences – the decline of both the human race and of property values."

As Ed Rendell rode in a limousine with the president of the United States, he had a choice of ways in which to reinforce the point of what federal policy had really done to the country's cities. He could point out, as he did so effectively, the inequity of taking jobs away from his own city, in which a quarter of the population was at the poverty level, and relocating them to an area in which 4 percent of the population was at the poverty level. As the motorcade made its way over [Interstate Highway] 95, he could have pointed to the navy yard and questioned the efficacy of silencing a place with 192 years of history and service to the country. To his left, he could have pointed out the malarial towers of Southwark and told the president how the federal government's answer to housing for the poor in the city, in the aftermath of slum removal (also know as Negro removal), had been these high-rise horrors that were doomed to fail. He could have mused aloud about what it meant that a federal Department of Agriculture had become a Cabinet-level department in 1889 whereas a federal Department of Housing and Urban Development had not been established until 1965.

Or he could have just given the president two maps: one would have been the map prepared by the federal HOLC appraisers in 1937, with its shades of green and blue and yellow and red; the other would have been a map of the city in 1993, similarly colored showing the areas with the greatest loss of population and vacant housing. Holding the maps side by side, looking at them for several seconds, the president would have discovered what anyone else would have discovered: despite a span of fifty-five years and eleven months, they were virtually the same in terms of what they revealed: they were mirrors of each other.

North Philadelphia had been painted red by the HOLC appraisers, meaning it wasn't worth mortgage investment. In the forty years between 1950 and 1990, North Philadelphia had lost 53 percent of its population; it now had a vacancy rate of 17 percent and contained 11,512 abandoned residential structures. South Philadelphia had been painted the red obsolescence by the HOLC appraisers. Subsequently, in the forty years between 1950 and 1990, South Philadelphia had lost 45 percent of its population: it now had a vacancy rate of 12 percent and contained 4,185 by the HOLC appraisers. In the forty years between 1950 and 1990, West Philadelphia had lost 33 percent of its population; it now had a vacancy rate of 12 percent and contained 4,118 abandoned structures. . . . Virtually all the suburbs favored by HOLC, on the other hand, ultimately grew into vibrant and steady residential areas.

Was it fate that had driven these cataclysmic changes? Or were they the result of the prophesies of those federal appraisers, who saw the city as doomed? In the outlines of those two maps, capable of being laid on top of each other like identical twins even though one was old and the other modern, the answer was apparent: by predicting the obsolescence of so much of the city, they guaranteed it; by promoting the promised land of the suburbs, they had guaranteed it.

[. . .]

In city after city, the changing dynamics of population coupled with a catastrophic loss in the industrial base made urban America more dependent on the federal government than ever before. But in the reality of politics, the reality in which money, public works projects, and certain policies flowed because they meant votes, the influence of cities in the national arena had never counted for less. Cities could deliver votes, but not in the way that had counted in the past, not in the way that had made Roosevelt and Kennedy and Johnson feel something of a political debt to them. Of all the facts that had been written about the 1992 presidential election, the one that was the most startling in what it revealed about the country – and was perhaps focused on the least – was that for the first time ever a majority of the country's voters lived in the suburbs.

In a piece in *The Atlantic Monthly* in July 1992 entitled "The Suburban Century Begins," William Schneider wrote that, "urban America is facing extreme economic pressure and the loss of political influence. The cities feel neglected, and with good reason: they are the declining sector of American life."

Schneider's exhaustive analysis showed that in metropolitan areas around the country the huge margins of urban votes that had swept Democrats into office in the past could now be outmatched by the margins of suburban votes: "The urban base doesn't have enough votes anymore. The Democrats have to break into the suburbs by proving that they understand something they have never made an effort to understand in the past – namely, the values and priorities of suburban America."

The clout of America's cities was little more than a whisper in the rant of retirees and health care providers and failed savings and loans and those who wanted their personal and property taxes lowered. The obsession was with the middle class, the same middle class that had begun to empty out of the cities forty years earlier, thanks in no small measure to the incentives of the federal government. Cities weren't ignored in the conference rooms and offices of the nation's lawmakers were treated with the faint whiff of patronization; the mayors who represented them were patted on the head, promised careful thought on the subject, and then quietly whisked away. The attitude toward cities could be measured in urban policy – more precisely, what urban policy – and it could be measured in reams of statistical and demographic data showing the mightiness of the suburbs.

For Ed Rendell, it could also be measured in the personal interactions of one day spent in Washington in May 1992. The riots in Los Angeles had occurred less than a month earlier, so there was now an impetus for Rendell and the nation's other big-city mayors to be there, beyond the usual hat-in-hand begging. The window of opportunity for cities had presumably never been opened wider, for the riots had given proof to the mayors' repeated warnings that it was only a matter of time before hopelessness and despair resulted in violence and lawlessness. In the reactive responses of Washington, riots were strangely good for cities; it was no accident that enough political support for a Department of Housing and Urban Development, which had been actively discussed and debated since the late 1950s, materialized only after the 1964 riots.

In a round of private meetings on that day in May, Rendell came face-to-face with the men who moved Washington, or who at least tried to move it despite the perpetual gridlock that now defined it – [Senators] Bradley, Cranston, Dole, Danforth, Durenberger, Kasten, Kerry, Lautenberg, Moynihan, Mitchell. Just as important, he came face-to-face with the men behind the men, the chiefs of staff and deputy chiefs of staff who knew how to drive and push policy. He also came face-to-face with other big-city mayors – Ray Flynn of Boston, David Dinkins of New York, Maynard Jackson of Atlanta, among others.

[. . .]

Rendell was hardly silent during the meetings. He too shared the fears of his fellow mayors. But the amount being requested seemed preposterous to him, not because the cities didn't deserve it but because he knew they would never get it, regardless of the impetus of the L.A. riots. He seemed more inclined to listen to the comments of Senator Bob Dole, who, unlike his colleagues, Democrat or Republican, saw little purpose in being obsequious with men who presided over something that the majority of Americans didn't care about. Given the nature of Washington politics, said Dole, there was a chance for the cities to get something, but the window was already closing. "We don't have a very long attention span. In my view, it's going to have to be done in the next thirty days. Even in the two weeks since L.A., there are already voters saying, 'What are we doing this for?' "

Rendell himself favored the tactics of compromise and conciliation that had worked so well in Philadelphia – not what should happen, but what could happen given the political reality. He advocated something far more modest than what the other mayors were asking for: $4 billion to $5 billion in urban aid, precisely what the Bush administration was giving Russia, and an extensive program of urban enterprise zones. The amount wasn't plucked from the air but contained a clever

bit of political blackmail: if President Bush wasn't willing to give such aid to cities, then the message was clear that he cared more about Soviet citizens than he did about the citizens of his own country. "The hardest thing to understand is that we're not Washington bashers," said Rendell to Dole's deputy chief of staff, Jim Wittinghill, in the privacy of Dole's Senate office. "The frustration out there and the hopelessness out there is enormous. Of all the emotions out there, lack of hope is the most tantamount. If we don't do something about that, cities are going to burn."

. . . Aware of how the suburban middle class ruled, he then posed the problem in a startlingly blunt way. "Even if you say, 'Screw the cities,' then you're going to have to pay a ton of police to encircle us and keep us in."

[. . .]

Several weeks later Congress and the president did come to an agreement on an urban-aid package. It wasn't the $35 billion that the U.S. Conference of Mayors believed was necessary, nor was it the far more modest $4 billion to $5 billion advocated by Rendell. It was $1 billion, the majority of which would go to Los Angeles for riot relief and to Chicago for relief from a downtown flood. A $5 billion urban-aid package was then passed by the House over the summer, this one containing fifty enterprise zones. The amount to be spent was to be spread over six years. But as a compromise, only twenty-five of the enterprise zones would actually be in urban areas; the rest would be in rural ones. The bill, said *The New York Times* in an editorial, "spreads money too thinly, in an obvious attempt to buy support." But it didn't matter. Shortly after his loss to Clinton, Bush vetoed the bill.

As America's newest president, Clinton might be different, or he might not be. Rendell was enormously fond of him personally, and the president at least talked about cities as if they had a place somewhere in American society. But Clinton was still subject to the rigors of a Congress focused on the suburban middle class, and although he seemed inclined to test the theory of enterprise zones in a way that was bold and might actually accomplish something, there was no national mandate for a far-reaching urban policy. Cities were not rioting, despite Maynard Jackson's prediction. Their neighborhoods were just continuing to fall apart, the schism between the poor who lived in the cities and the wealthy who lived in the hidden fringes, in palatial hills and gated communities, wider than ever. What a mayor could gain for his city, if he or she could gain anything, seemed largely dependent on the rapport established with the president. Appeals and fiery, impassioned rhetoric about America's moral imperative to save its cities seemed out of touch. To the contrary, the history of federal policy, with the exception of Lyndon Johnson's Great Society days of the middle 1960s, had proved that America felt no such moral imperative whatsoever. The best a mayor could hope for were stopgaps to staunch the bleeding every now and then – a conversation with the president not about saving the city, but about saving something within it.

At the end of June 1993, a month after Rendell made his pitch to the president to save those ten thousand defense jobs, the mayor gathered with a hundred politicians, workers, and reporters around an eighteen-inch television set in the City Hall Reception Room. They were there to witness live on C-SPAN the Base Closure and Realignment Commission's decision on the fate of those jobs, and it was an eerie moment, reminiscent of the one when the commission had decided to eliminate more than seventy-five hundred jobs at the navy yard.

The commission voted. . . . Not all 10,000 jobs had been saved, but 8,444 had been, and every politician and every worker who had gathered around that television set knew whom to thank: the mayor of the city, grinning from ear to ear. He had no direct proof, but he was convinced that his pitch to the president had made a significant difference in the outcome of the decision.

In a series of wonderful crescendos, it was yet another wonderful crescendo, another defying of the odds, and in the back of his mind, Ed Rendell wondered whether maybe, just maybe, there was a chance for something stunning and everlasting. Fuck the federal government. Fuck those bureaucrats whose contempt for cities could barely be concealed by the smarmy glint of their smiles. Fuck those senators whose idea of urban hardship was a lumpy pillow at the Four Seasons. He and his city would do it alone, if only they got just the slightest push of help every now and then, if only there wasn't some issue they could not control, if only confetti would fall from the sky forever.

"Regionalisms Old and New"

from *Place Matters: Metropolitics for the Twenty-first Century* (2004)

Peter Dreier, John Mollenkopf and Todd Swanstrom

Editors' Introduction

Duplication of services. Destructive competition for investment. Dwindling open space. The costs of political fragmentation can be easily enumerated. Metropolitan areas share a regional economy, co-exist in a common environment, share concerns about things like transportation and water quality. Would it not be sensible if political authorities could make decisions for the metropolitan area as a whole? Should not the common interests of a region be articulated in the political arena?

There are a number of arguments in favor of regional cooperation. Good government advocates as far back as the 1920s pushed for regional approaches to metropolitan government for efficiency reasons. When the metropolitan area is divided among hundreds of local governments, they argued, tax dollars are likely to be wasted on hundreds of local police chiefs and school superintendents. More recently, regionalists have focused less on efficiency and more on the economic and quality-of-life benefits promised by regional governance. Regional development partnerships can marshal a variety of resources to attract new investment, presenting a more appealing development environment than is found in areas in which localities are in competition. And it is on the regional, not the local level that planners can best address the congestion and strains on public services commonly labeled as and associated with "sprawl."

Although the idea of regional government – a layer of government above the municipal level, able to make policy for a central city and its suburbs – has long been popular among planning and public administration experts, but at least in the U.S. it has been, and continues to be, a political orphan. It is easy to identify and mobilize those who might fear regionalization measures. Better-off suburban residents may be reluctant to merge their political fortunes with cities which many have intentionally fled. They may fear an increase in property taxes, assuming that central-city residents will have greater service demands; they may be averse to marrying their fortunes to the urban political leaders who they see as corrupt, or sharing political space with urban constituencies that are racially and economically different. Likewise, urban voters may be concerned that regionalization will diminish their local voting power, leaving them with less access to decision-making. As Dreier, Mollenkopf, and Swanstrom note, this may have been the case in Memphis, where regionalization proposals were considered just at the moment in which the urban black electorate had managed to elect one of their own. Finally, local officials in city, suburban, and county governments may be loath to embrace a new regional entity that could usurp their power. As a result of all these potential opponents, those areas that have adopted regional governing institutions have often done so half-heartedly, and thereby doomed these institutions to failure.

The authors of *Place Matters* discuss two examples, the Minneapolis and Portland regional governments, as relative success stories. Consolidated city-county government (found in Miami, Charlotte, and Lexington, KY) represents another, somewhat less ambitious permutation of regionalism. In Canada, the government of Metro Toronto was created in 1954 and granted meaningful powers; it remains in existence today, and is considered a successful example of regionalization. In Europe, there have been many experiments with metropolitan governance: metropolitan county government was established in Britain in 1972 (and abolished in 1986); "Greater Rotterdam" was governed by a single authority from 1964 to 1985. German metropolitan regions have long-standing issue-specific cooperative mechanisms (*Zweckverbaende*), but Hannover and Stuttgart are among the few large cities to have a fuller set of regional governance institutions.

To learn more about metropolitan area governance, we recommend *Regional Politics: America in a Post-city Age*, edited by H.V. Savitch and Ronald Vogel (Thousand Oaks, CA: Sage, 1996), which includes chapters on a range of U.S. examples of regional governance models. Myron Orfield's *American Metropolitics: The New Suburban Reality* (New York: Brookings, 2002) makes a plea for a deepening of regionalism as a way to redress urban–suburban inequalities. L.J. Sharpe's edited volume, *The Government of World Cities: The Future of the Metro Model* (New York: Wiley, 1995) includes case studies from Europe, Asia, and North America.

"Regionalisms Old and New" is a chapter from Peter Dreier, John Mollenkopf and Todd Swanstrom's *Place Matters: Metropolitics for the Twenty-first Century* (Lawrence, KS: University Press of Kansas, 2004). *Place Matters* looks at the causes and consequences of patterns of segregation and inequality that mark urbanized areas in the U.S., and considers policies and political arrangements that can address these problems.

Hartford is the capital of Connecticut, the richest state in the nation ($42,706 per capita income in 2002), but it has one of the highest overall poverty rates in the country (30.6 percent) and contains many large areas of concentrated poverty. In 1969, Hartford's major employers formed a new organization, the Greater Hartford Process (GHP), to address the region's problems. Three years later, this elite group unveiled a plan to rebuild the city's low-income areas, create an entire new town in the suburbs, and develop new regional approaches to housing, transportation, health care, education, and social services. The plan would affect 670,000 persons in twenty-nine communities over a 750-square mile area.

The most dramatic impact was on Coventry, a town located fifteen miles from the city. GHP had quietly begun buying land in the all-white town of 8,500 residents in order to create, from scratch, a new town of 20,000; GHP planned to set aside 15 percent of the housing units for low-income families. When they learned of this, Coventry officials refused to cooperate with the plan, which was ultimately shelved. One suburban official noted, "I don't see the problems of the central city as my responsibility." Within a few years, GHP had ceased to exist.

[. . .]

Then, in 1996, the city council took a different approach: it placed a moratorium on opening new social service centers for the poor, such as soup kitchens, homeless shelters, and drug treatment centers. Within its 18.4 square miles, the city (population 140,000) already had 150 social service agencies, which city officials said attracted poor people to the city. Hartford city council members claimed that these programs were "ruining the climate for urban revitalization, business, shrinking the tax base, and scaring away the middle class." They also argued that the suburbs were not doing their share to address the needs of the poor – certainly an accurate statement. It was almost as if Hartford officials had decided that if the suburbs refused to cooperate, the city would stop accommodating the region's poor, potentially forcing them out to the suburbs.

Memphis is another central city with many poor, minority citizens that has an uneasy relationship with its suburbs. In 1971, suburban residents in Shelby County, Tennessee, voted two to one against merging with Memphis. In 1990, suburban opposition also thwarted an attempt to merge the largely black Memphis school system with the predominantly white county school system. In 1993, Mayor W.W. Herenton, the city's first African American mayor and a former school superintendent, again proposed merging the city with suburban Shelby County. Herenton's proposal was politically courageous, because if it had succeeded, it would have diluted the black electoral power so recently gained through his election.

Between 1940 and 1980, Memphis captured 54 percent of all population growth in the region by annexing suburbs. After 1980, suburban opposition ended Memphis's expansion, and during the 1980s, Memphis lost 6 percent of its population, while that of the surrounding Shelby County suburbs more than doubled. Many African American political leaders, who had recently gained six of thirteen seats on the city council and five on the nine-member school board, argued that a merger would dilute black voting strength and cost blacks "some of the political control we have fought for for so long." Herenton, who had been elected mayor in a highly racially polarized election, countered that "blacks gain little by controlling a city that is broke." Herenton garnered support from Memphis business leaders, who predicted that the merger would lessen their tax burden, but many suburbanites opposed the merger, fearing that suburban Shelby County (median household income $43,784) would face additional tax burdens when merged with Memphis (median household income $22,674). The proposal failed at the polls.

. . . The examples of Hartford and Memphis show how difficult it can be to craft and implement regional solutions to the problems of concentrated poverty and suburban sprawl. These are not the only examples . . . nor is suburban opposition to central-city initiatives the only obstacle such initiatives face. In this chapter, we examine regional initiatives over the past 100 years and take a closer look at the "new regionalism" that has emerged since the early 1990s. While these efforts are praiseworthy, state and federal rules of the game make it difficult to forge regional cooperation and restrict what they can do.

THE PROBLEM

The competition among metropolitan jurisdictions to attract higher-income residents and exclude the less well-off has been a powerful factor promoting the concentration of poor people in central cities. In the typical metropolitan area, dozens, sometimes hundreds, of suburban towns try to establish and maintain higher positions in the metropolitan pecking order. A "favored quarter" of suburbs houses upper-income people and select business activities that pay property taxes but do not demand many services. These places use large lot zoning, high housing prices, and even tacit discrimination to keep out the unwanted, or at least the less privileged. Elsewhere, less exclusive residential suburbs, suburban commercial and industrial areas, and aging inner-ring suburbs also seek to carve out their own places in the metropolitan hierarchy below these exclusive areas, but still above the central city. Even aging, economically declining inner suburbs often try to keep out inner-city residents. For those with limited means, living in the central city may be their only choice.

[. . .]

Federal policies made suburbs more attractive relative to central city neighborhoods by building freeways, underwriting suburban home ownership, and funding subsidized housing and social services in central cities that made them attractive to needy constituents. Suburbs often reinforce this arrangement by regulating land uses to maximize property values, by customizing their services for middle-class professional households, and by declining to build subsidized housing or sometimes even rental housing of any kind. Many central-city public officials have also contributed to this state of affairs by supporting the expansion of spending on social services. This wins favor from those who use them, enhances budgets, provides jobs for constituents, and builds political support. They are no more willing to give up responsibility for, and control over, these activities than exclusive suburbs are to embrace them. Defenders of this system draw on widely held beliefs that local autonomy, private property, and homogeneous communities are sacred parts of the American way of metropolitan living. . . .

Although this deeply embedded system may seem rational to suburban residents and urban public officials, it has produced dysfunctional consequences in the larger society. Metropolitan political fragmentation has encouraged unplanned costly sprawl on the urban fringe. . . . [I]t has imposed longer journeys on commuters, allowing them less time for family life. It has undermined the quality of life in older suburbs, hardened conflicts between suburbs and their central cities, hampered financing for regional public facilities such as mass transit, and encouraged disinvestment from central cities. Countries with strong national land-use regulation and regional governments have avoided or tempered many of these problems. Indeed, the United States could have avoided them if we had chosen a more intelligent path for metropolitan growth over the last one hundred years.

As these problems became increasingly evident over the last century, they drew criticism from scholars, planners, and good-government groups. These critics focused on how metropolitan political fragmentation undermines administrative efficiency, environmental quality, economic competitiveness, and social equity. As early as the 1920s administrative experts promoted regional solutions as ways to address the overlap, duplication, lack of coordination, and ante in the provision of public services. Concern today is becoming widespread. Even long-time suburban residents have expressed concern over the environmental costs of sprawl as they see their countryside being gobbled up by new development and find themselves stuck in traffic jams even while doing their Saturday morning shopping. They have made "smart growth" a hot-button issue across the country. Executives of large firms, transportation planners, and economic development officials most often express concern that fragmented metropolitan areas undermine the economic competitiveness of urban regions.

Finally, those who crusade for civil rights and racial desegregation, who care about the plight of the inner-city poor, and who champion greater civic participation have criticized how metropolitan fragmentation has strained the social and economic fabric of our communities. They favor fair housing and housing mobility programs, metropolitan administration of economic opportunity-programs, tax-base sharing, and metropolitan school districts. Robert Putnam's *Bowling Alone* documented the ways in which metropolitan fragmentation, sprawl, and inequality helped to drive the decline in community and civic participation. In the words of Frug, "The suspicion and fear that infect our metropolitan areas threaten to generate a self-reinforcing cycle of alienation: the more people withdraw from each other, the higher percentage of strangers that cause them anxiety, thereby producing further withdrawal."

[. . .]

[Sections discussing the historical precedents for the "New Regionalism" and traditional arguments in favor of regionalism have been omitted]

THE PRACTICE OF METROPOLITAN COOPERATION

Although support for new, regional approaches to metropolitan problems has grown steadily over the last several decades, the actual practice of metropolitan government has made less progress. Most regions have some elements of cooperation, but they vary considerably in terms of their institutional arrangements, the political constituencies they bring into play, and their capacity to address their region's social and economic challenges. A variety of constituencies has supported regional cooperation – business leaders seeking to enhance regional economic competitiveness, program administrators concerned with better coordination, community groups seeking regional equity, and suburban advocates of slowing growth. Their aims and interests obviously diverge on many points. In contrast to this diversity, the local interest in autonomy, particularly among suburban jurisdictions, is quite consistent. This has hindered the growth of metropolitan cooperation but may not prevent it in the long run.

[. . .]

THE LOCAL EXPERIENCE: AVOIDANCE AND CONFLICT

In the typical metropolitan area, a region's constituent towns and cities may realize that they belong to a common region, but they seek to retain their autonomy and continue to act independently or compete with one another for economic resources. . . .

The New York City consolidated metropolitan statistical area includes thirty-one counties in New York, New Jersey, Connecticut, and Pennsylvania and houses more than 21 million residents. It contains some 1,787 county, municipal, town, school district, and special district governments. The nonprofit, business-backed Regional Plan Association has advocated three regional plans since 1923, most recently calling for a new emphasis on concentrated development around regional transit nodes and greater emphasis on workforce development. These plans, however, have had more effect on thinking among academics and policy elites than on actual development patterns.

Efforts to create a true regional planning agency with significant authority have failed. . . . The Port Authority of New York and New Jersey, created in 1921, operates the region's seaport, its three airports, the Hudson River bridges and tunnels, and a commuter railroad. It also built the World Trade Center office complex and continues to own and oversee the site on which a memorial and new office buildings will be rebuilt in the wake of the destruction of the complex on September 11, 2001, which had a devastating impact on the organization. In recent decades, the Port Authority has responded primarily to the development agendas of the two governors who appoint its board. The agency has suffered from the need to balance any benefit to one state with an equivalent benefit to the other. It does not serve as an agent of the region's municipalities. Indeed, former Mayor Rudolph Giuliani of New York City campaigned to dissolve the Port Authority, or at least require it to return the New York airports to city control. Current New York Mayor Michael Bloomberg has little influence over how the Port Authority will rebuild lower Manhattan, despite its obvious importance to the city's future. . . .

Although New York City dominates the region and the borough of Manhattan draws more than 41 percent of the region's daily commuters, it does not dominate the region's highly fragmented politics. The surrounding states and municipalities compete vigorously to attract business investment away from the city. Since the 1970s, Westchester County and Stamford, Connecticut, have lured away many large corporate headquarters; New Jersey enticed the New York Giants and Jets football teams to the Meadowlands complex, and Jersey City has attracted many back-office operations. In response to "predatory moves" by other jurisdictions, New York City has granted large tax abatements and other concessions to attract corporations to Manhattan or to retain them. Immediately to the west of lower Manhattan, the Jersey City waterfront has used deep incentives to foster the construction of thirteen million square feet of office space over

the last decade, mostly tenanted by firms formerly located in New York.

Concerned about such bidding wars, the governors of New York, New Jersey, and Connecticut and New York City's mayor signed a "nonaggression" pact in 1991. They vowed to avoid negative advertising and the use of tax breaks and other incentives to steal investment from one another and agreed to cooperate on a regional development strategy. Within a year, however, the economic rivalry accelerated, all three states launched new business incentive and tax reduction programs, and the job wars continued as before. For example, New Jersey induced First Chicago Trust Company to move 1,000 jobs from lower Manhattan by subsidizing the company's office space. . . .

Similarly, the St. Louis region comprises twelve counties in Missouri and Illinois with a 2000 population of 2.6 million, divided by the Mississippi and Missouri Rivers. In 2002, it had 795 local governments, primarily special districts. The central cities of St. Louis (in Missouri) and East St. Louis (in Illinois) suffered massive depopulation while the suburban parts of the region grew. For example, the population of St. Louis fell from 857,000 in 1950 to 348,000 in 2000. The region's poor and black residents are concentrated in the central cities and some older suburbs. Per capita taxable property among the region's municipalities ranged from $2,178 to $143,285.

The St. Louis region created a regional agency responsible for sewers, junior colleges, zoos, museums, and a regional medical center. However, these serve only St. Louis County and the city of St. Louis, not the remainder of the region. In 1949, the two states created a seven-county Bi-State Development Agency (now called Metro), but it lacks taxing authority and focuses almost exclusively on transportation matters. Twice voters in St. Charles County, the most rapidly growing area of the region, rejected ballot measures to link to the region's public transit system. Efforts to promote broader regional cooperation failed in 1926, 1955, and 1959. In 1992, voters rejected ballot measures to create a metropolitan economic development commission (funded by a 2 percent tax on nonresidential utility service) and a metropolitan park commission (to be funded by property taxes). In 2000, voters in five counties approved the creation of a bistate Metropolitan Park and Recreation

District, with dedicated sales tax revenues to plan and implement a regional parks system.

[. . .]

New York [and] St. Louis . . . typify the experience of most metropolitan regions, especially the older ones. Regional cooperation is restricted to a few specific functions, parts of the region still compete for investment and advantage, and regional rivalries hamper regional planning for public improvements that would spur economic growth. Indeed, these regions have difficulty coordinating even simple functions, such as meshing the schedules of regional mass transit systems. Myriad local jurisdictions take their own approaches to federally funded activities such as the construction and management of subsidized housing, the distribution of housing vouchers, the creation of job training programs, and the like, all of which might be more effective if carried out on a regional basis.

EXPERIMENTS IN METROPOLITAN GOVERNANCE

Advocates of regional approaches consistently point to two places that have most fully developed the promise of regional government: the Twin Cities of Minneapolis and St. Paul, Minnesota, where an appointed metropolitan council carries out a number of functions and a regional tax-sharing scheme redistributes revenues from high-growth to low-growth areas; and Portland, Oregon, where an elected metropolitan government regulates suburban land development within a regional growth boundary.

The Twin Cities had a population in 2000 of 669,769 in a metropolitan area of 2.9 million people whose metropolitan per capita income is the sixteenth highest in the country. In 1967, the state legislature created a seven county Metropolitan Council to oversee planning for land use, housing, transit, sewage, and other metropolitan issues. Its members, appointed by the governor, set policy guidelines for local governments in these areas. In its first decade, the council solved a crisis in wastewater treatment. In the 1960s, the Federal Housing Administration threatened to stop insuring mortgages in burgeoning suburbs that lacked sewage treatment. In response, the state government created a Metropolitan Water Control Commission to finance and run treatment plants and build trunk sewer lines as well as creating a regional transportation system. By 1994, the Metropolitan Council not only guided planning for these functions, but had also absorbed the regional agencies for transportation and wastewater treatment. It had its own revenue streams from user fees, a small property tax, and federal grants that enabled it to undertake a substantial regional infrastructure capital program. It encouraged affluent suburbs like Golden Valley to develop low- and moderate-income housing and created regional parks where shopping centers might otherwise have been developed.

In 1971, Minnesota also adopted the Fiscal Disparities Act, which required metropolitan jurisdictions to pool 40 percent of the growth in their commercial and industrial tax bases and to allocate the proceeds according to population and level of tax capacity. This dampened competition for new development and reallocated resources from affluent, fast-growing suburbs to older, more urban parts of the region. Minneapolis and St. Paul began as beneficiaries of this system, but over time, the development of downtown Minneapolis converted it into a contributor, while St. Paul's improvement lessened its dependence on these funds. The formula only deals with revenue-raising capacity, which reflects the value of downtown construction, but does not consider spending needs generated by the relatively large poor populations of the two cities.

Despite its considerable successes, the Metropolitan Council has disappointed some early supporters. The existence of the Metropolitan Council and regional tax sharing did not prevent the Twin Cities from becoming one of the most sprawling metropolitan areas in the United States. The council did not have much impact on the major development issues of the 1980s, including the building of several sports complexes and the world's largest shopping mall. It continues to lack an independent political base (its members all serve at the pleasure of the governor, although they each represent home districts). Because the council cannot pursue policies that lack support from the governor, influential state legislators, or even some local officials, it has difficulty mobilizing support for its agenda. The council did not address neighborhood decay in the central cities or restrain the growth of central-city–suburban disparities. During the 1980s and 1990s, the two cities' population rose by 4.5 percent, but the seven-county metro area grew by 25 percent. By 2000, median family income in the Twin Cities was $48,750, compared with $65,665 in the metro area; the two cities' poverty rate was 16.4 percent, compared with 6.9 percent in the metro area. . . . Most job growth took place in the affluent southern and western suburbs, but many central-city residents (including almost half of black households) lacked cars or public transit access to these areas. This has perhaps contributed to what some have called "Minnesota malaise," and in November 2001, a newcomer displaced the incumbent Minneapolis mayor with the slogan, "I was born in a great city, but I don't want to die in a mediocre one."

Beginning in 1993, then-state legislator Myron Orfield of Minneapolis sought to give the Metropolitan Council more effective tools for addressing these disparities. He introduced a series of bills to elect the members of the Metropolitan Council, to mandate low- and moderate-income housing development goals for each suburb, to empower the council to deny sewer and highway funds to suburbs that failed to comply, to create an affordable housing trust fund by taxing residential construction valued in excess of $150,000 per home, and to give the council more power to control sprawl. After calculating that three-quarters of the region's suburbs would benefit from the housing fund and only one-quarter (including the wealthiest suburbs) would contribute, Orfield used color maps to convince city and inner suburban legislators that they had more in common than they thought, especially in terms of social problems and the distribution of state and regional funds.

The legislature passed Orfield's housing bills, but the Republican governor vetoed them; the bill to elect the council failed by one vote in the state house and five in the senate. But Orfield's efforts drew public attention to the council's potential and its weaknesses and shifted the political climate toward reform. A growing number of business, civic, and political leaders, including the daily newspapers, acknowledged the need to address the region's social disparities along the lines suggested by Orfield. After Reform Party candidate Jesse Ventura was elected governor in 1998, he continued to oppose Orfield's measures.

Between 2000 and 2030, the council expects the Twin Cities metropolitan area to add another million inhabitants and has developed a *2030 Regional Development Framework* that urges more compact development and denser residential development, but recognizes that successful implementation depends on cooperation from local municipalities and the rapidly growing parts of the metropolitan area outside the council's boundaries. In 2003, Republican governor Tim Pawlenty appointed his council members, who in turn chose the brother of his chief of staff as the council's top administrator. At this juncture, the Metropolitan Council may be considered as much a part of the governor's administration as it is a body representing the constituent jurisdictions.

Portland, Oregon, has the nation's only directly elected regional governing body, the Metro Council. It serves twenty-four cities and three counties. It has six nonpartisan members elected for four-year terms from districts containing about 200,000 constituents and a president, elected at large from the region. A chief operating office is responsible for daily management. The Metro Council formulates and implements policy on land use, growth management, solid waste, and parks and recreation and operates the region's zoo, convention center, performing arts center, and solid waste disposal system. It thus has a slightly narrower mandate than the Metropolitan Council in the Twin Cities, but it has home rule charter status and modest taxing abilities. The Oregon legislature authorized it in 1977 and the voters . . . approved it the following year. Its initial duties were designating the urban growth boundary limit (within which new development was to be confined), planning for municipal solid waste disposal, regional transportation planning, and operating the Washington Park Zoo.

Portland was a city of 382,619 in a three-county region of 878,676 in 1970. By 2000, the city had grown to 529,121 and the region to 1.9 million. The region's economy is dominated by shipping, electronics, and manufacturing. In contrast to many cities, the city of Portland's employment base grew significantly, even though job growth has been greater in the surrounding suburbs. Many believe that the region's planning efforts, which emphasized compact development and the development of open space amenities, have been a major factor in this outcome.

Until the 1970s, Portland seemed headed for decline. It lost its ability to annex suburban localities in 1906. By 1956, the three-county area had 176 governmental units, including many special districts. In 1960, the League of Women Voters published *A Tale of Three Counties*, which criticized uncoordinated services and wasteful spending and called for a new, more efficient, more accountable government structure. During the 1960s and 1970s, the state legislature and voters approved regional agencies for transportation, the zoo, and solid waste; the 1977 creation of Metro provided the occasion to consolidate these activities. In 1983, a large budget deficit in Multnomah County (which includes Portland) and the need to provide services to the unincorporated suburban areas east of Portland led the County Board of Supervisors to encourage Portland to annex some unincorporated areas, provide services to some suburbs, and create a sewer construction program for the county. These reforms created support for a metropolitan-wide governance structure. Nevertheless, heavily dependent on such troubled industries as paper and pulp, timber exports, shipbuilding, fishing, and metalworking, the Portland economy did not appear especially well positioned for future growth.

Following several blue-ribbon study commissions, the voters of the three counties approved the creation of an elected metropolitan government in May 1978. Its first task was to designate a metropolitan urban growth boundary under Oregon's land-use legislation, the nation's strongest. The state legislation authorizes Metro to compel local governments and counties to coordinate their land-use and development plans. As baby boomers moved to Portland to enjoy its environment, forest and farm product exports recovered from their slump in the 1980s, and employment surged in the region's major employers (including Intel, Tektronix, and Nike), the demand for new development grew steadily, as did the political base for managing this growth. A political alliance between Portland's business interests and residents of older neighborhoods sought to strengthen the downtown area against competition from suburban shopping malls while saving abandoned housing from the threat of large-scale land clearance and redevelopment. When Neal Goldschmidt was elected mayor in 1972, he incorporated neighborhood activists

into his administration (1972–79) and continued that policy after becoming governor (1987–90). Recently, Metro was reorganized to provide for an elected president separate from a chief operating officer serving alongside six council members elected from districts. . . .

Portland Metro largely achieved its goals for coordinating land-use planning and transportation policy, extending public transit, revitalizing older neighborhoods, and strengthening the downtown business district. Its efforts were bolstered by an urban growth boundary that preserved farmland and forests around Portland and directed urban growth into Portland and its neighboring areas. Although more office and retail development has taken place outside the city than within it, the downtown still accounts for 60 percent of the region's office space and half its upscale retail space. Several highway projects were abandoned in favor of buses and light-rail connections between the central city, outlying neighborhoods, and nearby suburbs. By the 1990s, the region's bus and rail system carried 43 percent of downtown commuters, compared with 20 percent in Phoenix, 17 percent in Salt Lake City, and 11 percent in Sacramento. Portland shows that metropolitan government can succeed. As Orfield observed, its "regional government has been more willing than the Twin Cities appointed Metropolitan Council to exercise its powers vis-à-vis competing authorities."

At the same time, Metro has narrowly defined powers. It does not provide most public services, does not build affordable housing, and does not transcend local zoning and land-use powers. Local school districts remain distinct, eighty single-purpose special districts still have the ability to tax, and twenty cities and three counties provide a broad range of services. Key activities that might help promote mobility of the poor out of the central city – subsidized housing construction and the administration of Section 8 vouchers – are administered by the Portland city housing authority and do not operate on a metrowide basis. Two metrowide agencies with appointed boards, Tri-Met (which runs the transit system) and the Port of Portland (which runs the port, industrial parks, and the airports), operate independently of Metro. Metro's most important function is to set ground rules for development and growth – for example, by siting and building the region's $65 million convention center. It has enhanced the region's ability to resist costly and inefficient sprawl, but it is not a full-service government.

Portland suffers less from the spatial concentration of poverty than most other central cities. Portland's poverty rate (14.5 percent in 1990 and 13.1 percent in 2000) was below most other large central cities. The area's low-income residents were spread out, not highly segregated, concentrated, and isolated. The income gap between the central city and the suburbs remains one of the smallest in the nation. Because Portland has a small minority population (6.5 percent African American, 6.8 percent Hispanic, and 6.6 percent Asian in 2002), the relationship between the city and its suburbs has not been racially charged. The state's "fair-share" housing mandate, along with Metro's planning efforts, have encouraged suburbs to develop more low-income housing than in other metropolitan areas. The region's housing prices did increase more than the national average during the early 1990s, but not in other periods over the last twenty years, and not as much as other growing western cities that lack Portland's growth management. Cooperation between low-income housing advocates and the region's real estate developers, together with the institutional strength of regional government, has led to discussions about creating a regional housing fund. If this does occur, it would put Portland at the forefront of regional equity, as well as environmental planning. Metro has not, however, made dispersing the inner-city minority poor to the surrounding area a priority.

EFFORTS TO PROMOTE REGIONAL EQUITY

As Scott Bollens observed, even at its best, "the current state of regional governance in the United States does not effectively address issues of concentrated poverty and social equity" because it tends to focus on infrastructure rather than people and because, lacking a broad political base, it takes a narrow, technical approach to its work. Recognizing the difficulty of changing this situation and the likelihood that it would not soon reduce the embedded patterns of regional inequity and constrained opportunities for the inner-city poor, many advocates have sought more direct regional solutions to the problems of concentrated urban poverty. Because the courts have generally upheld suburban jurisdictions' right to impose restrictive zoning and have found that federal law on housing discrimination does not apply to low-income households but applies only to particular groups such as racial minorities, these advocates have generally attacked the racial dimension of exclusion. They have, for example, sought to induce suburbs to take on their fair share of housing for lower-income residents and designed programs to connect the inner-city poor with suburban job opportunities, such as reverse commuting, portable Section 8 vouchers, and new institutions that train inner-city residents and place them in suburban job openings.

Because a "spatial mismatch" often prevents inner-city poor people from taking advantage of suburban job opportunities, advocates of regional equity have long sought to increase access to suburban housing. Because many urban poor are also members of federally protected minority groups, vigorous enforcement of federal fair housing and affirmative mortgage lending regulations seems to be an obvious way to provide better access. As Michael Schill has pointed out, this legal framework does not provide a way to roll back the overall framework of exclusionary zoning practices in the suburbs. Instead, it is most effective where least needed: for middle-class minority individuals who are obviously victims of discriminatory practices such as "steering" by real estate agents. Even in New Jersey, where the Mt. Laurel cases mandated suburban jurisdictions to do more to house the state's low-income residents, they can (and do) meet this obligation by financially supporting the construction of subsidized housing elsewhere. Shifting the financing of local schools away from the property tax and toward state budgets may reduce suburbanites' fiscal incentive to exclude the less well-off, but it is not likely to lessen the propensity of middle-class whites to flee districts when the percentage of minorities in their children's classrooms begins to rise significantly. Recent efforts to help central-city residents use Section 8 vouchers for suburban rental housing have shown promising results. But because these programs run up against exclusionary suburban zoning laws, it is difficult for them to help significant numbers of families.

A number of metropolitan regions have adopted fair-share housing policies to provide some affordable housing in otherwise expensive housing markets that exclude low- and even moderate-income residents. One of the oldest and most successful programs is in Montgomery County, Maryland, an affluent and fast-growing area of 497 square miles adjacent to Washington, DC, with a population in 2000 of 873,341 divided among fourteen incorporated municipalities. During the 1970s and 1980s, Montgomery County changed from a bedroom community of Washington to a large employment center. The population boomed, and housing costs spiraled. In the early 1970s, housing advocacy groups such as Suburban Maryland Fair Housing and the League of Women Voters began pushing to increase the supply of affordable housing. The elected county council responded by adopting an inclusionary zoning law.

Since 1974, Montgomery County has required that all new housing developments with fifty or more units include a percentage (now 12.5 to 15 percent) of moderate-priced units. In exchange, it provides developers with a "density bonus," permitting them to increase a project's density by 20 percent. The first units were built in 1976. To maintain the supply of affordable housing, the county limits the resale price for ten years and the rent for twenty years, and the county's Housing Opportunity Commission or nonprofit agencies can purchase up to 40 percent of these units. By 2004, the program had created more than ten thousand units, one-third of which may be purchased by the county's Housing Opportunity Commission (HOC), its public housing authority. HOC has purchased about 1,600 units for very low-income families. It also manages another 4,700 units of its own and administers 3,500 Section 8 vouchers.

Montgomery County requires developers to integrate these affordable units within market-rate housing rather than isolate them and create mini ghettos. The county insists on high standards of design and construction for these developments. They do not look like the stereotype of government-subsidized "projects." The program is limited to people who live or work in Montgomery County. Although there has been some opposition to particular developments, the programs have been generally well-accepted by developers and by neighbors. Montgomery's program is a successful effort to address the problem of economic segregation, but it is a drop in the bucket. Indeed, because the program has a long waiting list, the families are chosen by lottery.

[. . .]

So far, these experiments have produced modest but promising results. Several lessons can be drawn from them. First, they have been tried only on a pilot basis. Substantial additional investments will have to be made to bring them to full scale. It appears that with sufficient counseling of tenants and careful handling of landlords and suburban communities, the central-city poor can find better suburban housing opportunities without bringing on the social calamities and political opposition that defenders of suburban exclusion predict. It appears, therefore, that deconcentrating the central-city poor is feasible at a reasonable cost and will produce desirable results. Second, it is unlikely that physical juxtaposition or spatial mobility alone will dramatically improve the access of the urban poor to suburban job opportunities. Instead, the central-city poor require not only better skills (both "soft" and "hard") but also incorporation into networks of contact, reciprocity, and support with potential employers. In short, new labor market intermediaries may be needed. Finally, it is clear that the vast bulk of social programs are administered in a way that hinders them from functioning on a regional basis. That is, programs for subsidized housing, education, job training, and the like are typically administered by units of central-city governments that operate within those restricted jurisdictions. Even if they have the authority to operate outside central-city boundaries (as in the case of county public housing authorities), they generally do not. Suburban jurisdictions simply, unobtrusively, opt out. For such programs to take a regional approach to their work, they must be reorganized into a metropolitan-wide jurisdiction.

THE POLITICS OF REGIONALISM IN THE NEW MILLENNIUM

Historically, many forces have worked against a regional perspective in urban governance. Suburbs have been happy to benefit from being located in a large metropolitan area while excluding the less well-off and avoiding the payment of taxes to support services required by the urban poor. Central cities, for their part, have made a virtue of necessity by increasing spending on social services as a way of expanding the employment of central-city constituents. This spending has become an increasingly substantial part of municipal budgets and an important form of "new patronage" in city politics. As federal benefits have increasingly flowed to needy people, as opposed to needy places, it has helped to expand these functions. Legislators elected to represent areas where the minority poor are concentrated develop a stake in this state of affairs. Big-city mayors have also been reluctant to lose any of their powers within a broader metropolitan jurisdiction.

. . . [T]his dynamic is gradually but steadily shifting. The first wave of inner, working-class suburbs has long since been built out, their populations have aged, and their residents' incomes have stagnated since the early 1970s. Increasingly, suburbs have developed "urban" problems that they cannot solve on their own. Black and Hispanic central-city residents have increasingly moved to the suburbs. Although minority suburbs generally have significantly better conditions than inner-city minority neighborhoods, they still have higher rates of poverty and disadvantage than white suburbs and may face some of the same forces of decline that operate on inner cities. Even as metropolitan economic segregation has increased, metropolitan racial segregation has declined and suburban diversity has increased. As Myron Orfield has pointed out, these inner suburbs are coming to realize that they share significant interests with each other that are not all that dissimilar with those of their central cities. And as Juliet Gainsborough has shown, voters residing in more diverse suburbs are substantially more likely to vote like their urban neighbors.

[. . .]

Across the country, many powerful players, including corporations, foundations, unions, political leaders, and community organizations, have come to think that some form of regional collaboration is necessary to achieve their objectives, whether that be competitive advantage in the new global economy, more equitable access to housing and employment, or more sensible forms of land-use development that limit sprawl and preserve open space. Regions strongly divided against themselves are least able to undertake the necessary physical, human, and social capital investments. Metropolitan governance institutions – whether ad hoc regionalism focused on one objective like helping the inner-city poor to move to better suburban neighborhoods or full-blown governments like Metro Portland or Metro Louisville – are moving to carry out functions that make sense from economic, environmental, and equity perspectives. These include regional capital investments, transportation, land-use planning, economic development, job training, education, and tax-base sharing. Still, we do not yet know whether these efforts will succeed in integrating metropolitan areas by following a high-productivity, high-wage high road or will instead reflect a suburban predominance that stresses new economic investments, whatever the payoff to those at the low end of the labor market.

Which path they will take may depend on the extent to which metropolitan cooperation develops a broad, democratic base and the organizational capacity to articulate the common good, not merely to sum up the individual parts of the metropolis. Cooperation of the region's constituent elements must be secured through consent, not through unwanted mandates imposed on resistant local jurisdictions. To achieve this consent, the new regional form must provide tangible benefits to all or most of its constituent jurisdictions, not just an exclusive favored quarter.

"Are Europe's Cities Better?"

from *The Public Interest* (1999)

Pietro Nivola

Editors' Introduction

The distinction between urban development patterns in the United States and those in most of Europe becomes apparent to any airplane passenger. Flying over the populated parts of the United States, it is hard to know where one urban area ends and the other begins – development seems to sprawl out continuously, a maze of highways, malls, office parks, and suburban tract housing. Flying over Europe, in contrast, one can easily distinguish urban from rural. Areas of dense development are ringed by woods and farm land. These bird's-eye observations are confirmed on the ground: one can navigate most European cities by foot and mass transit, whereas one would need an automobile (and a full tank of gas!) to make one's way around most American cities.

In this article, as in his book, *Laws of the Landscape*, Pietro Nivola is interested in exploring the factors that account for these differences. After all, as we have noted in other sections of this Reader, there are many forces that could lead to a growing convergence between the policies of U.S. and European cities. Economic restructuring has imposed similar pressures throughout the industrialized world – would not urban form follow suit? Nivola, like other authors featured in this volume, is quick to note that political differences between nations lead to different policy outcomes even in countries faced with similar economic pressures.

In Nivola's view, urbanization patterns have been shaped by whole complexes of interwoven public policies and social practices that, together, encourage Europeans to settle in compact cities, and encourage Americans to buy large houses in far-flung suburbs. These include tax policies (which encourage Americans to spend, and in particular to spend on housing while encouraging Europeans to conserve, and in particular to conserve fuel), and fiscal revenue sharing practices (through which many European cities receive sufficient funds to maintain a high level of services). Social factors matter, too: high crime rates in U.S. cities push residents to the periphery. Geography also plays a role: Europeans simply have less room in which to spread out, so they have more incentive, as individuals and as societies, to learn to live "small."

Nivola understands that "Europe" is not monolithic, and that urban policy practices vary considerably even just within the European Union. In addition, he does not claim that European nations have all the right answers. He notes that some of the practices that have led to vibrant, compact cities (higher taxes, more stringent regulations, subsidies for farmers and small businesses) have some undesirable consequences.

A fuller examination of comparative planning practices may be found in Nivola's *Laws of the Landscape* (Washington, D.C.: Brookings Institution, 1999). H.V. Savitch's *Post-industrial Cities* (Princeton: Princeton University Press, 1988) compares development and planning practices in the U.S., Great Britain, and France by examining case studies of their largest cities. Susan Fainstein's *The City Builders* (University Press of Kansas, 2001) similarly examines the politics of downtown development in New York City and London. Few comparative policy studies focus on urban planning issues, but Arnold J. Heidenheimer *et al.*'s *Comparative Public Policy: The Politics of Social Choice in America, Europe, and Japan* (New York: St. Martin's Press, 1990) includes an excellent chapter on land-use policies.

Pietro Nivola is a vice-president of the Brookings Institution in Washington, D.C., and director of its Governance Studies Program. He has written numerous books and articles on a range of policy issues, including trade, energy, and land use. A few of his many publications are *Tense Commandments: Federal Prescriptions and City Problems* (Washington, D.C.: Brookings Institution, 2002); (ed.) *Comparative Disadvantages? Social Regulations and the Global Economy* (Washington, D.C.: Brookings Institution, 1997), *Regulating Unfair Trade* (Washington, D.C.: Brookings Institution, 1993), and *The Politics of Energy Conservation* (Washington, D.C.: Brookings Institution, 1986). His shorter essays on topical issues may frequently be found on the Brookings Institution website: www.brookings.edu.

Cities grow in three directions: in by crowding, up into multi-story buildings, or out toward the periphery. Although cities everywhere have developed in each of these ways at various times, nowhere in Europe do urban settlements sprawl as much as in the United States. Less than a quarter of the U.S. population lived in suburbia in 1930. Now well over half does. Why have most European cities remained compact compared to the hyper-extended American metropolis? At first glance, the answer seems elementary. The urban centers of Europe are older, and the populations of their countries did not increase as rapidly in the post-war period. In addition, stringent national land-use laws slowed exurban development, whereas the disjointed jurisdictions of U.S. metropolitan regions encouraged it.

But on closer inspection, this conventional wisdom does not suffice. It is true that the contours of most major urban areas in the United States were formed to a great extent by economic and demographic expansion after the Second World War. But the same was true in much of Europe, where entire cities were reduced to rubble by the war and had to be rebuilt from ground zero.

Consider Germany, whose cities were carpet bombed. Many German cities today are old in name only, and though the country's population as a whole grew less quickly than America's after 1950, West German cities experienced formidable economic growth and in-migrations. Yet the metropolitan population density of the United States is still about one-fourth that of Germany. New York, our densest city, has approximately one-third the number of inhabitants per square mile as Frankfurt.

Sprawl has continued apace even in places where the American population has grown little or not at all in recent decades. From 1970 to 1990, the Chicago area's population rose by only 4 percent, but the region's built-up land increased 46 percent. Metropolitan Cleveland's population actually declined by 8 percent, yet 33 percent more of the area's territory was developed.

The fragmented jurisdictional structure in U.S. metropolitan areas, wherein every suburban town or county has control over the use of land, does not adequately explain sprawl either. Since 1950, about half of America's central cities at least doubled their territory by annexing new suburbs. Houston covered 160 square miles in 1950. By 1980, exercising broad powers to annex its environs, it incorporated 556 square miles. In the same 30-year period, Jacksonville went from being a town of 30 square miles to a regional government enveloping 841 square miles – two-thirds the size of Rhode Island. True, the tri-state region of New York contains some 780 separate localities, some with zoning ordinances that permit only low-density subdivisions. But the urban region of Paris – Ile de France – comprises 1,300 municipalities, all of which have considerable discretion in the consignment of land for development.

To be sure, European central governments presumably oversee these local decisions through nationwide land-use statutes. But is this a telling distinction? The relationship of U.S. state governments to their local communities is roughly analogous to that of Europe's unitary regimes to their respective local entities. Not only are the governments of some of our states behemoths (New York State's annual expenditures, for example, approximate Sweden's entire national budget) but a significant number have enacted territorial planning legislation reminiscent of European guidelines. Indeed, from a legal standpoint, local governments in this country are mere "creatures" of the states, which can direct, modify, or even abolish their localities at will. Many European municipalities, with their ancient independent charters, are less subordinated.

The enforcement of land-use plans varies considerably in Europe. In Germany, as in America, some *Laender* (or states) are more restrictive than others. The Scandinavians, Dutch, and British take planning more seriously than, say, the Italians. The late Antonio Cederna, an astute journalist, wrote volumes about the egregious violations of building and development codes in and around Italy's historic centers. Critics who assume that land regulators in the United States are chronically permissive, whereas Europe's growth managers are always scrupulous and "smart," ought to contemplate, say, the unsightly new suburbs stretching across the northwestern plain of Florence toward Prato, and then visit Long Island's East End, where it is practically impossible to obtain a building permit along many miles of pristine coastline.

BIG, FAST, AND VIOLENT

The more important contrasts in urban development between America and Europe lie elsewhere. With three and half million square miles of territory, the United States has had much more space over which to spread its settlements. And on this vast expanse, decentralizing technologies took root and spread decades earlier than in other industrial countries. In 1928, for example, 78 percent of all the motor vehicles in the world were located in the United States. With incomes rising rapidly, and the costs of producing vehicles declining, 56 percent of American families owned an automobile by that time. No European country reached a comparable level of automobile ownership until well after the Second World War. America's motorized multitudes were able to begin commuting between suburban residences and workplaces decades before such an arrangement was imaginable in any other advanced nation.

A more perverse but also distinctive cause of urban sprawl in the United States has been the country's comparatively high level of violent crime. Why a person is ten times more likely to be murdered in America than in Japan, seven times more likely to be raped than in France, or almost four times more likely to be robbed at gun point than in the United Kingdom, is a complex question. But three things are known.

First, although criminal violence has declined markedly here in the past few years, America's cities have remained dangerous by international standards. New York's murder rate dropped by two-thirds between 1991 and 1997, yet there were still 767 homicides committed that year. London, a megacity of about the same size, had less than 130. Second, the rates of personal victimization, including murder, rape, assault, robbery, and personal theft, tend to be much higher within U.S. central cities than in their surroundings. In 1997, incidents of violent crime inside Washington, D.C., for instance, were six times more frequent than in the city's suburbs. Third, there is a strong correlation between city crime rates and the flight of households and businesses to safer jurisdictions. According to economists Julie Berry Cullen of the University of Michigan and Steven D. Levitt of the University of Chicago, between 1976 and 1993, a city typically lost one resident for every additional crime committed within it.

Opinion surveys regularly rank public safety as a leading consideration in the selection of residential locations. In 1992, when New Yorkers were asked to name "the most important reason" for moving out of town, the most common answer was "crime, lack of safety" (47.2 percent). All other reasons – including "high cost of living" (9.3 percent) and "not enough affordable housing" (5.3 percent) – lagged far behind. Two years ago, when the American Assembly weighed the main obstacles to business investments in the inner cities, it learned that businessmen identified lack of security as the principal impediment. In short, crime in America has further depopulated the cores of metropolitan areas, scattering their inhabitants and businesses.

THE NOT-SO-INVISIBLE HAND

In addition to these fundamental differences, the public agendas here and in major European countries have been miles apart. The important distinctions, moreover, have less to do with differing "urban" programs than with other national policies, the consequences of which are less understood.

For example, lavish agricultural subsidies in Europe have kept more farmers in business and dissuaded them from selling their land to developers. Per hectare of farmland, agricultural subventions are twelve times more generous in France than in the United States, a divergence that surely helps explain why small farms still surround Paris but not New York City.

Thanks to scant taxation of gasoline, the price of automotive fuel in the United States is almost a quarter of what it is in Italy. Is it any surprise that Italians would live closer to their urban centers, where they can more easily walk to work or rely on public transportation? On a per capita basis, residents of Milan make an average of 350 trips a year on public transportation; people in San Diego make an average of seventeen.

Gasoline is not the only form of energy that is much cheaper in the United States than in Europe. Rates for electric power and furnace fuels are too. The expense of heating the equivalent of an average detached U.S. suburban home, and of operating the gigantic home appliances (such as refrigerators and freezers) that substitute for neighborhood stores in many American residential communities, would be daunting to most households in large parts of Europe.

Systems of taxation make a profound difference. European tax structures penalize consumption. Why don't most of the Dutch and Danes vacate their compact towns and cities where many commuters ride bicycles, rather than drive sport-utility vehicles, to work? The sales tax on a new, medium-sized car in the Netherlands is approximately nine times higher than in the United States; in Denmark, 37 times higher. The U.S. tax code favors spending over saving (the latter is effectively taxed twice) and provides inducements to purchase particular goods – most notably houses, since the mortgage interest is deductible. The effect of such provisions is to lead most American families into the suburbs, where spacious dwellings are available and absorb much of the nation's personal savings pool.

Tax policy is not the only factor promoting home ownership in the United States. Federal Housing Administration and Veterans Administration mortgage guarantees financed more than a quarter of the suburban single-family homes built in the immediate postwar period. In Europe, the housing stocks of many countries were decimated by

the war. Governments responded to the emergency by erecting apartment buildings and extending rental subsidies to large segments of the population. America also built a good deal of publicly subsidized rental housing in the postwar years, but chiefly to accommodate the most impoverished city-dwellers. Unlike the mixed-income housing complexes scattered around London or Paris, U.S. public housing projects further concentrated the urban poor in the inner cities, turning the likes of Chicago's South Side into breeding grounds of social degradation and violence. Middle-class city-dwellers fled from these places to less perilous locations in the metropolitan fringe.

Few decisions are more consequential for the shape of cities than a society's investments in transportation infrastructure. Government at all levels in the United States has committed hundreds of billions to the construction and maintenance of highways, passenger railroads, and transit systems. What counts, however, is not just the magnitude of the commitment but the distribution of the public expenditures among modes of transportation. In the United States, where the share claimed by roads has dwarfed that of alternatives by about six to one, an unrelenting increase in automobile travel and a steady decline in transit usage – however heavily subsidized – was inevitable.

Dense cities dissipate without relatively intensive use of mass transit. In 1945, transit accounted for approximately 35 percent of urban passenger miles traveled in the United States. By 1994, the figure had dwindled to less than 3 percent – or roughly one-fifth the average in Western Europe. If early on, American transportation planners had followed the British or French budgetary practice of allocating between 40 and 80 percent of their transport outlays to passenger railroads and mass transit systems, instead of nearly 85 percent for highways, there is little question that many U.S. cities would be more compressed today.

Dense cities also require a vibrant economy of neighborhood shops and services. (Why live in town if performing life's simplest everyday functions, like picking up fresh groceries for supper, requires driving to distant vendors?) But local shopkeepers cannot compete with the regional megastores that are proliferating in America's metropolitan shopping centers and strip malls. Multiple restrictions on the penetration and predatory pricing practices of large retailers in various European countries protect small urban businesses. The costs to consumers are high, but the convenience and intimacy of London's "high streets" or of the corner markets in virtually every Parisian *arrondissement* are preserved.

"SHIFT AND SHAFT" FEDERALISM

Europe's cities retain their merchants and inhabitants for yet another reason: European municipalities typically do not face the same fiscal liabilities as U.S. cities. Local governments in Germany derive less than one-third of their income from local revenues; higher levels of government transfer the rest. For a wide range of basic functions – including educational institutions, hospitals, prisons, courts, utilities, and so on – the national treasury funds as much as 80 percent of the expense incurred by England's local councils. Localities in Italy and the Netherlands raise only about 10 percent of their budgets locally. In contrast, U.S. urban governments must largely support themselves: They collect two-thirds of their revenues from local sources.

In principle, self-sufficiency is a virtue; municipal taxpayers ought to pay directly for the essential services they use. But in practice, these taxpayers are also being asked to finance plenty of other costly projects, many of which are mandated, but underfunded, by the federal government. Affluent jurisdictions may be able to absorb this added burden, but communities strapped for revenues often cannot. To satisfy the federal government's paternalistic commands, many old cities have been forced to raise taxes and cut the services that local residents need or value most. In response, businesses and middle-class households flee to the suburbs.

America's public schools are perhaps the clearest example of a crucial local service that is tottering under the weight of unfunded federal directives. Few nations, if any, devote as large a share of their total public education expenditures to nonteaching personnel. There may be several excuses for this lopsided administrative overhead, but one explanation is almost certainly the growth of government regulation and the armies of academic administrators needed to handle the red tape.

Schools are required, among other things, to test drinking water, remove asbestos, perform recycling, insure "gender equity," and provide something called "special education." The latter program alone forces local authorities to set aside upwards of $30 billion a year to meet the needs of students with disabilities.

Meanwhile, according to a 1992 report by the U.S. Advisory Commission on Intergovernmental Relations, the federal government reimburses a paltry 8 percent of the expense. Compliance costs for urban school districts, where the concentrations of learning-disabled pupils are high and the means to support them low, can be particularly onerous. Out of a total $850 million of local funds budgeted for 77,000 students in the District of Columbia, for instance, $170 million has been earmarked for approximately 8,000 students receiving "special education."

Wretched schools are among the reasons why most American families have fled the cities for greener pastures. It is hard enough for distressed school systems like the District's, which struggle to impart even rudimentary literacy, to compete with their wealthier suburban counterparts. The difficulty is compounded by federal laws that, without adequate recompense, divert scarce educational resources from serving the overwhelming majority of students.

Schools are but one of many municipal services straining to defray centrally dictated expenses. Consider the plight of urban mass transit in the United States. Its empty seats and colossal operating deficits are no secret. Less acknowledged are the significant financial obligations imposed by Section 504 of the Rehabilitation Act and subsequent legislation. To comply with the Department of Transportation's rules for retrofitting public buses and subways, New York City estimated in 1980 that it would need to spend more than $1 billion in capital improvements on top of $50 million in recurring annual operating costs. As the city's mayor, Edward I. Koch, said at the time, "It would be cheaper for us to provide every severely disabled person with taxi service than make 255 of our subway stations accessible."

Although the Reagan administration later lowered these costs, passage of the Americans with Disabilities Act in 1990 led to a new round of pricey special accommodations in New York and other cities with established transit systems. Never mind that the Washington Metro is the nation's most modern and well-designed subway system. It has been ordered to tear up forty-five stations and install bumpy tiles along platform edges to accommodate the sight impaired, a multi-million dollar effort. At issue here, as in the Individuals with Disabilities Education Act, is not whether provisions for the handicapped are desirable and just. Rather, the puzzle is how Congress can sincerely claim to champion these causes if it scarcely appropriates the money to advance them.

Nearly two decades ago, Mayor Koch detailed in *The Public Interest* what he called the "millstone" of some 47 unfunded mandates [Koch 1980]. The tally of national statutes encumbering U.S. local governments since then has surpassed at least one hundred. And this does not count the hundreds of federal court orders and agency rulings that micromanage, and often drain, local resources. By 1994, Los Angeles estimated that federally mandated programs were costing the city approximately $840 million a year. Erasing that debit from the city's revenue requirements, either by meeting it with federal and state aid or by substantial recisions, would be tantamount to reducing city taxes as much as 20 percent. A windfall that large could do more to reclaim the city's slums, and halt the hollowing out of core communities, than would all of the region's planned "empowerment zones," "smart growth" initiatives, and "livability" bond issues.

FOLLOW EUROPE?

To conclude that greater fiscal burden sharing and a wide range of other public policies help sustain Europe's concentrated cities is not to say, of course, that all those policies have enhanced the welfare of Europeans – and hence, that the United States ought to emulate them. The central governments of Western Europe may assume more financial responsibilities instead of bucking them down to the local level, but these top-heavy regimes also levy much higher taxes. Fully funding all of Washington's many social mandates with national tax dollars would mean, as in much of Europe, a more centralized and bloated welfare state.

Most households are not better off when farmers are heavily subsidized, or when anticompetitive practices protect microbusinesses at the expense of larger, more efficient firms. Nor would most

consumers gain greater satisfaction from housing strategies that encourage renter occupancy but not homeownership, or from gas taxes and transportation policies that force people out of their cars and onto buses, trains, or bicycles.

In fact, these sorts of public biases have exacted an economic toll in various Western European countries, and certainly in Japan, while the United States has prospered in part because its economy is less regulated, and its metropolitan areas have been allowed to decompress. So suffocating is the extreme concentration of people and functions in the Tokyo area that government planners now view decentralization as a top economic priority. Parts of the British economy, too, seem squeezed by development controls. A recent report by McKinsey and Company attributes lagging productivity in key sectors to Britain's land-use restrictions that hinder entry and expansion of the most productive firms.

The densely settled cities of Europe teem with small shops. But the magnetic small-business presence reflects, at least in part, a heavily regulated labor market that stifles entrepreneurs who wish to expand and thus employ more workers. As the *Economist* noted in a review of the Italian economy, "Italy's plethora of small firms is as much an indictment of its economy as a triumph: many seem to lack either the will or the capital to keep growing." The lack of will is not surprising; moving from small to midsize or large means taking on employees who are nearly impossible to lay off when times turn bad, and it means saddling a company with costly mandated payroll benefits. Italy may have succeeded in conserving clusters of small businesses in its old cities and towns, but perhaps at the price of abetting double-digit unemployment in its economy as a whole.

STRIKING A BALANCE

America's strewn-out cities are not without their own inefficiencies. The sprawling conurbations demand, for one thing, virtually complete reliance on automotive travel, thereby raising per capita consumption of motor fuel to four times the average of cities in Europe. That extraordinary level of fossil-fuel combustion complicates U.S. efforts to lower this country's considerable contribution to the buildup of greenhouse gases. Our seemingly unbounded suburbanization has also blighted central cities that possess irreplaceable architectural and historic assets. A form of metropolitan growth that displaces only bleak and obsolescent urban relics, increasingly discarded by almost everyone, may actually be welfare-enhancing. A growth process that also blights and abandons a nation's important civic and cultural centers, however, is rightfully grounds for concern.

Still, proposals to reconfigure urban development in the United States need to shed several misconceptions. As research by Helen Ladd of Duke University has shown, the costs of delivering services in high-density settlements frequently increase, not decrease. Traffic congestion at central nodes also tends to worsen with density, and more people may be exposed to hazardous levels of soot and smog. (The inhabitants of Manhattan drive fewer vehicle miles per capita than persons who inhabit New York's low-density suburbs. Nevertheless, Manhattan's air is often less healthy because the borough's traffic is unremittingly thick and seldom free-flowing, and more people live amid the fumes.) Growth boundaries, such as those circumscribing Portland, Oregon, raise real estate values, so housing inside the boundaries becomes

less, not more, "affordable." Even the preservation of farmland, a high priority of managed growth plans, should be placed in proper perspective. The United States is the world's most productive agricultural producer, with ample capacity to spare. Propping up marginal farms in urbanizing areas may not put this acreage to uses most valued by society.

In sum, the diffuse pattern of urban growth in the United States is partly a consequence of particular geographic conditions, cultural characteristics, and raw market forces, but also an accidental outcome of certain government policies. Several of these formative influences differ fundamentally from those that have shaped European cities. Critics of the low-density American cityscape may admire the European model, but they would do well to recognize the full breadth of hard policy choices, and tough tradeoffs, that would have to be made before the constraints on sprawl in this country could even faintly begin to resemble Europe's.

REFERENCE FROM THE READING

Edward I. Koch. "The Mandate Millstone," *The Public Interest.* Number 61, Fall 1980.

"The American City in the Age of Terror: A Preliminary Assessment of the Effects of September 11"

from *Urban Affairs Review* (2004)

Peter Eisinger

Editors' Introduction

"Tuesday the Senate also voted to continue distributing a significant portion of security dollars equally among the states, rather then by likelihood of attack. Bad news for [places like New York City]. But good news for smaller states like Wyoming, which only has one high risk target – the popular tourist attraction 'The World's Largest Pile of Homeland Security Money.'"

(Jon Stewart, the *Daily Show*)

Terrorism finds its causes in countless conflicts reflecting regional, national, and global tensions. But it has its impact almost exclusively on cities. Most terrorist attacks seek to destabilize nations by causing disruption and instilling fear. With their concentrations of people, landmarks, and media, cities make disturbingly attractive targets for those wishing to inflict extensive damage. Whatever source of discord drives extremist groups, they will carry out their attacks in places like Madrid or Baghdad, London or Oklahoma City, Nairobi or New York.

Security concerns are not new to cities. Indeed, the impenetrable walls fortifying medieval towns make it clear that defensive requirements have long shaped urban form. Centuries later, early Cold War era planners advocated the dispersal of population and industry out of central cities to minimize damage from a Soviet attack (although suburbanization would no doubt have occurred even in the absence of these concerns). Terrorist attacks, which occur without warning or provocation and prey on people as they go about their everyday lives, pose particular challenges for cities. There is no question that policing practices, the use of public space, and building design have been shaped by the threat of terrorism.

Contemporary urbanists have recently begun to consider the more profound impacts that terrorism and the fear of terrorism could have upon cities. Writing after the 2001 hijackings and attacks in New York and Washington, D.C. (but before the Madrid and London bombings), Peter Eisinger reflects on how the September 11 attacks have affected, and could affect, New York City and other U.S. cities. He identifies how heightened security concerns have affected three aspects of urban governance and planning. The first is public policy. One might assume that the U.S. federal government would be eager to aid the specific cities that had suffered the brunt of the attacks, both to help their recovery and to protect them against future attacks. But Congress has been slow to approve new aid, and new funds that have been approved have been dispersed across the country with little effort to focus them on the most vulnerable cities. (Comedian Jon Stewart's quip, quoted above, alludes to the distribution of Homeland Security funds between regions.)

Eisinger concludes that cities have had to bear the costs of higher security on their own. Surveys of city governments undertaken by the U.S. Conference of Mayors provide support for that view (see their website, www.usmayors.org, for the most recent Homeland Security report).

Eisinger finds little evidence that the September 11 attacks left long-term scars on the New York City economy, and it would be difficult to discern any security-related movement out of cities given the many factors that determine where businesses and people locate. And although one finds a few more barricades, the continued proliferation of security cameras, and baggage searches in new places, cities have not, he concludes, become fortresses. Press reports from London and Madrid suggest that there, as well, the openness of city life has not changed.

Peter Eisinger is the Henry Cohen Professor of Urban Affairs at the Milano Graduate School at New School University. Over his career, he has written on many aspects of state and urban policies. His books include *The Rise of the Entrepreneurial State* (Madison: University of Wisconsin Press, 1988); *Federal Policy Toward the Cities* (New York: HarperCollins, 1995), and *Toward an End to Hunger in America* (Washington, D.C.: Brookings Institution, 1998).

Those interested in understanding security concerns as a factor in U.S. urban development since World War II should consult Jennifer S. Light's *From Warfare to Welfare: Defense Intellectuals and Urban Problems in Cold War America* (Baltimore, MD: Johns Hopkins University Press, 2003). More specific information on post-September 11 homeland security measures are discussed at length in periodicals such as *Governing* and *American City and County*. A trio of edited volumes published by the Russell Sage Foundation in 2005 track the impact of 9/11 on New York City: *Resilient City*, edited by Howard Chernick, *Wounded City*, edited by Nancy Foner, and *Contentious City*, edited by John Mollenkopf.

The terror attacks of September 11, 2001, on New York and Washington, D.C., were fundamentally challenges to American values, optimism, and global economic dominance, but they must also be seen as assaults on cities as urban places. In targeting these open and unprotected places, populated by a large, socially diverse workforce engaged in the knowledge-intensive occupations of the new cosmopolitan economy in signature buildings that had come to represent some of the most powerful symbols of modern urban achievement, the terrorists took aim at the very essence of American cities. It would hardly be surprising under the circumstances, therefore, if Americans did not begin to wonder in the aftermath of the attacks about the security, role, and importance of urban life and forms in modern society. Yet as the so-called war on terror proceeds, the enduring impacts on cities of those terrible events of September 11 remain unclear.

In the immediate period after the attacks, certain commentators were quick to predict the end of American urban life as we know it. Some believed that fearful cities would respond primarily through strategies of repressive fortification. Peter Marcuse (2002) was among the most emphatic of these pessimists, warning of the erosion of urban democracy, the closing of public spaces, and the emergence of the citadel city, a fortress protected by "pervasive surveillance." Others, however, although just as convinced that there would be deep and lasting effects of the attacks, offered a polar scenario to the so-called hardening of the city, arguing that terrorism would simply hasten patterns of business and residential diffusion that had long been in evidence. No one would build tall buildings anymore, for no one would wish to work or live in them; firms would increasingly move their headquarters and back office functions to suburban locations, joining their out-migrating labor force; business travel and tourism investment would diminish. The terror attacks would simply accelerate the stampede to the far suburbs. . . .

Not everyone believes that American cities were forever altered in profound ways by the 2001 terror attacks. Many observers have long been convinced that the enduring feature of great American urban centers is their resilience. Over half a century ago, E.B. White wrote of New York City that it "is peculiarly constructed to absorb almost anything that comes along" and that New Yorkers survive the myriad challenges and threats of urban life with "a sort of perpetual muddling through" ([1949]1999, 23, 33). That view is shared by many contemporary social scientists. . . . Hank Savitch (2003) acknowledges both short-term economic costs of the attacks in New York and the breadth of fear of future terror among the American public, but in the end, he argues, the impact on cities has been negligible. The American city has not been transformed into a garrison state.

James Harrigan and Phillipe Martin (2002) make a more explicit argument for the resilience of cities in the face of catastrophic events, including terror attacks: Cities exist in the first place because of agglomeration economies. Applying separate simulation models based on labor pooling (where employers derive economic benefits from the proximity of a large and diverse workforce) and core/periphery assumptions (which take account of the desire of firms to be near customers and suppliers to minimize transportation costs), Harrigan and Martin show that the economic advantages derived from concentration of resources are powerful enough to overcome the costs or "tax" of terror events, not only in New York but in other large cities as well.

The effort to assess the impacts of terror on American cities is an on-going enterprise. No examination so far has marshaled much empirical evidence on the matter, however. It is appropriate, therefore, to take stock periodically and to do so from a variety of different angles. This note examines some of the evidence to date relating to the impact on U.S. cities in three broad areas: government and policy, the economy, and what may be called simply the *texture* of city life.

GOVERNMENT AND POLICY

Within days of the collapse of the World Trade Center towers, the National League of Cities (NLC) polled 456 of its member cities to find out how they were responding to the appearance of massive terror on American soil. Communities of all sizes and in all parts of the country, it turned out, had set about immediately to secure water supplies, assign guards to critical transportation facilities and government buildings, alert hospitals and public health departments to stand by, and convene officials to discuss emergency plans. Mobilization of local public safety resources was not unexpected: Not only did a majority of communities already have terrorism contingency plans in place, but as New Haven Mayor John DeStefano has pointed out,

when disaster strikes, "We're the people who show up" (Peirce 2003).

These early responses were simply a foretaste of what have come to be significant new, continuing, and costly burdens borne by municipal governments as they have sought to respond to public demands for security. The emerging institutionalization of these new responsibilities is clear from another survey conducted nearly a year after the attacks by the U.S. Conference of Mayors (USCM). To deal with their concerns about terror attacks, local officials, not just in New York and Washington but in cities of every size in every region, began routinely to reassign police officers to guard public buildings and public utilities, conduct vulnerability assessments of likely targets, expand biological and chemical surveillance through increased testing of water supplies and installation of hardware detection devices, increase training of first responders, and switch over to interoperable communications systems among fire and police departments.

Because Congress did not appropriate any funds to help state or local governments defray the costs of these new responsibilities until March 2003, cities were reliant entirely on their own resources for nearly two years after September 11. None of this new activity came cheaply. Surveys conducted by the USCM estimate that cities had to spend $2.1 billion of their own funds in 2002 alone for first-responder overtime wages, new equipment, and additional personnel. When the Department of Homeland Security raised the terror alert warning from yellow to orange as the war in Iraq began in March 2003, large and small cities all across the country increased their security spending by an estimated nationwide total of $70 million per week. These expenditures came on top of existing homeland security outlays. One result of these added burdens is that 42 percent of the cities contacted in the summer 2002 NLC survey reported that they would be less able to meet their financial needs as a result of the additional security costs after September 11.

By the time Congress finally began to pass federal programs to help defray these locally borne costs, city leaders were already deeply frustrated by the slow pace at which Washington had come to their aid. In the immediate period of shock and resolve after the attacks, there had been an expectation that federal, state, and local governments would come together quickly in a new and tighter partnership to provide security and share the financial burdens. But what has emerged is a nexus of new intergovernmental relationships where the rhetoric of partnership is burdened by feelings of disappointment and neglect.

Mayors complained not only about the complete absence of federal aid but also that the Department of Homeland Security provided no guidelines about how to respond to the terror alert system when it changes from yellow to orange. Until January 2004, there were no federal funds to offset these particular added costs, nor are there yet federal guidelines pertaining to different levels of risk that different cities might face and the corresponding steps they might take.

On a broader level, most local officials believed, at least in the early period, that intergovernmental coordination had not improved or had improved only slightly since the attacks. Only a minority of local officials polled by the NLC believed that their local government had increased its coordination with the federal government "a great deal" (8 percent) or "a good amount" (20 percent) (Baldassare and Hoene 2002).

At the heart of the new intergovernmental partnership is the issue of federal aid to defray the local costs of increased security. Although President Bush asked Congress in January 2002 for $3.5 billion for training and equipment for local first-responders, no money was actually appropriated until 18 months later, when Congress provided $750 million in direct grants to local firefighters (see Table 1). Shortly thereafter, Congress passed a host of programs to finance equipment acquisition, training, personnel costs, and planning by local governments, including the Urban Area Security Initiative, the State Homeland Security Program, and a Wartime Supplemental Appropriation/Critical Infrastructure program.

Except for the Assistance to Firefighters, all these programs provide money to local governments – municipalities, counties, and special authorities – through the states. In most cases, Congress required the states to pass through 80 percent of their allocation to local government applicants within 45 or 60 days, depending on the program. If local political leaders have been angry at the federal government for its perceived tardiness in providing aid, they are even more unhappy with their respective states. They claim in USCM surveys that states routinely miss the pass-through deadlines. Furthermore, mayors contend that they have little or no voice in influencing the allocation process at the state level and that the states favor counties and special authorities over municipalities for emergency response funding.

In sum, the first governmental impact of the terror attacks on cities has been that municipal governments have taken on new and costly security responsibilities in an intergovernmental environment in which the state and federal partners have often been perceived as both dilatory and

Table 1 Federal funding to state and local governments for homeland security (in millions)

	$ for fiscal year 2003	$ for fiscal year 2004
Direct aid to local governments		
Assistance of firefighters	750	750
Pass-through aid to local governments[a]		
Urban Area Security Initiative	100	725
Urban Area Security Initiative supplement, FY 2003 budget	700[b]	
State Homeland Security Program	566	1,685
Wartime supplemental appropriation, federal first responder/critical infrastructure	1,500	
Law enforcement terrorism prevention		500
Citizen Corps	35	

Notes

a. In all but one of the cases, 80% of the allocation of the pass-through funds must, by law, be in the hands of local governments within 45 days to 60 days, depending on the program. The one exception is the Critical Infrastructure Program, in which only 50% must be passed through. The state retains the other half.

b. Of this supplemental sum, $500 million was to be allocated to 30 selected cities, $75 million was for port security, and $65 million was for mass-transit security. Where ports or mass-transit systems are run by special authorities, those jurisdictions received the bulk of the funds.

unresponsive. Far from drawing the local, state, and federal levels closer together in a common effort to ensure against further terrorist attacks, the intergovernmental funding nexus created new burdens, frustrations, and resentments in cities across the nation.

ECONOMIC EFFECTS

In the months immediately following the September 11 catastrophe, New York City lost more than 100,000 jobs, an estimated $3 billion in tax revenues for fiscal year (FY) 2002 and 2003, and approximately $4.5 billion in income – all from diminished business activity and travel. Nearly 8,000 jobs in the theater industry alone were lost, as were approximately 12,000 restaurant jobs. The securities industry and retail trade experienced even larger employment losses. Altogether, the attacks in New York cost the city about $83 billion in lost output, wages, business closings, and spending reductions, although these losses are being offset to some degree by roughly $67 billion in insurance and federal assistance, as well as the increased economic activity that resulted from the destruction, such as clean-up and new construction.

Economic impacts have extended far beyond the borders of New York and Washington, D.C., of course, affecting the national and global economies. A review conducted by the Government Accounting Office of studies of the domestic economic impacts of the terror attacks reports a cost to the nation's 315 metropolitan areas in 2001 alone of $191 billion from diminished economic activity. Various analysts have predicted a variety of other economic effects in urban areas around the country, including drastic reductions in tourism and business travel, an end to unsubsidized office tower construction, and accelerated rates of business and residential dispersal to the suburbs, and reduced demand for central city office space. The rising cost of terrorism insurance coverage may also have a chilling effect on financing for high-profile buildings and on business profits.

An analysis of the economic impacts on cities more than two years after the attacks suggests a more nuanced picture than the vision of urban economic desertification. The analysis is complicated by factors in the national economy that have no connection to the events of September 11, including the hangover from the bursting of the Internet stock market bubble and the movement of jobs offshore. Also, the economic downturn that was in process when the terror events occurred, as well as the recovery that has set in since that time have made it difficult to isolate the independent effects of terrorism. Nevertheless, some of the data that follow are suggestive: they indicate on balance that cities – not just the attack sites but all cities – suffered initially but that economic recovery in the subsequent two years has been substantial and widespread. Severe lasting economic effects on cities that can be traced to fear of future terrorist attacks are difficult to discern.

Consider hotel occupancy rates. Table 2 reports data on fluctuations in hotel occupancy in the top twenty-five largest U.S. metropolitan hotel markets for the years 1999 through 2003. The figures in the table report average occupancy rates for October of each year, first for the five metropolitan areas that might be considered at particularly high risk of terror attacks and then for the remaining twenty lower-risk largest hotel markets. Note that average occupancy rates were highest for both categories of cities in October 2000 and then fell precipitously in October 2001 in the wake of the attacks. Although the drop in occupancy in the high-risk cities was greater than elsewhere (on average, 23 percent vs. 16 percent), recovery has been swifter (18.5 percent vs. 9.5 percent). Occupancy rates have not returned to their pre-2001 levels in any market, but they have significantly rebounded, particularly among the higher-risk metropolitan areas.

A different set of economic consequences of the terror attacks has to do with business location decisions. In particular, analysts predicted that businesses would be reluctant to seek new locations in densely concentrated downtown buildings and that departures of city firms for suburban sites would accelerate. If business reluctance to locate or stay in the central city has increased significantly, then one should see increasing office vacancy rates in central cities and signs of pressure on suburban office markets, reflected in decreasing vacancy rates or increasing office supply in the form of new construction.

In fact, however, office vacancy rates in central cities and suburbs have both been increasing since the middle of 2000, a trend related not to fear of terror but to speculative building in the 1990s. Furthermore, national data, reported quarterly, show

Table 2 Hotel occupancy rates in the twenty-five largest hotel markets, 1999 to 2003 (October)

Description	% 1999	% 2000	% 2001	% 2002	% 2003
High-risk metro areas					
New York	90.8	86.7	70.9	80.2	84.1
Los Angeles	71.6	76.0	59.3	69.0	72.8
Chicago	77.1	76.7	62.6	66.4	68.1
San Francisco	85.7	85.8	57.0	67.8	69.1
Washington, D.C.	80.4	81.4	62.9	70.0	76.6
Average	81.1	81.3	62.5	70.7	74.1
Lower-risk metro areas					
Anaheim	63.4	71.3	53.0	61.8	65.3
Atlanta	70.4	67.4	59.2	59.6	59.7
Boston	86.8	87.9	66.9	77.1	74.8
Dallas	68.7	72.4	54.4	58.0	59.3
Denver	66.2	67.1	57.1	58.6	59.1
Detroit	67.5	69.7	59.8	59.3	55.6
Houston	65.4	64.3	63.1	62.6	58.4
Miami	69.6	69.4	49.7	60.6	64.7
Minneapolis	73.6	74.0	61.7	66.3	67.4
Nashville	69.1	67.1	59.2	62.2	62.4
New Orleans	78.5	75.3	64.6	72.7	65.6
Norfolk	60.0	57.8	56.6	58.2	71.8
Oahu	72.6	75.9	54.0	69.4	73.6
Orlando	74.4	70.6	52.9	58.9	62.8
Philadelphia	74.9	74.0	67.5	70.5	73.0
Phoenix	67.7	66.2	57.7	60.2	62.6
San Diego	71.0	73.4	61.5	64.3	68.0
Seattle	67.2	70.3	57.6	63.7	63.0
St. Louis	65.3	67.1	63.4	64.2	63.2
Tampa	59.4	60.4	54.6	56.9	56.5
Average	69.6	70.1	58.7	63.3	64.3

Source: Cornell University Schools of Hotel Administration (unpublished data, 2003).

that suburban office vacancy rates were consistently higher than central city rates between the third quarter of 2000 and the third quarter of 2002, that is, from one year before to one year after the terror attacks. Between the month of the attacks, September 2001 and the end of September 2002, office vacancy rates in central city downtowns went from 10.4 percent to 12.9 percent, an increase of 24 percent. By contrast, suburban vacancy rates went from 13 percent to 16.5 percent, an increase of 26.9 percent. Notably, midtown Manhattan and Washington, D.C., were among the five places in the nation with the lowest downtown office vacancy rates at the end of 2002. It is true that hundreds of firms in lower Manhattan were displaced by the destruction of the World Trade Center. The city lost about 13 million square feet of office space.

Many firms relocated in the immediate aftermath to midtown and many more went to the suburbs.

But the New York City Partnership has concluded that most businesses that were dislocated had actually returned to lower Manhattan a year later. Of the fifty companies that had occupied the most office space in and around the World Trade Center at the time of the attack, 54 percent had returned to downtown locations by September 2002. Another 26 percent had relocated to midtown Manhattan. These data are consistent with the conclusion of the U.S. General Accounting Office that many of the jobs that left New York City after the attack had returned by May 2002.

National data on new office construction show declining numbers of new office starts beginning in mid-2001, before the terror attacks, as developers responded to the office surplus. In each successive quarter of 2002, new starts declined both for central city and suburban sites after September 11. If there had been a mass business exodus from dense downtown office concentrations, one should have seen signs of pressure or responses in suburban markets. By 2003, however, PricewaterhouseCoopers saw instead a glut of empty suburban office buildings. They conclude that although many businesses now believe that it is unwise to concentrate all key decision makers in one place, "no sea change has occurred. Suburbs haven't benefited conspicuously at the expense of cities."

If suburbs have not gained significantly in the competition with central cities for business investment, neither do they appear to have gained as locations for federal facilities. The Government Services Administration, responsible for the planning and construction of federal courthouses and office buildings, continues its Good Neighbor program. Begun in 1996, the program promotes, among other things, the use of downtown locations for federal facilities. Periodic e-news bulletins on the GSA Web site indicate that new construction, building renovation, and the hosting of public events in federal buildings continue to take place in downtown locations. A list of capital projects funded by Congress in the period from 2000–2003 shows no tendency to choose suburban rather than downtown sites in metropolitan areas.

Downtown commercial districts nationwide also seem to be thriving since the September 11 attacks. Each year, the National Trust for Historic Preservation surveys approximately 1,400 cities and towns to assess the economic health of historic and older downtown and neighborhood retail areas. Survey respondents were asked in 2002 to assess changes over the previous year in a variety of economic indicators. About three-quarters of the 370 respondents that answered the survey reported that the attacks had not had significant economic effects on businesses in their so-called Main Street districts. Specifically, almost half said that retail sales had increased over figures in 2001, nearly two-thirds reported an increase in property values, and over four-fifths said that attendance at festivals and special events had increased over the prior year. The only negative economic effects that respondents traced to September 11 were a fall-off in business at local travel agencies and greater caution among would-be small business entrepreneurs as they contemplated opening a business.

As the events of September 11 recede in time, other economic forces loom as larger influences on the fortunes of U.S. cities. This is certainly the con-

clusion of Berube and Rivlin, who point to immigration, globalization, and technological innovation as the more crucial economic factors in urban growth and development (2002, 3, 22). Similarly, the GAO observes that stronger than anticipated economic growth as the nation emerged from recession has probably mitigated the economic impacts of the terror attacks on cities.

CITY LIFE

In an article titled "Is Density Dangerous?" architect David Dixon worries that the war against terrorism might become "a war against the livability of American cities" (2002, 1). Measures to protect buildings and people, he suggests, could compromise cities' vitality, sense of community, and civic quality. These are hard to measure, of course, but Dixon's concerns evoke visions of closed-off public spaces, fortress-like buildings, constant security screenings in office buildings and mass transit stations, a forest of surveillance hardware, a heavy and constant police presence, and a general sense of dread, constraint, and foreboding.

As early as the mid-1990s, the federal government had begun to undertake some of these measures to protect its property in Washington, D.C. The result, according to the Task Force of the National Capital Planning Commission, was "an unsightly jumble of fences and barriers [that made us] look like a nation in fear." The task force called for an integrated design for the capital's monumental core, including landscaping, building setbacks, decorative street furniture, and various traffic and parking modifications. The message of the task force was that Americans must resist the impulse to build garrison cities but rather develop unobtrusive and aesthetically pleasing security measures, while maintaining an open and accessible public environment reflective of democratic values.

In Washington, D.C., more security-related construction is under way. In a few vulnerable neighborhoods of New York City, there is also an unusual police presence and a scattering of concrete Jersey barriers. But the casual pedestrian in the American city, including in most parts of Washington, D.C., and New York, in fact, rarely encounters security measures designed to thwart murderous terrorism. . . . [A]lthough visitors to office buildings in New York must pass through increased security since September 11, so-called "closed buildings" are rare in other parts of the country. Even in Washington, D.C., most private commercial buildings have free access, except when there are specific terror alerts.

When people gather in public places at high-profile events – a championship football game, New Year's Eve in Times Square – they are searched, and there is a heavy police presence, but people gather, nevertheless. Public political assemblies still take place as they did before terrorism struck American cities. And according to the National Restaurant Association, restaurant patronage has risen nationally every year for twelve straight years and was predicted to rise again in 2004.

Survey data do show signs of public anxiety, it is true, particularly in New York, but it is diminishing. There is little indication that fear has changed the ordinary habits of daily life and entertainment in cities. In a survey conducted by the *New York Times* and CBS News in September 2003, 68 percent of the respondents said they were "personally very concerned about another terrorist attack in New York City". This was down from 74 percent in October 2001. But two-thirds (67 percent) said their daily routine had gotten back to normal, up from slightly over half (52 percent) the respondents who gave the same answer in 2001.

In an earlier poll of a national sample of slightly under 1,000 people, this one by Gallup in March 2002, only 8 percent of people reported being "very worried" and 31 percent said they were "somewhat worried" that they or members of their family might become victims of a terrorist attack. Gallup also polled samples of 500 people each in New York, Washington, D.C., and Oklahoma City to compare levels of fear in those cities with national patterns. Proportions saying they were very or somewhat worried in Washington, D.C., and Oklahoma City were similar to national levels, but New Yorkers were slightly more afraid.

[. . .]

If the texture and pace of city life are clouded somewhat by public anxieties about terror, the

Plate 7 Homeland security billboard, New York City.

actual changes urban dwellers encounter in their daily lives in most places in the country and at most times are small and relatively unobtrusive. Most cities have not invested heavily in camera surveillance hardware. Retail and entertainment remain vibrant in most places, affected by economic fluctuations to all appearances more than by fears of terror. The urban streetscape has hardly become fortified, except in the official precincts of the nation's capital, nor is the police presence oppressive in these places. There may be threats to civil liberties in the law enforcement methods permitted under the federal Patriot Act, but measures undertaken by cities themselves do not seem to dampen political expression or public assembly, either by design or coincidence.

CONCLUSIONS

In certain respects, the September 11 attacks changed America deeply, but the fundamental character of American cities remains much as it was before that day. It would not have been unreasonable to expect that Americans might radically rethink the openness of urban places after such assaults, that security concerns would trump the free-wheeling, increasingly cosmopolitan values that make U.S. cities magnets for people from all over the world who seek opportunity. But the city-as-garrison-state has not taken form. People come and go to the cities, start businesses, stay in hotels and eat in restaurants, rent space and buy property, attend public political and entertainment events, and immigrate from abroad much as they once did before the world changed. Perhaps the biggest lesson so far about the cities in the aftermath of the terror attacks is that they are resilient.

Indeed, the cities seem, if not impervious to change, then at least highly resistant to external shocks like periodic terror events. Cities are shaped instead by much larger change agents. Immigration, global trade and travel, the business cycle, racial competition and conflict, changing tastes for urban living, and the regional and overseas cost of labor all have more to do with the character of cities in America than the fear of future attacks.

This is not to say that there have been no observable effects on cities of the experience of deadly assaults on American soil. A new, but frustrating set of intergovernmental relations is just beginning to take shape, posing new challenges to be navigated by public officials at all levels of the federal system. Local governments face new responsibilities and must develop new capacities, particularly in the area of emergency planning and risk assessment. Local law enforcement personnel must learn new habits of vigilance. Certain economic sectors were badly hurt after September 11 and are only now beginning to recover, particularly those in the travel and hospitality industry, although there is clear evidence that, barring other terrorist attacks, the worst is over. Finally, there is a new appreciation of the need to use architecture, planning, and landscaping not simply in service to efficiency and aesthetic considerations but also for security purposes. But nearly three years after the events of that September, the fundamental character of the American city – open, free-wheeling, striving, competing, diverse – is very much as it was before.

REFERENCES FROM THE READING

Baldassare, Mark, and Christopher Hoene. 2002. Coping with Homeland Security: Perceptions of city officials in California and the United States. Paper presented at the Congress of Cities and Exposition, National League of Cities, Salt Lake City, Utah.

Berube, Alan, and Alice Rivlin. 2002. *The potential impacts of recession and terrorism on U.S. cities.* Washington, DC: Brookings Institution.

Dixon, David. 2002. "Is Density Dangerous?" In *Perspectives on preparedness.* Vol. 12. Cambridge, MA: Harvard University Press.

Harrigan, James, and Phillipe Martin. 2002. Terrorism and the resilience of cities. *Economic Policy Review* 8 (November): 97–116.

Marcuse, Peter. 2002. Urban form and globalization after September 11th: The view from New York. *International Journal of Urban and Regional Research* 26 (September): 596–606.

Peirce, Neal. 2003. Cities and homeland security spending: As war unfolds, what *now? Government Finance Review* 19 (April): 78–79.

Savitch, H.V. 2003. Does 9/11 portend a new paradigm for cities? *Urban Affairs Review* 39 (September): 103–127.

White, E.B. [1949] 1999. *Here is New York.* New York: The Little Book Room.

S
I
X

"A Progressive Agenda for Metropolitan America"

from *What We Stand For* (2004)

Bruce Katz

Editors' Introduction

Bruce Katz's essay provides an apt conclusion to this Reader. In it, he raises a variety of issues that have been addressed throughout these pages, and urges us to see them as interconnected. Economic change is linked to population movement. Federal policies subsidizing highways and large single-family homes leads to such congestion that the average worker spends twenty-six hours a year stuck in traffic. Racial and class stratification hurt public services in poor areas most directly, but also discourage economic growth in better-off suburbs.

Moreover, he notes, we must accept the "new spatial geography of work and opportunity," in which cities and suburbs, linked together as metropolitan areas, acknowledge their shared interests and work together. Work and population have decentralized, and those trends are unlikely to change. Rather than wax nostalgic for an age in which central cities were the leaders in population and employment, we are best advised to accept the regional nature of our economies and espouse policies that maximize success for the various components of the region.

Out of this analysis, Katz develops a policy agenda that he feels will "span jurisdictional, ideological and party lines." Focusing on reforms in the funding and implementation of transportation and housing policies, his agenda fosters greater accessibility and integration across class and race lines, and reduces suburban sprawl, with its high environmental price tag.

Katz's analysis draws on census data, most notably those found in his three-volume co-edited series, *Redefining Urban and Suburban America* (with Alan Berube and William Frey, published by the Brookings Institution). He also refers to other work that draws on census data, including Richard Florida's *The Rise of the Creative Class.* (New York: Basic Books, 2002), in which Florida argues that cities which have been successful in attracting creative, innovative workers have experienced higher economic growth.

Bruce Katz is Vice-president of the Brookings Institution and founding director of its Center on Urban and Metropolitan Policy. From 1993 to 1996, he served as Chief-of-staff to Henry G. Cisneros, Secretary of Housing and Urban Development. Previously, Mr. Katz served as Senior Counsel and then Staff Director of the United States Senate Subcommittee on Housing and Urban Affairs.

[...]

[We are in the midst of] a period of profound change in the United States, comparable in scale and complexity to the latter part of the nineteenth century. Broad *demographic forces* – population growth, immigration, domestic migration, aging – are sweeping the nation and affecting settlement patterns, lifestyle choices and consumption trends. Substantial *economic forces* – globalization, deindustrialization, technological innovation – are restructuring our economy, altering what Americans do and where they do it.

Together, these complex and inter-related forces are reshaping the metropolitan communities that drive and dominate the national and even global economy. Cities – while still the disproportionate home to poor, struggling families – are re-emerging as key engines of regional growth, fueled by the presence of educational and health care institutions, vibrant downtowns, and distinctive neighborhoods. Suburbs, meanwhile, are growing more diverse in terms of demographic composition, economic function and fiscal vitality. In many respects, the differences between cities and suburbs are becoming less important than their similarities and their interdependence.

The nation's grab bag of "urban" policies – subsidized housing, community reinvestment, community development, empowerment zones – does not address or even recognize the challenges emerging from this new metropolitan reality. The almost exclusive focus of these policies on central cities ignores the fact that an entire generation of suburbs now faces city-like challenges and limits the potential political coalition for change. Renewing city neighborhoods in isolation disregards the metropolitan nature of employment and educational opportunities and inhibits the access of low-income families to good schools and quality jobs. Furthermore, principally focusing on the "deficits" of communities fails to recognize that cities and older places have assets and amenities (e.g., entrepreneurs, educational institutions, density, waterfronts, historic districts) that are highly valued by our changing economy. In general, national "urban" policies largely ignore the broader market forces and other federal policies that grow economies, shape communities and influence people's lives.

In shaping solutions by 2010, this chapter will contend that federal policies need to grow up and reflect the Metropolitan America that is rather than the urban America that was. It will argue that, after better than a half century of sprawl, that "urban" means "metropolitan" – central cities, their surrounding older suburbs, and the larger economic regions described by their effective labor market. Finally, it will put forward a progressive agenda to respond to the pressing economic, fiscal and social challenges faced by Metropolitan America.

I. THE PROBLEM

According to the 2000 census, eight in ten Americans and 95 percent of the foreign-born population live in the nation's nearly 300 metropolitan areas. Together, these regions produce more than 85 percent of the nation's economic output, generate 84 percent of America's jobs and produce virtually all the nation's wealth.

Metropolitan America is also at the vanguard of our changing economy, leading the transition to an economy based on ideas and innovation. As metropolitan experts Robert Atkinson and Paul Gottlieb have shown, the 114 largest metropolitan areas account for 67 percent of all jobs, but 81 percent of high tech employment and 91 percent of Internet domain names. According to Richard Florida, author of *The Rise of the Creative Class*, even fewer metropolitan areas are winning the competition for the young, talented, educated workers who form the nucleus of our entrepreneurial economy.

More and more, how Metropolitan America is organized and governed determines how most Americans do in life, and how we do as a nation. Yet while the indicators cited above tell a story of economic strength and productivity, America's metropolitan areas are growing in unbalanced ways that pose significant competitive, fiscal, and social challenges that require federal attention and action.

Despite clear signs of renewal in many central cities, a close examination of the 2000 census and other market data shows that the decentralization of economic and residential life remains the dominant growth pattern in the United States. As Brookings researcher Alan Berube has shown, rapidly developing new suburbs – built since the 1970s on the outer fringes of metropolitan areas –

are capturing the lion's share of employment and population growth. In the largest metropolitan areas, the rate of population growth for suburbs from 1990 to 2000 was twice that of central cities – 18 percent versus 9 percent. Suburban growth outpaced city growth irrespective of whether a city's population was falling like Baltimore or staying stable like Kansas City or rising rapidly like Denver. Even Sun Belt cities like Phoenix, Dallas and Houston grew more slowly than their suburbs.

Suburbs dominate employment growth as well as population growth. As economists Edward Glaeser and Matthew Kahn have demonstrated, employment decentralization has become the norm in American metropolitan areas. Across the largest 100 metro areas, on average only 22 percent of people work within three miles of the city center and more than 35 percent work more than ten miles from the central core. In cities like Chicago, Atlanta and Detroit, employment patterns have radically altered, with more than 60 percent of the regional employment now located more than 10 miles from the city center. The American economy is essentially becoming an "exit ramp economy," with new office, commercial, and retail facilities increasingly located along suburban freeways.

With suburbs taking on a greater share of the country's population and employment, they are beginning to look more and more like traditional urban areas. In many metropolitan areas, the explosive growth in immigrants in the past decade skipped the cities and went directly to the suburbs. As demographer William Frey has illustrated, every minority group grew at faster rates in the suburbs during the past decade; as a consequence, racial and ethnic minorities now make up more than a quarter (27 percent) of suburban populations, up from 19 percent in 1990.

Even with these profound changes, most metropolitan areas in the United States remain sharply divided along racial, ethnic, and class lines. America's central cities became majority minority for the first time in the nation's history during the 1990s and, while generally improving, have poverty rates that are almost double those of suburban communities. As metropolitan scholar Myron Orfield has shown, suburban diversity also tends to be uneven, with many minorities and new immigrants settling in older suburbs that are experiencing central city-like challenges – aging infrastructure, deteriorating schools and commercial corridors, and inadequate housing stock.

These patterns – of racial, ethnic, and class stratification, of extensive growth in some communities and significantly less growth in others – are all inextricably linked. Poor schools in one jurisdiction push out families and lead to overcrowded schools in other places. A lack of affordable housing in thriving job centers leads to long commutes on crowded freeways for a region's working families. Expensive housing – out of the reach of most households – in many close-in neighborhoods creates pressures to pave over and build on open space in outlying areas, as people decide that they have to move outwards to build a future.

The cumulative impact of these unbalanced growth patterns has enormous economic, fiscal, and social implications for the nation that deserve and require federal attention.

Unbalanced growth undermines the economic efficiency of metropolitan markets. Some of this is fairly obvious in metropolitan areas that are literally "stuck in traffic." Traffic congestion – a product in large part of growth patterns that are low density and decentralizing – has become the bane of daily existence in most major metropolitan areas. Such congestion places enormous burdens on employers and employees alike and substantially reduces the efficiency of labor and supplier markets. A recent study by the Texas Transportation Institute of 75 urban areas in the US found that the average annual delay per person was 26 hours or the equivalent of about three full work days of lost time.

Some economic consequences of unbalanced growth reflect the lost opportunities of cities and older communities that never reach their true potential. As *Business Week* has noted, "cities still seem best able to provide business with access to skilled workers, specialized high-value services, and the kind of innovation and learning growth that is facilitated by close contact between diverse individuals." Indeed, as Harvard economist Edward Glaeser has argued, the density of cities offers the perfect milieu for the driving forces of the new economy: idea fermentation and technological innovation. These broader theories on human capital formation and metropolitan growth help explain why metropolitan areas without strong central cities – Detroit, St. Louis, Cleveland, Milwaukee – are having so much difficulty making the transition to a higher road economy.

The fiscal costs of unbalanced growth are also enormous. Low-density development increases demand for new infrastructure (e.g., schools, roads, sewer, and water extensions) and increases the costs of key services like police, fire and emergency medical. Then there is the substantial impact of abandonment in older communities on the property values of nearby homes as well as the implications of concentrated poverty for additional municipal services in the schools and on the streets. Ultimately, these factors lead to reduced revenues, higher taxes and over-stressed services for older communities.

Finally, unbalanced growth imposes enormous social and economic costs on low-income minority families. As economies and opportunity decentralize and low-income minorities continue to reside principally in central cities and older suburbs, a wide spatial gap has arisen between low-income minorities and quality educational and employment opportunities. Poor children growing up in neighborhoods of poverty are consigned to inner city schools where less than a quarter of the students achieve "basic" levels in reading compared to nearly two thirds of suburban children. Similarly, inner city residents are cut off from regional labor markets where entry-level jobs in manufacturing, wholesale trade and retailing (that offer opportunities for people with limited education and skills) are abundant.

Federal 'anti-metropolitan' policies

The metropolitan growth patterns described above are the product of many factors. Population growth, consumer housing preferences and lifestyle choices have fueled suburbanization. Market restructuring and technological change have altered the location patterns of manufacturing, retail and other key employment sectors. Yet the shape and extent of decentralization in America are not inevitable. Since the middle of the twentieth century, broad federal policies – the policies often ignored by "urban" initiatives – have contributed substantially to unbalanced growth patterns in metropolitan areas.

First, and foremost, federal polices taken together set "rules of the development game" that encourage the decentralization of the economy and the concentration of urban poverty. Federal transportation policies generally support the expansion of road capacity at the fringe of metropolitan areas and beyond, enabling people and businesses to live miles from urban centers but still benefit from metropolitan life. The deductibility of federal income taxes for mortgage interest and property taxes appears spatially neutral but in practice favors suburban communities, particularly those with higher income residents. Federal and state environmental policies have made the redevelopment of polluted "brownfield" sites prohibitively expensive and cumbersome, increasing the attraction of suburban land.

Other federal policies have concentrated poverty rather than enhancing access to opportunity. Until recently, federal public housing catered almost exclusively to the very poor by housing them in special units concentrated in isolated neighborhoods. According to housing scholar Margery Turner, more than half of public housing residents still live in high poverty neighborhoods; only 7 percent live in low poverty neighborhoods where fewer than 10 percent of residents are poor. Even newer federal efforts – for example, the low-income housing tax credit program – are generally targeted to areas of distress and poverty, not to areas of growing employment. We now know that concentrating poor families in a few square blocks undermines almost every other program designed to aid the poor – making it harder for the poor to find jobs and placing extraordinary burdens on the schools and teachers that serve poor children.

The effect of all these policies: they lower the costs – to individuals and firms – of living and working outside or on the outer fringes of our metro regions, while increasing the costs of living and working in the core. They push investment out of high-tax, low-service urban areas and into low-tax, high-service favored suburban quarters, while concentrating poverty in the central city core.

The second major flaw of federal policies is that they rely on states and localities to "deliver the goods." Federal policies have not recognized the primacy of metropolitan areas and have been slow to align federal programs to the geography of regional economies, commuting patterns, and social reality.

Despite the fact that the bulk of the funds for transportation programs are raised in metropolitan areas, federal law currently empowers state departments of transportation to make most transportation decisions. These powerful bureaucracies are principally the domain of traffic engineers and are notorious for disproportionately spending transportation funds raised in metropolitan areas in rural counties. Incredibly, metropolitan areas make decisions on only about 10 cents of every dollar they generate even though local governments within metropolitan areas own and maintain the vast majority of the transportation infrastructure.

Despite the metropolitan nature of residential markets, the federal government has devolved responsibility for housing voucher programs to thousands of local public housing authorities. The Detroit metropolitan area, for example, has more than 30 separate public housing authorities, greatly limiting the residential mobility of poor families. The hyper-fragmentation of governance makes it difficult for low-income recipients to know about suburban housing vacancies, let alone exercise choice in the metropolitan marketplace.

Progress during the 1990s

During the 1990s, the federal government began to recognize the importance of metropolitan areas (and cities) to national wealth and prosperity as well as the costs and consequences of unbalanced growth patterns. A series of reform efforts in the transportation and housing arenas sought to "level the playing field" between older and newer communities and devolve more responsibility and flexibility to metropolitan decision-makers.

Federal transportation laws in the early and late 1990s, for example, devolved greater responsibility for planning and implementation to metropolitan planning organizations ("MPOs"), thus giving these areas some ability to tailor transportation plans to their distinct markets. The laws also introduced greater flexibility in the spending of federal highway and transit funds, giving state transportation departments and MPOs the ability to "flex" funding between different modes. Finally,

the laws directly funded special efforts to address metropolitan challenges such as congestion and air quality, job access for low-income workers, and the linkage between transportation and land use planning.

The changes in housing policy were equally ambitious. Public housing reforms mandated the demolition of the nation's most troubled projects and supported (through the multi-billion dollar HOPE VI program) the development of a new form of public housing – smaller scale, economically integrated, well constructed, and better designed. Other housing reforms enhanced the ability of low-income residents to move to areas of growing employment and high performing schools. The rules governing housing vouchers (now the nation's largest affordable housing program) were streamlined, making this rental assistance tool more attractive to private sector landlords. Regional counseling efforts were initiated to provide voucher recipients with the kind of assistance they need to make smart neighborhood choices.

These transportation and housing reforms, while still relatively new, have already shown some positive results. Federal money spent on transit almost doubled during the 1990s and new light rail systems are being constructed in metropolitan areas as diverse as Salt Lake City, Denver, Dallas, Charlotte, and San Diego. For the first time since World War II, growth in transit ridership has outpaced the growth in driving for five straight years. The public housing reforms became the catalyst for urban regeneration as cities like Atlanta, Louisville and St. Louis leveraged the HOPE VI funding with other private and public investments to modernize local schools, stimulate neighborhood markets and rebuild local infrastructure, parks, and libraries. The public housing reforms also contributed to one of the real success stories of the 1990s – the precipitous decline in the number of neighborhoods with poverty rates of 40 percent or higher and the number of people living in those neighborhoods.

These changes happened in the course of one decade and illustrate the kind of substantial impact a sustained course of federal action could have.

[...]

II. SOLUTIONS

It is time to develop a federal metropolitan agenda that takes account of the new spatial geography of work and opportunity in America. A progressive metropolitan agenda is necessary to help shape growth patterns that are economically efficient, fiscally responsible and environmentally sustainable. It is also necessary to revitalize central cities and older suburbs and to connect low-income families to broader educational and employment opportunities.

A federal metropolitan agenda should cover many aspects of domestic policy, ranging from workforce development to economic development to homeland security. It should also be developed in close coordination with traditional urban policies as well as major federal policies on immigration, working families and the environment. Reform of current transportation and housing policies, however, is at the core of the new metropolitan agenda.

A new transportation agenda for Metropolitan America

Metropolitan America faces a daunting set of transportation challenges – increasing congestion, deteriorating air quality, crumbling infrastructure, spatial mismatches in the labor market – that threaten to undermine their competitive edge in the global economy. Three reform ideas stand out for federal attention and action.

The federal government should continue to expand the responsibility and capacity of metropolitan transportation entities. These institutions are, after all, in the best position to integrate transportation decisions with local and regional decisions on land use, housing and economic development. At the same time, states should be required to tie their decisions more closely to the demographic and market realities of metropolitan areas. Both states and metropolitan areas should be encouraged to work together on major commercial corridors and to knit together what are now separate air, rail and surface transportation policies.

Besides governance reform, metropolitan areas also need access to broader tools and policies. A "Metropolitan Transportation Fund" should be created to provide metropolitan areas with the predictability of resources required for long term planning and the flexibility necessary to tailor transportation solutions to individual markets. The fund and all other federal programs should treat highway and transit projects equally in terms of financing and regulatory oversight. New resources, including tax credits, should be made available to stimulate development around existing light rail and other rail projects. At the same time, transportation reform should encourage the greater use of market mechanisms – such as tolls and value pricing – to ease congestion on major thoroughfares at peak traffic times. London's recent experimentation with congestion pricing, in particular, offers lessons for large American cities and metropolitan areas.

Finally, a metropolitan transportation agenda should hold all recipients of federal funding to a high standard of managerial efficiency, programmatic effectiveness, and fiscal responsibility. To that end, transportation reform should establish a framework for accountability that includes tighter disclosure requirements, improved performance measures, and rewards for exceptional performance. Transportation reform should also increase the practical opportunities for citizen and business participation in transportation decision making. States and metropolitan areas should be provided the funding to experiment with state-of-the-art technologies for engaging citizens in public debates.

A new housing agenda for Metropolitan America

Federal housing policy must also be recast to fit the new metropolitan reality. As discussed above, the uneven residential patterns in most metropolitan areas are placing special burdens on older communities and limiting the educational and employment opportunities of a wide cross-section of families.

A new federal housing agenda must expand housing opportunities for moderate- and middle-class families in the cities and close-in suburbs while creating more affordable, "workforce" housing near job centers. Ideally, federal policies should help regional elected leaders balance their housing markets through zoning changes, subsidies and tax incentives so that all families – both middle class and low income – have more choice about where they live and how to be closer to quality jobs and good schools. A new federal housing agenda can build on the replicable models of balanced housing policies that are already emerging in the metropolitan areas of Minneapolis, Portland, Seattle, and Washington, D.C.

To achieve these ends, federal tax incentives should be expanded to boost homeownership in places where homeownership rates are exceedingly low. Incentives could include a tax credit that goes directly to first time homebuyers (as in Washington, D.C.) and a tax benefit that entices developers to construct or renovate affordable homes (like the existing tax credit for rental housing). Such incentives would enhance the ability of working families to accumulate wealth and contribute to the stability of neighborhoods by lowering the costs of homeownership.

In addition, the federal government should continue its efforts to demolish and redevelop distressed public housing and promote economic integration in federally-assisted housing. The successful HOPE VI program should be renewed for another decade of investment and its reach should be extended beyond public housing to distressed housing projects financed by the federal government. The federal government should also make it easier in all housing programs to serve families with a broader range of incomes, particularly in neighborhoods with high concentrations of poverty.

To enhance housing choice, the federal government should invest more substantially in vouchers. A national goal of a million more vouchers over the next decade sets an ambitious, but achievable, target. Vouchers have consistently proven to be the most cost effective and market-oriented of federal housing programs and, more than any other housing program, enable low-income parents to base their housing decisions on the performance of local schools.

Besides these additional investments, more substantial governance and statutory reforms will be necessary to promote greater housing choice for low-income families. The federal government should, for example, shift governance of the housing voucher program to the metropolitan level. As previously described, the federal voucher program is administered by thousands of separate public housing bureaucracies operating in parochial jurisdictions. Competitions should be held in dozens of metropolitan areas to determine what kind of entity – public, for-profit, nonprofit, or a combination thereof – is best suited to administer the program.

The federal government should also make it easier to allocate low-income housing tax credits to areas of growing employment, not only to areas of distress and poverty. And existing funds should be invested in creating a network of regional housing corporations to develop and preserve affordable housing in suburban areas. A national network of regional housing corporations can build on the achievements of community development corporations, many of which can naturally graduate to operate at the metropolitan level.

The most important action, however, will be the hardest. Many wealthy communities will only open up their communities if they are denied something they want. To this end, the federal government should prohibit lucrative federal highway investments in communities that have been found in violation of federal civil rights laws or otherwise have engaged in exclusionary housing practices.

The metropolitan agenda described above could have a transforming affect on the physical and social landscape of metropolitan areas. For example, obsolescent freeways that currently block access to urban waterfronts and other valuable real estate can be removed, as in Milwaukee, Boston, and Portland. At the same time, new, dense residential communities can emerge along commuter rail, light rail, and rapid bus lines (as in Dallas and Arlington, Virginia), giving commuters greater residential and transportation choices and responding more adequately to the changing demographics of the country.

Providing affordable housing throughout a region will also produce substantial benefits. It should help workers live closer to suburban areas of employment and reduce congestion on roadways. It should help reduce the concentration of poverty, thereby making school reform and educational achievement real possibilities. It should help cities and older suburbs create mixed-income communities, thereby revitalizing neighborhoods and generating markets. By strengthening older communities, it will take the pressure off of sprawl, thereby improving the quality of life in outer exurban areas.

Some of the reforms described above are feasible in the current political environment and should be enacted in the near term. The homeownership tax credit idea, for example, has already received broad bipartisan support.

Yet other reforms and investments will take longer to accomplish. State departments of transportation will oppose the further devolution of responsibility to metropolitan entities as well as greater levels of federal oversight and accountability. Some neighborhood advocates will oppose further efforts to demolish distressed housing and provide low-income residents with greater choice in the metropolitan marketplace. Some low-income housing advocates will oppose efforts to promote economic integration in federally assisted housing. Many suburban areas will surely resist the production of affordable housing. In general, the constrained fiscal environment created by Bush policies will make any new housing investments extremely difficult.

This new metropolitan agenda, therefore, will require not just new policy ideas but new political coalitions that span jurisdictional, ideological and party lines. Existing local constituencies will have to think differently about metropolitan issues and make connections between policies – housing, workforce, education, transportation – that are now kept separate and distinct.

To a large extent, this change is inevitable. Urban policy in America can no longer be exclusively about cities or neighborhoods. It must be about the new metropolitan reality that defines our economy and society and the larger government rules that help shape that reality. The next administration has an historic opportunity to design and implement a metropolitan agenda that promotes balanced growth, stimulates investment in cities and older suburbs and connects low-income families to employment and educational opportunities.

SIX

COPYRIGHT INFORMATION

1 THE SOCIAL AND ECONOMIC CONTEXT OF URBAN POLITICS

2 THE ROOTS OF URBAN POLITICS

3 UNDERSTANDING URBAN POWER

4 THE POLITICAL ECONOMY OF CITIES AND COMMUNITIES

5 THE POLITICS OF RACE, ETHNICITY, AND GENDER

6 CITIES, REGIONS, AND NATIONS

Index

CPSIA information can be obtained
at www.ICGtesting.com
Printed in the USA
FSOW04n1329310316
18675FS